MW00620591

Dissipative Structures and Chaos

Springer
Berlin
Heidelberg
New York
Barcelona
Budapest
Hong Kong
London
Milan
Paris
Santa Clara
Singapore
Tokyo

Hazime Mori   Yoshiki Kuramoto

Dissipative Structures and Chaos

Translated by Glenn C. Paquette

With 142 Figures and 5 Tables

Springer

Professor Hazime Mori
Professor Emeritus, Kyushu University
Private address:
5-6-27 Kasumigaoka
Higashi-ku
Fukuoka 813-0003, Japan

Professor Dr. Yoshiki Kuramoto
Graduate School of Sciences
Kyoto University
Kyoto 606-8502, Japan
e-mail:
kuramoto@ton.scphys.kyoto-u.ac.jp

Translator

Dr. Glenn C. Paquette
1-9 Yokonawate-cho Matsugasaki
Sakyo-ku
Kyoto 606-0965, Japan
e-mail: vye03010@niftyserve.or.jp

The cover picture shows spatio-temporal chaos displayed by solutions to the KS equation as discussed in Chap. 4 (by permission of Toshio Aoyagi)

Title of the original edition:
Sanitsu Kozo to Kaosu (Dissipative Structures and Chaos)
by Hazime Mori and Yoshiki Kuramoto

Originally published in Japanese by Iwanami Shoten, Publishers, Tokyo 1994

Library of Congress Cataloging-in-Publication Data. Mori, Hazime, 1926– Dissipative structures and chaos / Hazime Mori, Yoshiki Kuramoto; translated by G. C. Paquette. p. cm. Translation of: Sanitsu kozo to kaosu. Includes bibliographical references and index. ISBN 3-540-62744-8 (hard: alk. paper) 1. Chaotic behavior in systems. 2. Fluid dynamics. 3. Self-organizing systems. I. Kuramoto, Yoshiki. II. Title. Q172.5.C45M67 1998 003'.857–dc21 98-3125

ISBN 3-540-62744-8 Springer-Verlag Berlin Heidelberg New York

Typesetting: Data conversion by Springer-Verlag
Cover design: *design & production* GmbH, Heidelberg

SPIN 10493881 56/3144 – 5 4 3 2 1 0 – Printed on acid-free paper

Preface to the English Edition

This book consists of two parts, the first dealing with dissipative structures and the second with the structure and physics of chaos. The first part was written by Y. Kuramoto and the second part by H. Mori.

Throughout the book, emphasis is laid on fundamental concepts and methods rather than applications, which are too numerous to be treated here. Typical physical examples, however, including nonlinear forced oscillators, chemical reactions with diffusion, and Bénard convection in horizontal fluid layers, are discussed explicitly.

Our consideration of dissipative structures is based on a phenomenological reduction theory in which universal aspects of the phenomena under consideration are emphasized, while the theory of chaos is developed to treat transport phenomena, such as the mixing and diffusion of chaotic orbits, from the viewpoint of the geometrical phase space structure of chaos.

The title of the original, Japanese version of the book is *Sanitsu Kozo to Kaosu* (Dissipative Structures and Chaos). It is part of the *Iwanami-Koza Gendai no Butsurigaku* (Iwanami Series on Modern Physics). The first Japanese edition was published in March 1994 and the second in August 1997.

We are pleased that this book has been translated into English and that it can now have an audience outside of Japan.

We would like to express our gratitude to Glenn Paquette for his English translation, which has made this book more understandable than the original in many respects.

Fukuoka and Kyoto
January 1998

Hazime Mori
Yoshiki Kuramoto

Preface to the Second Japanese Edition

Roughly four years have passed since the first edition of this book was completed. During this time, there have been a number of important discoveries, and the significance of certain previously discovered phenomena has come to be newly understood. In this second edition of the book, we have added two supplementary chapters in which some of these new developments are discussed.

The ever-changing natural world that surrounds us is an extremely unstable system, containing 'internal discord' manifested in many forms. This system, infinitely rich in change, exhibits an unlimited variety of structures existing on multiple levels. We would like briefly to discuss here the formation of this multilevel structure in the context of dissipative structures and chaos.

In the study of dissipative systems, the concept of *symmetry breaking* is very important for the purpose of understanding the formation of multilevel structure. This concept has its origin in the study of phase transition phenomena, but it later came to be one of the most important concepts in high-energy physics as well. Now it plays a central role in all fields of physics, as the principle at work in the appearance of variety in physical systems.

In the context of dissipative structures, symmetry breaking appears through bifurcation phenomena. For example, when periodic structure appears in an originally uniform field, the spatial symmetry of the system is reduced. Such a reduction in symmetry can then result in the appearance of even more complicated patterns. The emergence of oscillation constitutes the breaking of time translational symmetry. A succession of such reductions of symmetry leads to a chaotic state. The development of oscillation and the appearance of periodic structure, found in infinitely extended physical systems, imply the *breaking of continuous symmetry.* It is well known that, as the result of such symmetry breaking, there appear degrees of freedom of extremely long time scales. In the case of dissipative structures, these correspond to the phase degrees of freedom. The exceedingly 'soft' spatio-temporal behavior characteristic of dissipative structures is due to the presence of phase modes of very weak stability.

Dynamical system chaos displays two fundamentally different types of behavior, which are observed on different time scales. Short time scales are characterized by deterministic, predictable motion, while long time scales are characterized by random, unpredictable motion. This two-layered structure of chaos provides a foundation with which we can understand the myriad macroscopic structures displayed by nature. For example, note that motion in nature on a macroscopic scale entails the dissipation of kinetic energy into microscopic thermal motion and is thus irreversible. However, the transport coefficients that correspond to the processes responsible for this energy dissipation are determined by microscopic, reversible molecular motion subject to Onsager's reciprocity theorem (1931). The existence of these two levels – those of microscopic, reversible motion and macroscopic, irreversible motion – results from the two-layered structure of chaos appearing in conservative dynamical systems of many degrees of freedom. This macroscopic irreversibility accompanies the *breaking of time-reversal symmetry* that results from the instability of chaotic orbits.

Large-scale motion in the Earth's atmosphere, often spanning hundreds of kilometers, is subject to energy dissipation, as energy is carried through the turbulent flow of intermediate-sized vortices to vortices existing on successively smaller scales. Turbulent viscosity arises as the mechanism tending to impose uniformity on this large-scale flow. This coexistence of large-scale and intermediate-scale flow can be understood in terms of the two-layered structure existing in the chaos of dissipative dynamical systems.

Fukuoka and Kyoto
June 1997

Hazime Mori
Yoshiki Kuramoto

Preface to the First Japanese Edition

Every day, we observe fluid motion in nature – in the atmosphere, in the ocean, in rivers. From the ruffling of the leaves on a tree to the welling of summer storm clouds, the raging of hurricanes and monsoons, and the dynamics of massive cloud formations seen from weather satellites, a myriad of natural phenomena, consisting of exquisitely self-organized motion, are played out in open systems far from equilibrium, presenting to us nature's limitless variation. Whether our purpose is to maintain our coexistence with Nature or to control and utilize it, a deep understanding of these phenomena is indispensable. However, since most of this behavior is nonlinear and nonequilibrium in nature, it is very difficult to treat, and for the first half of the 20th century, it was not regarded as an important subject in physics. Since the 1960s, though, with developments in the theory of dynamical systems and computer experimental methods, rapid progress has been made in the study of these systems, and as a result many of nature's processes have become newly understood.

A clear example of these recent developments is the discovery of the many varieties of dissipative structures and chaotic behavior existing in nature. Through this research, new concepts in nonlinear dynamics, used to probe the source of nature's complexity and variety, have been formed. In particular, because the phase space trajectories associated with chaotic behavior are random, chaotic systems display unpredictable motion, and their study has introduced a new manner of thinking with regard to the concepts of chance and inevitability in nonlinear dynamics. These developments have brought into question the deterministic and mechanistic view of Nature set forth by Laplace in 1776, which has characterized scientific thinking for roughly 200 years.

At present, however, it is not possible to develop the study of these phenomena as a whole from a mechanics point of view. In this book we consider, as representative constituent subsystems of mechanical systems appearing in nature, nonlinear forced oscillators, fluids exhibiting convective motion, and reaction–diffusion systems. With a foundation in the geometric, qualitative theories regarding the solutions to the equations of motion describing these phenomena, we describe the recently developed physical and mathematical methods used in the study of dissipative structures and chaos.

This book, written in two parts, is intended for readers at or above the undergraduate level in science and engineering fields. In Part I, *Dissipative Structures*, with regard to nonequilibrium open systems, we consider questions concerning the nature of systems from which macroscopic structure can arise, the stability of this structure, the way in which its instability leads to complexity, and the manner in which spontaneous turbulence arises. The importance of focusing on the universal aspects of the systems in question – that is, those aspects which are independent of each system's underlying material nature – is emphasized throughout Part I. As a mathematical method to facilitate a qualitative understanding of this type of phenomenon, the *theory of reduction* is developed. In Part II, *The Structure and Physics of Chaos*, questions concerning the nature of the generation of the many varieties of chaotic behavior displayed by mechanical systems, the forms and structures that it assumes, the manner in which it is described, and the nature of the statistical laws to which it is subject are considered. In particular, with a combination of geometric and statistical descriptions of chaotic orbits, we construct the statistical physics of chaos. While giving this statistical characterization of the form and structure of chaos, our purpose is to elucidate the nature of mixing and diffusion, energy dissipation fluctuations, and other types of transport phenomena exhibited by chaotic systems.

A macroscopic system far from thermal equilibrium will irreversibly approach an attractor with the passing of time. In such a situation, the system's energy will be dissipated (for example, by friction), but if energy is continuously added to the system at some fixed rate, it will converge to an attractor representing a state of thermal nonequilibrium. In many cases, such a system will possess a spatio-temporal pattern referred to as a *dissipative structure.* A well-known example of this is the so-called Bénard convective roll structure arising in the horizontal layers of a fluid heated from below. Dynamical systems are inherently coherent, but in the case of dissipative structure formation, this coherence is in reference to a specific macro-scale spatio-temporal pattern. This type of macroscopic system is referred to as a nonequilibrium open system. The nature of the dissipative structure that is formed depends on the nature of the system in question, as well as on the values of the *control parameters* or *bifurcation parameters*, which constitute a measure of the distance by which the system is separated from equilibrium. As the value of such a parameter is increased, upon reaching some particular value (the bifurcation point), the structure of the solution to the equation of motion describing the system will change qualitatively. This change is called a *bifurcation.* In other words, at a bifurcation point, a system becomes structurally unstable, and as a result, new structure appears.

If the bifurcation parameter is not too much larger than its value at a given bifurcation point, only degrees of freedom corresponding to slow motion are excited, and the attractor will correspond to macroscopic motion consisting of only a few degrees of freedom. As the bifurcation parameter is increased further, the dissipative structure will, as a result of bifurcation, undergo further qualitative change and become gradually more complex. Usually, the nature of this structure undergoes change in the following manner. As the parameter increases, first a spatial pattern develops. Next, the pattern begins to undergo periodic oscillation in time. As the parameter becomes even larger, aperiodic undulation appears and eventually random, unpredictable chaos arises. Then, this chaotic state itself may experience a series of bifurcations, resulting in the appearance of various forms of chaos. However, in describing the motion of the fluid, depending on whether we use the Euler picture, in which we focus on each fixed point in space and consider the fluctuations in the current as seen there, or the Lagrange picture, in which we follow the trajectory of each fluid particle, we find the observed character of the motion to differ. For example, even if the flow at each point is regular and laminar, if the streamlines oscillate in time, when viewed in the Lagrange picture, chaotic behavior arises in which the trajectories of the fluid particles are random.[1] This phenomenon is known as Lagrangian turbulence and differs from the more familiar Eulerian turbulence.

The term 'dissipative structure' is used in reference to not only the spatio-temporal structure associated with an attractor but also the transient spatio-temporal patterns that appear as the system approaches the attractor. In the neighborhood of an attractor, the form of this transient behavior contains important information with regard to the dissipative structure of the attractor. For this reason, it is necessary to develop a qualitative dynamical theory of nonlinear, nonequilibrium systems which includes both the dynamics of dissipative structure on attractors and also the dynamics of structure arising out of the transient processes leading to attractors. As a basis for this development, we pick out only the slow modes of a given system's degrees of freedom and set out to describe a reduction theory that seeks the evolution equations describing the slow-motion time development associated with these degrees of freedom. In particular, in the neighborhood of a bifurcation point at which a stable solution destabilizes, the evolution equation describing the system becomes greatly reduced and comes to assume a universal form.

To this point, we have discussed dissipative dynamical systems; that is, those systems accompanied by energy dissipation. However, there is also a great variety of motion occurring in nature that can be described by conservative dynamical systems possessing fixed Hamiltonians. These systems are symmetric with respect to time reversal but can nonetheless exhibit chaotic behavior. The chaotic region in the phase space corresponding to such a system will contain islands of tori of many various sizes – a structure qualitatively different from that appearing in the case of dissipative systems. However, even in this conservative case, the mixing, diffusion, and other irreversible transport phenomena associated with chaotic orbits exist. In general,

---

[1] Aref, H. (1984) Stirring by chaotic advection. J. Fluid Mech. **143**, 1–21

chaos destroys time-reversal symmetry on large time scales and introduces irreversible behavior.

The thermal motion of molecules is an example of microscopic chaotic behavior. The physical quantity that characterizes this behavior is the Boltzmann entropy. What are the physical quantities that characterize the chaos of nonequilibrium open systems? At this level, we can ignore thermal fluctuations, but we cannot ignore the fluctuations associated with macroscopic chaos. Moreover, these fluctuations arise in the context of a variety of chaotic orbits and exhibit a diversity of geometric forms in phase space. The quantities used to characterize chaos must be capable of characterizing and classifying this diversity. The investigation of such physical quantities is one of the fundamental themes of this book.

Part I, *Dissipative Structures*, comprises Chaps. 1–5. In Chap. 1 we survey two systems that have been used extensively in research on dissipative structures, Bénard thermal convection and the Belousov–Zhabotinskii (BZ) reaction. We discuss the manner in which dissipative structures have been studied, as represented by the construction and analysis of the models that describe these natural phenomena, and, in particular, we consider the origin of behavior whose phenomenological and qualitative nature is described by such models. We also discuss the reason this study of dissipative structures has produced important results. This chapter provides preparation necessary for the topics developed in later chapters.

Chapter 2 presents an outline of the so-called amplitude equation approach, which is applicable to a system in the neighborhood of a bifurcation point at which dissipative structure arises. Several representative amplitude equations corresponding to a variety of physical situations are derived phenomenologically. Then we consider, among other topics, the stability of stationary propagating periodic patterns and the structure and motion of defects that form therein, as arising from these equations. These nonequilibrium patterns embody the most fundamental constituent elements of dissipative structure.

In Chap. 3 we treat the subject of interface dynamics using a theoretical approach involving 'interfaces', which is based on a manner of thinking that is completely different from that used in the previous chapter. An interface is a localized structure that represents an abrupt change in state occurring in a small region of space. Particularly in the case of reaction–diffusion systems, it is useful to consider the dynamics of interfaces, and from this point of view we study the structure and motion of the characteristic waves seen in the BZ reaction, including solitary waves, wave pulse trains, and spiral waves, and we outline the pattern formation resulting from interface instability.

Chapter 4 consists of a discussion of the method of phase dynamics, which, like the approaches based on amplitude equations, constitutes a representative example of a reduction method. Phase dynamics is a general reduction approach that can be applied in the situation where a pre-existing pattern breaks the continuous spatial symmetry associated with the system's constituent species. Using the equation of motion for the phase field derived using this method, we can treat the many types of motion exhibited by ordered patterns deformed under the influence of a variety of physical effects.

We pursue the fundamental nature of reduction in Chap. 5. The basic idea underlying the discussion of this chapter is that there exists a single clear universal structure forming the basis of the reductive construction. In making this point clear, we re-examine from a single, unified point of view, each of the various perturbative theories that have to that point been developed separately. This unified approach provides strong support for the phenomenological reduction theories developed in the previous chapters.

Part II, *The Structure and Physics of Chaos*, includes Chaps. 6–10. In Chap. 6 we give an overview of the method of description used for chaos and consider the necessity of the statistical physics approach. This point concerns the reproducibility problem as it relates to the system's statistical stability. The orbits of a chaotic system are unstable with respect to small disturbances, and thus the actual trajectory followed in any case is not reproducible. However, the long-time average of physical quantities measured over such a trajectory is stable and reproducible. The new physical quantity that we use to characterize chaos can be considered as an extension of the concept of an eigenvalue of the Poincaré map in the neighborhood of a periodic point. This quantity is the coarse-grained local expansion rate of the distance between neighboring orbits. Studying the statistical structure of the fluctuations of this quantity along chaotic orbits, we give a geometric characterization of the structure of chaos while describing its mixing nature.

Considering low-dimensional dissipative dynamical systems, in Chap. 7 we present an outline of attractor bifurcation phenomena and consider the geometric structure of bifurcations in chaos. This study provides the foundation for the analysis developed in later chapters, in which we elucidate the qualitative changes in the statistical structure of chaotic systems that occur at bifurcation points. Furthermore, we investigate similarity and renormalization transformations of bifurcation cascades in the contexts of period doubling, band splitting and attractor merging, thereby obtaining a description of the quasi-periodicity route to the emergence of chaos.

Chaotic orbits and the concept of chance inherent in their nature are addressed from a statistical point of view in Chap. 8. Here, we consider the framework needed in constructing the statistical physics of aperiodic motion. The physical quantities we use for this purpose are the coarse-grained local expansion rate between neighboring orbits and the local dimension describing the self-similar nested structure of strange attractors. As concrete examples, we treat, in the context of low-dimensional maps, the multifractal structure of critical attractors that appear at the emergence point of chaos and the qualitative change in a system's statistical structure corresponding to several universal chaotic bifurcations.

In Chap. 9 we consider the phenomena of band-merging and attractor-merging crises in two different types of forced oscillators, and the changes suffered at the bifurcation points at which these phenomena occur in the two-dimensional phase spaces representing such systems. We also investigate the nature of the statistical structure that appears here. Then, as examples of self-organized criticality, we investigate the self-similar time series of the coarse-grained orbital expansion rate for the case of a critical attractor, the dynamical self-similarity of a chaotic band attractor, and the form realized in the disappearance of two-dimensional fractality in the neighborhood of the emergence point of chaos.

In Chap. 10 we treat the mixing of chaotic orbits in the widespread chaotic sea of a conservative dynamical system. In addition, we consider the anomalous diffusion caused by accelerator modes in such systems. First, with regard to the two-dimensional standard map, we discuss both numerical and theoretical studies of the statistical structure of mixing and diffusion. We then show that the instability of chaotic orbits destroys the time-reversal symmetry of conservative dynamical systems. Next, as an application, with regard to Lagrangian turbulence appearing in a Bénard convection system with sufficiently large aspect ratio, we demonstrate the existence of accelerator mode tori and treat the mixing and diffusion of fluid particles.

Part I was written by Y. Kuramoto and Part II by H. Mori. Both authors would like to pay tribute to the individuals who contributed to the theory and experiments described in this book. We would also like to express our gratitude to Yasuji Sawada, Hidetsugu Sakaguchi and Shin-ichi Sasa for their enlightening influence on the content and presentation of Part I, and to Hiroki Hata and Takehiro Horita, who carefully read the manuscript of the original, Japanese version of Part II and offered a great deal of advice.

A set of references appears at the end of the book. This includes selected works containing simple explanations of material related to the content of the book.

Fukuoka and Kyoto
August 1993

Hazime Mori
Yoshiki Kuramoto

Contents

Part I. Dissipative Structures

Introduction 3

1. A Representative Example of Dissipative Structure 5
   1.1 Bénard Convection 5
   1.2 The Belousov–Zhabotinskii Reaction 14

2. Amplitude Equations and Their Applications 21
   2.1 The Newell–Whitehead Equation and the Stability of Periodic Solutions 21
   2.2 Anisotropic Fluids and the Ginzburg–Pitaevskii Equation 26
   2.3 Topological Defects and Their Motion 29
   2.4 The Amplitude Equation of an Oscillating Field 34
   2.5 The Properties of the Complex Ginzburg–Landau Equation 36
      2.5.1 Propagating Planar Wave Solutions and Their Stability 36
      2.5.2 Rotating Spiral Waves 38
      2.5.3 Hole Solutions and Disordered Patterns 39

3. Reaction–Diffusion Systems and Interface Dynamics 43
   3.1 Interfaces in Single-Component Bistable Systems 43
   3.2 Solitary Wave Pulses and Periodic Wave Pulse Trains in Excitable Systems 47
   3.3 Spiral Waves in Excitable Systems 52
   3.4 Multiple Spiral Waves and the Turing Pattern 58
      3.4.1 Compound Spiral Rotation 58
      3.4.2 The Turing Pattern 59
   3.5 The Instability of Interfaces and Formation of Structure 62

4. Phase Dynamics 69
   4.1 Weak Turbulence of Periodic Structures and the Phase Equation 69
   4.2 Phase Waves and Phase Turbulence of Oscillating Fields 76
      4.2.1 The Phase Equation of an Oscillating Field and Its Applications 76

4.2.2 Phase Waves and the Target Pattern ... 78
4.2.3 Phase Turbulence ... 82
4.3 The Phase Dynamics of Interfaces ... 84
4.4 Multiple Field Dynamics ... 86

5. Foundations of Reduction Theory ... 93
5.1 Two Simple Examples ... 93
5.2 The Destabilization of Stationary Solutions ... 98
5.3 Foundations of the Amplitude Equation ... 100
5.4 The Introduction of Continuous Spatial Degrees of Freedom ... 107
5.4.1 The Hopf Bifurcation ... 107
5.4.2 The Turing Instability ... 109
5.5 Fundamentals of Phase Dynamics ... 112
5.5.1 Phase Dynamics in a Uniform Oscillating Field ... 113
5.5.2 Phase Dynamics for a System with Periodic Structure ... 114
5.5.3 Interface Dynamics in a Two-Dimensional Medium ... 115

Supplement I:
Dynamics of Coupled Oscillator Systems ... 119
SI.1 The Phase Dynamics of a Collection of Oscillators ... 119
SI.2 Synchronization Phenomena ... 121

Part II. The Structure and Physics of Chaos

Introduction ... 127

6. A Physical Approach to Chaos ... 129
6.1 The Phase Space Structure of Dissipative Dynamical Systems ... 129
6.2 The Phase Space Structure of Conservative Dynamical Systems ... 134
6.3 Orbital Instability and the Mixing Nature of Chaos ... 139
6.3.1 The Liapunov Number ... 139
6.3.2 The Expansion Rate of Nearby Orbits, $\lambda_1(X_t)$ ... 141
6.3.3 Mixing and Memory Loss ... 144
6.4 The Statistical Description of Chaos ... 145
6.4.1 The Statistical Stability of Chaos ... 145
6.4.2 Time Coarse-Graining and the Spectrum $\psi(\Lambda)$ ... 146
6.4.3 The Statistical Structure of Chaos ... 148

7. Bifurcation Phenomena of Dissipative Dynamical Systems 151
7.1 Band Chaos of the Hénon Map ... 151
7.2 The Derivation of Several Low-Dimensional Maps ... 155
7.2.1 The Hénon Map ... 157

7.2.2 The Annulus Map ... 157
7.2.3 The Standard Map ($J = 1$) ... 159
7.2.4 One-Dimensional Maps ($J = 0$) ... 160
7.3 Bifurcations of the One-Dimensional Quadratic Map ... 161
7.3.1 2^n-Bifurcations and 2^n-Band Bifurcations ... 161
7.3.2 The Self-Similarity and Renormalization Transformation of 2^n-Bifurcations ... 166
7.3.3 The Similarity of 2^n-Band Bifurcations ... 170
7.4 Bifurcations of the One-Dimensional Circle Map ... 174
7.4.1 Phase-Locked Band Chaos ... 174
7.4.2 Phase-Unlocked Fully Extended Chaos ... 177

8. The Statistical Physics of Aperiodic Motion ... 183
8.1 The Statistical Structure Functions of the Coarse-Grained Orbital Expansion Rate ... 183
8.1.1 The Baker Transformation ... 185
8.1.2 Attractor Destruction in the Quadratic Map ... 187
8.1.3 Attractor Merging in the Circle Map ... 189
8.1.4 Bifurcations of the Hénon Map ... 192
8.1.5 The Slopes s_α and s_β of $\psi(\Lambda)$ ... 193
8.2 The Singularity Spectrum $f(\alpha)$ ... 195
8.2.1 The Multifractal Dimension $D(q)$ ... 198
8.2.2 Partial Local Dimensions $\alpha_1(X)$ and $\alpha_2(X)$ ... 198
8.2.3 $f(\alpha)$ Spectra of Critical Attractors ... 199
8.3 Theory Regarding the Slope of $\psi(\Lambda)$ ... 203
8.3.1 The Slope s_α Due to the Folding of W^{u} for Tangency Structure ... 203
8.3.2 The Slope s_β Resulting from Collision with the Saddle S ... 205
8.4 The Relation Between $f(\alpha)$ and $\psi(\Lambda)$ ... 209
8.4.1 The Linear Segment of $f(\alpha)$ Resulting from the Folding of W^{u} in the Presence of Tangency Structure ... 212
8.4.2 The Linear Segment of $f(\alpha)$ Caused by Bifurcation ... 214

9. Chaotic Bifurcations and Critical Phenomena ... 217
9.1 Crisis and Energy Dissipation in the Forced Pendulum ... 217
9.1.1 The Slope s_δ Induced by the Cantor Repellor ... 218
9.1.2 The Spectrum $\psi(W)$ of the Energy Dissipation Rate ... 222
9.1.3 The Formation of the Attractor Form in Figure 6.1 ... 224
9.2 Fully-Extended Chaos That Exists After Attractor Merging ... 224
9.2.1 Attractor Merging in the Annulus Map ... 225
9.2.2 Attractor Merging in the Forced Pendulum ... 227

9.3 Critical Phenomena and Dynamical Similarity of Chaos 230
9.3.1 The Self-Similar Time Series of Critical Attractors 230
9.3.2 The Algebraic Structure Functions of the Critical Attractor 233
9.3.3 The Internal Similarity of Bands for the Spectrum $\psi(\Lambda)$ 235
9.3.4 The Form Characterizing the Disappearance of Two-Dimensional Fractality 238

10. Mixing and Diffusion in Chaos of Conservative Systems 241
10.1 The Dynamical Self-Similarity of the Last KAM Torus 241
10.1.1 The Self-Similar F_m Time Series 242
10.1.2 The Symmetric Spectrum $\psi_\beta(\beta)$ 242
10.2 The Mixing of Widespread Chaos 243
10.2.1 The Form of $\psi(\Lambda)$ and the Breaking of Time-Reversal Symmetry 245
10.2.2 The Appearance of Anomalous Scaling Laws for Mixing 247
10.3 Anomalous Diffusion Due to Islands of Accelerator Mode Tori 250
10.3.1 Accelerator Mode Periodic Orbits 250
10.3.2 Long-Time Velocity Correlation 251
10.3.3 The Anomalous Nature of the Statistical Structure of the Coarse-Grained Velocity 252
10.4 Diffusion and Mixing of Fluids as a Result of Oscillation of Laminar Flow 254
10.4.1 Islands of Accelerator Mode Tori Existing Within Turnstiles 254
10.4.2 Anomalous Mixing Due to Long-Time Correlation 257

Supplement II: On the Structure of Chaos 261
SII.1 On–Off Intermittency 261
SII.2 Anomalous Diffusion Induced by an Externally Applied Force 262
SII.3 Transport Coefficients and the Liapunov Spectrum 263

Summary of Part II 265

A. Appendix 269
A.1 Periodic Points of Conservative Maps and Their Neighborhoods 269
A.2 Variance and the Time Correlation Function 271
A.3 The Cantor Repellor of Intermittent Chaos 271

Bibliography 279

Index 295

Part I

Dissipative Structures

Introduction

Dissipative structure is peculiar to nonequilibrium open systems. It is maintained by the balance of the influx of energy and matter with dissipation, and in many cases it appears on a macroscopic scale. For this reason, the most natural theoretical description of these phenomena should begin with a consideration of macro-level, nonlinear evolution equations such as the Navier–Stokes equation. Each chapter in Part I is based on this consideration. Of course, one can argue that there exists the problem of determining the microscopic physical source of macroscopic dissipative structure. However, in the end, this problem is equivalent to that of determining the statistical mechanical basis of the behavior of the macroscopic evolution equations themselves. Following this line of reasoning, questions regarding the microscopic physical source of dissipative phenomena can be separated from the study of dissipative phenomena themselves. Such microscopic considerations are beyond the scope of this book.

Dissipative structures and other nonequilibrium patterns have become the subject of serious study in physics only recently. Earnest research in this field did not begin until the 1970s. The approach used in the study of nonequilibrium patterns employs a phenomenological/qualitative manner of thinking which represents a bold departure from the physics that existed prior to this study, and as such, this approach represents an important success. Through this approach, a new theoretical framework has developed, while new terms have arisen for the purpose of describing complex natural phenomena.

There is no all-powerful theoretical method to treat the phenomena of nonequilibrium patterns. However, one standard and fairly well-established method consists of an approach based on reduced equations. The validity of this approach is now widely accepted. These reduced equations are produced through the application of a particular procedure whose purpose is to extract the most important features of the nonlinear equations describing the evolution of patterns. In Part I, concrete discussion of nonequilibrium patterns is based on and formed from study of reduced equations obtained phenomenologically through very simple considerations. The foundations of reduction theory are presented in Chap. 5.

1. A Representative Example of Dissipative Structure

The examples of dissipative structures in nature are too numerous to count. In most cases, in addition to the complexity arising from the nonlinear nature of the governing laws, there are various external factors whose combined influence causes the behavior to become even more complex. For this reason, it is very important in an attempt to understand dissipative behavior that we carefully select a few model examples from the multitude and concentrate our study on such examples by employing well-controlled conditions. In so doing, it becomes possible to apprehend the general features of dissipative structures that transcend the specific features of individual systems. In this sense, the examples discussed in this chapter, Bénard convection and the Belousov–Zhabotinskii reaction, are representative, and it is perhaps not inaccurate to state that the development seen in the research of nonlinear nonequilibrium systems in the context of chaos and dissipative structures has been made largely through the study of these two examples. In this chapter we discuss the prominent characteristics of the method used to study nonlinear nonequilibrium systems through the application of this method to these two specific problems. In so doing, we make several preliminary observations in preparation for later chapters.

1.1 Bénard Convection

Interest in convective flow is very old, dating to the time of Archimedes. However, this type of flow was first studied in a well-controlled experimental situation, as the subject of scientific inquiry, by H. Bénard in 1900. Applying heat from below to a fluid contained within a horizontally oriented container, Bénard induced convective motion in the fluid. Most of the convection experiments performed even presently for the purpose of studying dissipative structures and chaos use conditions that are essentially the same as those used in this original study. The type of phenomenon that Bénard observed is known as Rayleigh–Bénard convection, named in honor of Bénard and Lord Rayleigh, who first theoretically described the mechanism responsible for this behavior.

In a typical convection experiment, a thin layer of fluid is contained between two horizontally oriented plates of good thermal conductivity. The temperatures of the top and bottom plates, T_t and T_b ($T_b > T_t$) are maintained at constant values. The difference between these two temperatures, $\Delta T = T_b - T_t$, is the most important quantity in determining the system's behavior, but from a physics point of view, the dimensionless parameter R obtained from ΔT is more meaningful:

$$R = \frac{\alpha g d^3}{\kappa \nu} \Delta T \,. \tag{1.1}$$

Here, α is the coefficient of thermal expansion, g is the acceleration due to gravity, d is the thickness of the fluid layer, κ is the coefficient of thermal diffusion, and ν is the coefficient of kinematic viscosity. It is convenient to define the additional parameter $P = \nu/\kappa$, the so-called Prandtl number. The value of the Prandtl number is essentially determined by the type of fluid in question and, unlike the Rayleigh number, cannot be changed freely. The set of parameters appearing in the fluid mechanical equation used to describe convection that we introduce shortly is in essence equivalent to the Rayleigh and Prandtl numbers.

At $R = 0$ the system is in thermal equilibrium, while for some finite R it constitutes a nonequilibrium open system. However, below some critical value of R, R_c, the fluid remains quiescent. In this case, the temperature behaves as $T = T_0(z) = T_b - \beta z$ ($\beta = \Delta T/d$), possessing a constant gradient in the vertical direction, and heat is conducted steadily, in accordance with Fourier's Law. However, with such a temperature dependence, the fluid in the upper portion of the system will be heavier than that in the lower portion (as governed by α). This is an easily destabilized distribution. The reason why at sufficiently small values of R, the system can exist in such a state without flow developing to carry the system into a state of lower potential energy follows from the fact that, for such values of R, the rate at which the potential energy can be decreased due to such flow is less than the rate at which kinetic energy would be dissipated through viscosity as a result of this flow. However, when R exceeds R_c this is no longer the case, and it becomes energetically favorable for the flow to develop. Here, the fluid which has undergone expansion due to the heating rises. In so doing, it cools, thereby contracting, and thus descends. In this way, a continuous circulation develops.

The above-described convection appears from the time of its outbreak as a steady flow. As schematically depicted in Fig. 1.1, the fluid forms a flow pattern consisting of a set of regularly aligned rolls. This formation is referred to as a *roll pattern.* The flow directions of adjacent rolls are oppositely directed, and the distance between their centers is on the order of d. Usually the rolls are aligned parallel to the shorter of the container walls. This type of convective pattern is well-known as a typical example of dissipative structure. If R is increased further, these steady rolls become unstable. However, the resulting changes in the form of the convective pattern depend strongly on the shape of the container and the Prandtl number. A particularly important shape parameter is the aspect ratio $\Gamma = L/d$, where L is a characteristic length, representing the extent of the system in the horizontal plane. The quantity L can be thought of as the longer horizontal dimension of the container. In the case $L \sim O(d)$, following the above discussion, it would appear that at most a single pair of rolls could appear. Thus in this case, when R is not too large, only the relatively small number of modes possessing wavelengths on the order of L and satisfying the boundary conditions represent meaningful mechanical degrees of freedom. Because the large number of degrees of freedom corresponding to spatial fluctuations on a finer scale are quickly damped, these modes cannot play a role in the actual dynamics exhibited by the system. For this reason, Bénard convection systems possessing small aspect ratios are well-suited for the study of chaos in dynamical systems with few degrees of freedom, and they have acted as important subjects of study in this field (see Part II).

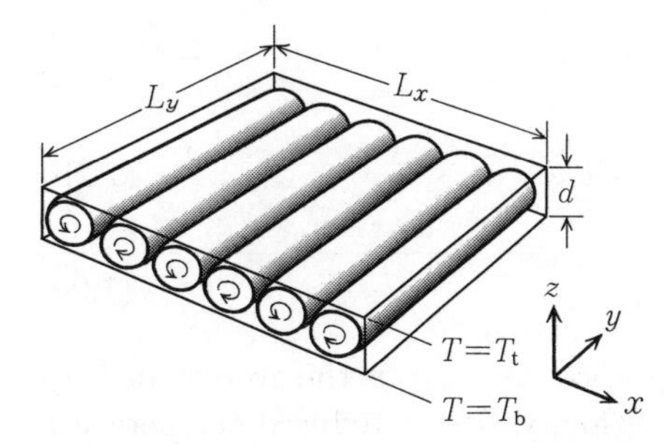

Fig. 1.1. The roll pattern of Bénard convection

In the study of chaos, information concerning details of spatial flow patterns is lost in the idealized mathematical treatment we employ, and our study becomes cast in terms of the time development described by the abstract model used to represent the system in question. In particular, interest becomes focused on the asymptotic behavior of phase space trajectories, and qualitative phenomena, such as the successive changes seen in the nature of attractors in response to changes applied to the bifurcation parameter (for example, R in the Bénard convection system) and the order hidden in the complex structure of chaotic orbits, become the objects of investigation.

Considering a different regime, thermal convection systems with large aspect ratios ($\Gamma \gg 1$) have provided important subject matter for the study of nonequilibrium patterns. In this case the effect of the container walls is negligible, and in the idealization of such a system, it can be treated as a nonequilibrium field of infinite two-dimensional extent.

It should be possible to understand the appearance of convection and the change undergone by the resulting flow through the investigation of the solutions to fluid dynamical equations satisfying appropriate boundary conditions. Fortunately, such equations have a firm physical foundation, and a vast amount of experience provides confirmation of their universal validity. However, when considered as models of physical behavior, fluid dynamical equations have the disadvantage of adhering too closely to the specific physical nature of fluids. If we are interested in studying the development of convective flow, it is not necessary to investigate a model that is able to describe more complex behavior. In some cases it is useful to take one step away from such physically realistic models (i.e., models capable of capturing detailed physical intricacies). Properly carrying out such an idealization, we expect that the universal nature of the phenomena displayed by a system under study – those phenomena which do not depend on the details of its actual material composition – will be made clear. This point of view is effective in the study of both chaos and dissipative structures. In what follows, we make clear the meaning of this point of view in the context of the dissipative structures.

The following coupled partial differential equation is fundamental in the description of convection:

$$\mathrm{div}\, \boldsymbol{u} = 0 , \tag{1.2a}$$

$$\frac{\partial \boldsymbol{u}}{\partial t} = \nu \nabla^2 \boldsymbol{u} - \rho_0^{-1} \nabla \pi - (\boldsymbol{u}\nabla)\boldsymbol{u} + \alpha g \hat{z} \theta , \tag{1.2b}$$

$$\frac{\partial \theta}{\partial t} = \kappa \nabla^2 \theta - \boldsymbol{u} \nabla \theta + \beta w . \tag{1.2c}$$

This is referred to as the Boussinesq equation. Here, ρ_0 is the average density of the fluid, $\boldsymbol{u} = (u, v, w)$ is the flow velocity, θ is a reduced temperature $[\theta = T - T_0(z)]$, π is the fluctuation in pressure as measured with respect to a quiescent fluid, and $\hat{z} = (0, 0, 1)$ is the unit vector in the vertical direction. In addition, $-(\boldsymbol{u}\nabla)\boldsymbol{u}$ and $-\boldsymbol{u}\nabla\theta + \beta w$ $(= -\boldsymbol{u}\nabla T)$ are the inertial terms. The nonlinear nature of fluid motion as modeled by this equation arises from these two inertial terms. In addition, $-\rho_0^{-1}\nabla\pi$ represents the acceleration of the fluid resulting from pressure fluctuations, and $g\hat{z}\theta$ corresponds to the buoyant force created by temperature fluctuations. The Boussinesq equation is derived as an approximation of a system consisting of the Navier–Stokes equation and the heat equation, and its validity with regard to the study of many problems has been duly demonstrated. The only approximation used in deriving this equation is that, except for considering the effect of the thermal expansion of the fluid in terms of a buoyant force, the fluid is treated as incompressible.

In order to investigate the linear stability of the quiescent state [the trivial solution of (1.2), $(\boldsymbol{u}, \pi, \theta) \equiv (0, 0, 0)$], we ignore the nonlinear terms in (1.2b) and (1.2c), $\boldsymbol{u}\nabla\boldsymbol{u}$ and $\boldsymbol{u}\nabla\theta$. The pressure term in (1.2b) vanishes when operated on by $\nabla\times$. Applying this operator to (1.2b) twice, we obtain

$$\frac{\partial}{\partial t} \nabla^2 w = \nu \nabla^4 w + \alpha g \left(\frac{\partial^2}{\partial x^2} + \frac{\partial^2}{\partial y^2} \right) \theta . \tag{1.3}$$

Here we have used the equality $\nabla \times (\nabla \times \boldsymbol{A}) = \nabla(\nabla \boldsymbol{A}) - \nabla^2 \boldsymbol{A}$ and the incompressibility condition (1.2a). Linearizing (1.3) and (1.2c), we obtain

$$\partial\theta/\partial t = \kappa\nabla^2\theta + \beta w \,. \tag{1.4}$$

Equations (1.3) and (1.4) constitute a closed system in w and θ. We now make several assumptions with regard to the boundary conditions. First, we assume that the horizontal plane extends to infinity. At the parallel plates ($z = 0, d$), we stipulate that $w = 0$. In addition, we can assume that the temperature of each plate is fixed, and therefore that here $\theta = 0$. Thus, if there is no slip at the plates, the equality $u = v = 0$ is also satisfied for these values of z. Then, following Rayleigh, we assume that the tangential stress vanishes (i.e., $\partial u/\partial z = \partial v/\partial z = 0$). While this is not necessarily a realistic assumption, it results in the simplest case mathematically, and we thus enforce this condition in the following. From the incompressibility condition, we also have $\partial^2 w/\partial z^2 = 0$ in this case. Solutions to (1.3) and (1.4) can be decomposed into normal modes satisfying these boundary conditions. These normal modes can be written as

$$\begin{aligned}(w(x,y,z,t), \theta(x,y,z,t)) \quad &= \quad (w_{m,\boldsymbol{k}}(t), \theta_{m,\boldsymbol{k}}(t)) \sin\frac{m\pi z}{d} \\ &\quad\times \exp(\mathrm{i}(k_x x + k_y y)) \quad (m \text{ integer})\,, \end{aligned} \tag{1.5}$$

with the wave vector $(m, \boldsymbol{k})$. Writing the time dependence of the mode amplitudes $w_{m,\boldsymbol{k}}$ and $\theta_{m,\boldsymbol{k}}$ as $\exp(\lambda t)$, the expansion rate λ obeys the relation

$$\{\tilde{\lambda} + (m\pi)^2 + \tilde{k}^2\}[\tilde{\lambda}\{(m\pi)^2 + \tilde{k}^2\} + P\{(m\pi)^2 + \tilde{k}^2\}^2] - PR\tilde{k}^2 = 0\,, \tag{1.6}$$

where $\tilde{\lambda}$ and $\tilde{k}$ represent the dimensionless expansion rate and the component of the dimensionless wave vector projected on the horizontal plane, respectively. These are defined as $\tilde{\lambda} = d^2\lambda/\kappa$ and $\tilde{k} = |\boldsymbol{k}|d$. If the real part of $\tilde{\lambda}$ is a decreasing function of time for all m and $\tilde{k}$, the stationary state is stable. If this condition is not satisfied, the stationary state is unstable. (Note that $\tilde{\lambda}$ cannot be an imaginary number.)

The limiting value of the Rayleigh number $R_c(m, \tilde{k})$ for the stability of a given mode characterized by $(m, \tilde{k})$ is obtained by stipulating that $\lambda = 0$:

$$R_c(m, \tilde{k}) = \frac{(m^2\pi^2 + \tilde{k}^2)^3}{\tilde{k}^2} \,. \tag{1.7}$$

The corresponding stability diagram is given in Fig. 1.2. As R is increased, the first eigenmode to become unstable is that corresponding to $m = 1$, $k = \pi/\sqrt{2}d \equiv k_c$. For this mode, $R_c(m, \tilde{k}) = 27\pi^4/4 \equiv R_c$. This is referred to as the critical value of the Rayleigh number. If R exceeds this value, the mode corresponding to k_c begins to grow, but because of the isotropic nature of the horizontal plane, the direction of the wave vector of the mode which appears is arbitrary. For this reason, it is possible for multiple modes with different wave vectors to appear simultaneously. However, in most cases some unique orientation characterizing the physically realized wave vector becomes selected. From this point, we will assume this to be the case.

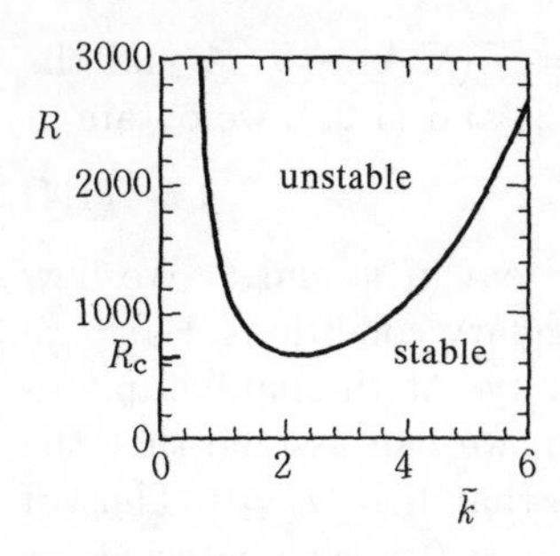

Fig. 1.2. The stability diagram for Bénard convection. The curve represents the Rayleigh number at which the system becomes unstable with respect to a disturbance of dimensionless wave number $\tilde{k}$

Let us choose the direction of the selected wave vector to be along the x-axis. With this assumption, the first pair of modes to grow as R is increased are those represented by $\pm\boldsymbol{k}_c = (\pm k_c, 0)$. Exactly at the critical point, these two modes neither grow nor decay, while all other modes decay with time. Using this fact, within the scope of linear analysis, as $t \to \infty$ we have

$$w(x, y, z, t) = (Z + \bar{Z}) \sin \frac{\pi z}{d}, \qquad Z = W\mathrm{e}^{ik_c x}, \tag{1.8}$$

where Z and $\overline{Z}$ are complex conjugates, and W is an arbitrary complex number. Here, there is a unique relation between θ and w, and θ can be written in a form similar to (1.8):

$$\theta(x, y, z, t) = -\frac{\beta d^2}{\kappa}(\pi^2 + k_c^2 d^2)^{-1}(Z + \bar{Z}) \sin \frac{\pi z}{d}. \tag{1.9}$$

Both u and π can now also be expressed uniquely in terms of w (and therefore in terms of θ), and thus all of the functions describing the state of the system can be represented by the single arbitrary parameter W. Referring to (1.2a) and (1.2b), in the present context, it is clear that u and π are proportional to $(Z - \overline{Z})\cos(\pi z/d)$ and $(Z + \overline{Z})\cos(\pi z/d)$, respectively. Of course, we also have $v = 0$ by the above assumption. The solution to the linearized system corresponding to the critical mode (1.8) is called a *neutral solution.* This solution possesses a one-dimensional periodic structure in the horizontal plane. Then, also considering the z dependence of u and w, we see that the neutral solution can be thought of as representing the roll pattern discussed above, where a single pair of rolls corresponds to one period of this solution's one-dimensional periodic structure along the x-axis.

For the discussion given below, let us consider two types of transformations, $W \to \mathrm{e}^{i\psi}$ and $W \to \overline{W}$. With respect to the neutral solution defined by (1.8), (1.9), and corresponding expressions for u, π and v, the first of these transformations is equivalent to the translation $x \to x + k_c^{-1}\psi$ along the direction of the wave vector of the roll pattern, while the second of these is equivalent to spatial inversion, $x \to -x$, along the same direction. This implies that the neutral solution is invariant under each of the following simultaneous transformations:

(i) $W \to W\exp(\mathrm{i}\psi)$, $x \to x - k_{\mathrm{c}}^{-1}\psi$ (ψ arbitrary) ;

(ii) $W \to \overline{W}$, $x \to -x$.

Let us consider the situation in which R differs slightly from R_{c}. In this case, if we consider W as some appropriate slowly varying function of time, we expect the neutral solution discussed above to represent an approximate solution of the system. Let us define the small parameter μ by the expression $(R - R_{\mathrm{c}})/R_{\mathrm{c}} = \mu$. We refer to μ as the *bifurcation parameter.* We are considering the case $|\mu| \ll 1$, and thus the eigenvalue of the critical mode $\lambda(k_{\mathrm{c}})$ can be assumed to depend linearly on μ; that is, we write

$$\lambda(\boldsymbol{k}_{\mathrm{c}}) = \mu\lambda_1 \qquad (\lambda_1 > 0) . \tag{1.10}$$

In the linear treatment implied by this approximation, we have $\dot{Z} = \mu\lambda_1 Z$, and thus $\dot{W} = \mu\lambda_1 W$, implying that W grows or decays exponentially with time, depending on the sign of μ.

Let us consider the case $\mu > 0$ – i.e., that of exponential growth of W, as given by its linear evolution equation. Then, if we can realize the appropriate inclusion of nonlinear terms into this equation, we will obtain a meaningful result. We write

$$\partial W/\partial t = \mu\lambda_1 W + F(W, \bar{W}) . \tag{1.11}$$

Here, F represents the nonlinear terms in question. In general, F will contain terms with many different powers of W and $\overline{W}$. However, since $|\mu|$ is small, we expect $|W|$ also to be small, and thus we need only consider low-order terms. Now, the state of this system that is represented approximately by the neutral solution is invariant with respect to transformations of the form (i) and (ii). It follows that the equation of motion for W must be invariant with respect to the same transformations. In general, we could consider the three possible types of second-order terms in F, those proportional to W^2, $\overline{W}^2$ and $|W|^2$, but these terms do not satisfy the condition of invariance with respect to transformation (i). The only third-order term satisfying this condition is $W|W|^2$. Thus the lowest-order nonlinear evolution equation for W must be

$$\mathrm{d}W/\mathrm{d}t = \mu\lambda_1 W - g|W|^2 W , \tag{1.12}$$

where g is a real constant, as implied by invariance of this equation under (ii). The value of this parameter should be expressible in terms of the parameters in the Boussinesq equation but, in the phenomenological treatment given here, it is not possible to obtain this relationship. However, using the reduction theory presented in Chap. 5, such a relationship can be obtained. In this analysis it is found that g is positive.

It is often the case that in the neighborhood of an instability point of a given system, the nonlinear evolution equation for the amplitude of the corresponding critical mode takes a form similar to (1.12). Equations of this type are termed *small-amplitude equations*, or simply *amplitude equations.*

The amplitude equation (1.12) possesses the trivial solution $W = 0$. This represents the quiescent state. In the $\mu < 0$ case, the trivial solution is the unique stationary solution. For $\mu > 0$, the trivial solution is unstable, and if $g > 0$, a stable stationary solution with small but finite amplitude $|W| = (\mu\lambda_1/g)^{1/2}$ exists. This is called the 'nontrivial' or 'bifurcating' solution. The concept of bifurcation is central to the study of dissipative structures, and it will appear repeatedly in later chapters.

In the present problem, g is a positive number, but in some cases it can be negative. When this occurs, the bifurcating solution of (1.12) appears for $\mu < 0$, but it is unstable (while the trivial solution is stable). There is no stable solution for $\mu > 0$. In this case, the description of any state of the physical system appearing in the region $\mu > 0$, be it stationary or otherwise, is beyond that provided by the amplitude equation (1.12). The case in which a bifurcating solution appears in the parameter region in which the trivial solution is unstable is referred to as a *normal bifurcation* or *supercritical bifurcation*, while that in which the bifurcating solution appears together with a stable trivial solution is referred to as an *inverted bifurcation* or *subcritical bifurcation.*

In the discussion given above, the critical mode is considered as the only effective degree of freedom in the neighborhood of the bifurcation point. However, strictly speaking, this assumption is not justified. This is because there is an uncountably infinite number of modes corresponding to wave numbers arbitrarily close to k_c (in other words, with eigenvalues arbitrarily close to 0). The critical eigenvalue is not isolated but, rather, exists as a single point in a continuous distribution of eigenvalues. For this reason, it is more correct to think in terms of each member of a set of roll patterns, each corresponding to some value of k in the neighborhood of k_c, as representing equally important degrees of freedom. This is equivalent to allowing for the possibility that the roll pattern will exhibit some gentle spatial modulation. With this interpretation, our model allows for the description of a much wider range of phenomena.

Even with this broader interpretation concerning wave numbers, it is natural to think of the neutral solution as representing an approximate description of the state of the system. However, in this case it is necessary to think of W as possessing spatial variation on a very long length scale corresponding to the gentle modulation mentioned above. Let us now consider the linear equation of motion for W. In this case, we can assume that the eigenvalue of a mode in the neighborhood of the critical mode will take the form $\lambda(\boldsymbol{k}) = \mu\lambda_1 - D(k - k_c)^2$, with D a positive constant. Then, noting that in the neighborhood of the selected critical wave vector $\boldsymbol{k}_c = (k_c, 0)$, the quantity k is given approximately by $k = k_x + (k_y^2/2k_c)$, we find

$$\lambda(\boldsymbol{k}) = \mu\lambda_1 - D\left(k_x - k_c + \frac{k_y^2}{2k_c}\right)^2 . \tag{1.13}$$

From this, we have the corresponding linearized equation of motion for Z,

$$\frac{\partial Z}{\partial t} = \mu\lambda_1 Z + D\left(\frac{\partial}{\partial x} - \mathrm{i}k_\mathrm{c} - \frac{\mathrm{i}}{2k_\mathrm{c}}\frac{\partial^2}{\partial y^2}\right)^2 Z\,, \tag{1.14}$$

which implies the relation

$$\frac{\partial W}{\partial t} = \mu\lambda_1 W + D\left(\frac{\partial}{\partial x} - \frac{i}{2k_\mathrm{c}}\frac{\partial^2}{\partial y^2}\right)^2 W \tag{1.15}$$

for W. Note that in going from (1.14) to (1.15) we have used the correspondences $\partial/\partial t \leftrightarrow \lambda$, $\partial/\partial \leftrightarrow ik_x$ and $\partial/\partial y \leftrightarrow ik_y$. With regard to nonlinear terms, we can use the same reasoning as that used above when W was considered as a function only of t. In this way we again find it sufficient to include the single term $-g|W|^2W$. We thus obtain

$$\frac{\partial W}{\partial t} = \mu\lambda_1 W + D\left(\frac{\partial}{\partial x} - \frac{i}{2k_\mathrm{c}}\frac{\partial^2}{\partial y^2}\right)^2 W - g|W|^2 W\,. \tag{1.16}$$

This is the Newell–Whitehead (NW) equation, derived by Newell and Whitehead (1969) from the Boussinesq equation.

We leave the investigation of the nature of solutions to amplitude equations similar to (1.16) for Chap. 2. At this time we discuss the characteristics of the various approaches that are effective in studying the dynamics of dissipative structures. As can be seen from the above discussion given in the context of Bénard convection, in general, when we consider a system in the neighborhood of the point at which dissipative structure first appears (the bifurcation point), the system's dynamics are greatly reduced. The equation describing these reduced dynamics allows for a much more detailed analysis than does the original equation of motion. In the neighborhood of the bifurcation point, the only important degrees of freedom are those corresponding to the relatively very small number of modes with respect to which the trivial solution is unstable. This is because in the neighborhood of the bifurcation point, the large number of stable degrees of freedom are adiabatically eliminated. This type of reduction in the number of relevant degrees of freedom was named the *slaving principle* by H. Haken.

Reducing the number of degrees of freedom is important because it results in the simplification of the equation of motion, as discussed above. This simplification can be thought of as the *universalization* of the equation of motion. For example, (1.12) possesses a universal form, meaningful in a much larger context than the convection system for which it was derived. By reinterpreting the meaning of the function W, this equation can be considered as embodying the extraction of the universal structure of a wide variety of systems in the neighborhood of bifurcation points at which instability arises. The partial differential equation (1.16) is such a universal equation, which can in fact be applied to describe a large number of physical situations in which one-dimensional periodic structure appears in a two-dimensional system.

The universal nature of a given equation implies that its analysis will provide information concerning a large number of actual systems simultaneously. Additionally, the fact that the physical foundation of the reduced, universal equation may be insufficient simply implies that, to the extent that we are interested only in universal features of the system, this foundation is not an important issue. The logic concerning the reduction in degrees of freedom and universality finds justification in the success obtained using phenomenological theories in the study of nonlinear dynamics.

It may be thought that the fact that amplitude equations arise only in the neighborhood of bifurcation points is a severe limitation. However, if our interest is limited to universal qualitative characteristics of systems, this is not necessarily the case. If we think of a bifurcation point as a threshold beyond which a system undergoes a qualitative change, we are led to conclude that this qualitative nature does not change between bifurcation points. Therefore, with our interest limited to universal qualitative features, we can achieve our purpose to a great extent by focusing on the neighborhood of bifurcation points.

1.2 The Belousov–Zhabotinskii Reaction

Like Bénard convection, the Belousov–Zhabotinskii (BZ) reaction has played a large role in research concerned with dissipative structures and chaos. This chemical reaction was first studied by the Russian biochemist B.P. Belousov. It is characterized by the continual oscillation of an intermediate compound. Belousov was interested in the metabolic cycle found in living systems, known as the Krebs cycle, or the citric acid cycle. In tests concerning similar cycles in inorganic chemical reactions, he discovered the BZ reaction. At the beginning of 1960, Zhabotinskii became aware of Belousov's discovery, which until that time had gone almost completely unnoticed, and carried out a series of detailed tests, in an attempt to clarify the reaction's structure. The prescription for a particular case in which the reaction is observed is given in Table 1.1. In reference to the reagents appearing in Table 1.1, malonic acid $CH_2(COOH)_2$ is oxidized by sodium bromate $NaBrO_3$, producing bromomalonic acid $BrCH(COOH)_2$. In this process the cerium ions Ce^{3+} and Ce^{4+} act as catalysts. If 0.025 M of ferroin is added as an oxidation-reduction indicator, the change in the concentration of the intermediate compound is reflected by a change in color from red to blue and blue to red.

If we express only the original and final species in the reaction, we have

$$2BrO_3^- + 3CH_2(COOH)_2 + 2H^+ \rightarrow 2BrCH(COOH)_2 + 3CO_2 + 4H_2O. \quad (1.17)$$

However, this overall reaction is composed of a number of constituent reactions, in which a number of intermediate compounds appear. R.J. Field, E. Körös, and R.M. Noyes studied the details of this reaction. They found that

Table 1.1. Quantities of the various participating reagents in a particular case for which the Belousov–Zhabotinskii reaction is realized

150 ml	1M H_2SO_4	
0.175 g	$Ce(NO_3)_6(NH_4)_2$	0.002 M
4.292 g	$CH_2(COOH)_2$	0.28 M
1.415 g	$NaBrO_3$	0.063 M

(1.17) consists of the fundamental steps displayed in Table 1.2 and they evaluated the rate of the reaction corresponding to each step. This set of steps is known as the FKN mechanism.

Table 1.2. Fundamental steps composing the BZ reaction

$HOBr + Br^- + H^+$	$\rightleftharpoons$	$Br_2 + H_2O$	(R1)
$HBrO_2 + Br^- + H^+$	$\rightarrow$	$2HOBr$	(R2)
$BrO_3{}^- + Br^- + 2H^+$	$\rightarrow$	$HBrO_2 + HOBr$	(R3)
$2HBrO_2$	$\rightarrow$	$BrO_3{}^- + HOBr + H^+$	(R4)
$BrO_3{}^- + HBrO_2 + H^+$	$\rightleftharpoons$	$2BrO_2 + H_2O$	(R5)
$BrO_2 + Ce^{3+} + H^+$	$\rightleftharpoons$	$HBrO_2 + Ce^{4+}$	(R6)
$Br_2 + CH_2(COOH)_2$	$\rightarrow$	$BrCH(COOH)_2 + Br^- + H^+$	(R7)
$6Ce^{4+} + CH_2(COOH)_2 + 2H_2O$	$\rightarrow$	$6Ce^{3+} + HCOOH + 2CO_2 + 6H^+$	(R8)
$4Ce^{4+} + BrCH(COOH)_2 + 2H_2O$	$\rightarrow$	$4Ce^{3+} + HCOOH + 2CO_2 + 5H^+ + Br^-$	(R9)
$Br_2 + HCOOH$	$\rightarrow$	$2Br^- + CO_2 + 2H^+$	(R10)

From these ten steps, Field, Körös, and Noyes identified (R2), (R3), (R4), (R10), and (R5) + 2(R6) as being of particular importance. The step (R5) + 2(R6) is given by

$$2Ce^{3+} + BrO_3{}^- + HBrO_2 + 3H^+ \rightarrow 2Ce^{4+} + 2HBrO_2 + H_2O. \qquad (Q)$$

The rate-determining process for (Q) is (R5). To construct a dynamical model based on these five reactions, we represent them as

$$A + Y \rightarrow X + P\,, \qquad (S1)$$

$$X + Y \rightarrow 2P\,, \qquad (S2)$$

$$A + X \rightarrow 2X + 2Z\,, \qquad (S3)$$

$$2X \rightarrow A + P\,, \qquad (S4)$$

$$B + X \rightarrow hY\,, \qquad (S5)$$

where A = BrO_3^-, $BrCH(COOH)_2$, P = HOBr, X = $HBrO_2$, Y = Br^-, and Z = Ce^{4+}. Here, we have used the fact that changes in $[H^+]$ can be effectively ignored. Clearly, (S1), (S2), and (S4) correspond, respectively, to (R3), (R2), and (R4). Furthermore, (S3) represents (Q), where we have taken into account the fact that (R5) is the rate-determining process. Finally, (S5) corresponds to (R9), but there are a number of unclear points with regard to the latter process, and therefore in (S5) we account for the consumption of Ce^{4+} and the production of Br^- in a rough manner, using the unknown chemical quantity h. In what follows, we represent the molar concentration of chemical A by the symbol A, that of B by B, etc. Field et al. assumed that A and B are essentially constant and that the important functions of time are X, Y and Z only. If we consider (S1) through (S5) to be the fundamental reactions, we obtain the following dynamical model consisting of these three functions:

$$\mathrm{d}X/\mathrm{d}t = k_1AY - k_2XY + k_3AX - 2k_4X^2 \,, \tag{1.18a}$$

$$\mathrm{d}Y/\mathrm{d}t = -k_1AY - k_2XY + hk_5BZ \,, \tag{1.18b}$$

$$\mathrm{d}Z/\mathrm{d}t = 2k_3AX - k_5BZ \,. \tag{1.18c}$$

Here k_1 through k_5 are constants representing the effective reaction rates of the corresponding reactions. The value of each of these parameters has been determined experimentally. The above model is known as the ‘Oregonator’. The behavior of this well-known model can be investigated to a certain extent analytically, and despite its relatively simple nature, it describes the BZ reaction quite well, even in a quantitative sense. If we rewrite (1.18a)–(1.18c) using the scaled concentration functions $u = 2k_4X/k_3A$, $v = k_4k_5BZ/(k_3A)^2$ and $w = k_2Y/k_3A$ and the scaled time $t' = k_5Bt$, and then replace t' by t, we obtain the concise system

$$\epsilon\frac{\mathrm{d}u}{\mathrm{d}t} = aw - uw + u - u^2 \,, \tag{1.19a}$$

$$\mathrm{d}v/\mathrm{d}t = u - v \,, \tag{1.19b}$$

$$\epsilon'\frac{\mathrm{d}w}{\mathrm{d}t} = -aw - uw + bv \,, \tag{1.19c}$$

where $a = 2k_1k_4/k_2k_3$, $b = 2h$, $\epsilon = k_5B/k_3A$, and $\epsilon' = 2k_4k_5B/k_2k_3A$.

In the situation discussed above, it has been assumed that the system is not operated on by any external influence for the time during which the reaction takes place. Thus the system will eventually reach chemical equilibrium. However, in the three-function model appearing above, constant-valued parameters are used in several cases to represent material concentrations which in fact should be slowly varying functions of time, and thus it appears as if a nonequilibrium state is maintained indefinitely. In fact, however, in order to maintain a nonequilibrium state for an indefinitely long period of time, reagents must be added to the system indefinitely. But, as a result of such a process, the concentrations of the various chemicals are very easily made inhomogeneous, and it is therefore necessary to stir the mixture as the reagents are added. Unfortunately, this has the effect of causing the chemical reaction to take place homogeneously throughout the system, precluding the formation of spatial patterns. Thus, in this type of experiment, rather than the spatial patterns of the various chemical concentrations, the time dependence of a spatially uniform state – and in particular the chaotic nature of this dependence – becomes the focus of investigation. We can think of the Oregonator model as describing the dynamics at a single point in the system or, because of the homogeneous nature of the system resulting from the continuous mixing, the state of the entire system. But in order to describe this situation in which there exists some net flux of chemicals participating in the reaction, corresponding terms must be added to the equations. In any case, it is expected that the homogenized BZ system can be modeled using a dynamical system with a small number of degrees of freedom. In recent years both this system and the Bénard convection system have provided important subject matter for the study of chaos in dynamical systems. In this context, a control parameter representing the flux rate of reacting chemicals corresponds to the Rayleigh number in Bénard convection.

In the study of spatial patterns arising in systems undergoing the BZ reaction, maintenance of the nonequilibrium system in the absence of stirring has long been regarded as a very difficult problem. Thus, since stirring causes the destruction of these spatial patterns, research on such systems has been advanced by studying relatively long-lived nonequilibrium transient states appearing in closed systems. However, in recent years an experimental method has been developed employing molecular diffusion which allows for the uniform infusion of reacting chemicals. This development offers great prospects for the study of dissipative structures. This method is discussed further in Chap. 3.

We will discuss the problem of patterns appearing in systems undergoing the BZ reaction in Chap. 3, but let us here show how the three degrees of freedom in the Oregonator (1.19) can be reduced to two. We will then consider the qualitative nature of the behavior of this reduced model. According to the results of experiments, it is known that the four parameters appearing in the Oregonator model satisfy the following relations: $\epsilon' \ll \epsilon \ll 1$, $a \ll 1$, and $b \approx 1$. The first of these relations implies that w is the most rapidly varying and v is the most slowly varying among the three functions of time. For this reason, it should be possible to adiabatically eliminate w. This is realized by replacing the left-hand side of (1.19c) by 0 and replacing w in (1.19a) and (1.19b) by the resulting relation, $w = bv/(u+a)$. This yields the two-component system

$$\epsilon \frac{\mathrm{d}u}{\mathrm{d}t} = u(1-u) - \frac{bv(u-a)}{u+a} \tag{1.20a}$$

$$\mathrm{d}v/\mathrm{d}t = u - v\ . \tag{1.20b}$$

The fundamentals of the theory concerning the reduction method based on this kind of adiabatic elimination are given in Chap. 5.

The two-component system given in (1.20a) and (1.20b) is referred to as the Keener–Tyson (KT) model. In order to understand the qualitative nature of this model, let us consider the two curves Σ and Γ (known as nullclines) drawn in the u–v coordinate system displayed in Fig. 1.3. These correspond to the cases $\dot{u} = 0$ and $\dot{v} = 0$, respectively. When the parameter a is small, these curves take on the forms shown in Fig. 1.3a. Figures 1.3b and 1.3c are schematic representations of Fig. 1.3a, qualitatively depicting the ways in which Γ and Σ cross. (In the following discussion, we refer to the cases represented by Figs. 1.3b and 1.3c as 'case (b)' and 'case (c)', respectively.) The curve Σ is nonmonotonic, possessing a maximum value at the point P and a minimum value at Q. As determined by P and Q, Σ is separated into three branches. Let us call these Σ_-, Σ_0, and Σ_+. In the case depicted by Fig. 1.3b, the intersection point of Σ and Γ (the stationary state **S**) lies to the left of the point at which the minimum value is realized, and in Fig. 1.3c it lies to the right of this value. The parameter ϵ is very small, and therefore u changes much more quickly than v. For this reason, the velocity vector at any point in this plane, except very close to the nullcline, is nearly horizontal. In other words, starting from arbitrary initial conditions, the system first approaches Σ with the value of v remaining nearly constant. It then proceeds to move slowly along Σ. As can easily be seen by inspecting the directions of the velocity vectors, Σ_- and Σ_+ are stable branches, attracting the point representing the system state, while Σ_0 is an unstable branch, repelling this point. The system point rises as it follows along Σ_+ and falls as it follows along Σ_-. When the system point reaches either of the branch endpoints, P or Q, it jumps to the opposite stable branch.

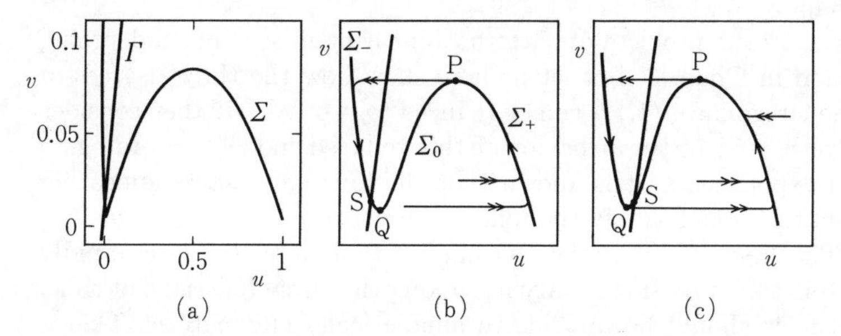

Fig. 1.3. (**a**) Nullclines of the KT model (1.20), describing the BZ reaction. Here, $a = 2 \times 10^{-4}$ and $b = 3$. (**b**) schematic representation of (**a**) for the case in which the stationary point S is to the left of Q. In this case, for sufficiently small ϵ, no limit cycle orbit exists, and the system displays transient excited behavior. (**c**) As the value of b decreases, S appears to the right of Q. Here there exists a stable limit cycle orbit

From the above considerations, the behavior in cases (b) and (c) is clear. In case (c), the stationary state S is unstable, and the system follows a closed trajectory (we call this trajectory C) enclosing S. No other closed trajectory in the neighborhood of C exists. This kind of isolated closed trajectory corresponds to oscillations of chemical concentrations in the BZ reaction. In case (b), the point S is locally stable, but when this point is situated very near Q, even when the initial conditions place the system very close to S, it is possible for the system first to follow along a trajectory closely resembling the limit cycles existing in case (c) before finally monotonically approaching S. This feature, characteristic of this type of system, is called *excitability*. When an excitable system is in a quiescent state, it can be made to suffer a large (but transient) change when subjected to even a small perturbation. This kind of excitability is well-known in the context of biological systems, appearing in certain living membranes. The BZ reaction can easily be made to exhibit excitability, and because this excitability very closely resembles that seen in living systems, it has attracted attention in a wide variety of fields.

The Hodgkin–Huxley equation is a realistic model of excitability in living membranes and has been the subject of a good deal of study. This is a system of ordinary differential equations in four functions whose behavior is similar to that of the Oregonator. The following simplification of the Hodgkin–Huxley equation is known as the FitzHugh–Nagumo (FN) equation:

$$\epsilon \frac{\mathrm{d}u}{\mathrm{d}t} = u(1-u)(u-a) - v \,, \tag{1.21a}$$

$$\mathrm{d}v/\mathrm{d}t = u \,. \tag{1.21b}$$

The nullclines of the FN equation are very similar to those shown in Fig. 1.3. For this reason, the arguments put forward concerning the qualitative nature of the KT system in the $\epsilon \to 0$ limit can also be made for the FN model. This model also displays limit cycle oscillations and excitable behavior for certain values of a. From the above discussion, it should be clear that a reduction of dynamics due to the elimination of nonessential degrees of freedom should be quite effective in the study of the BZ reaction, as it is in the case of Bénard convection.

Let us now consider a BZ system in the absence of stirring that displays some nontrivial pattern. (We can think of a nonequilibrium state which is maintained by the homogeneous infusion of reagents or one which is decaying slowly to equilibrium.) The spatial pattern appearing in such a homogeneous BZ reaction can be considered as simply arising from the interplay between the reaction itself and diffusion. Thus in the model used to describe such a system, a diffusive term should be added to the evolution equation of each component. The resulting system of coupled partial differential equations is referred to as a reaction–diffusion system. For example, the reaction–diffusion system corresponding to (1.20a) and (1.20b) is

$$\epsilon \frac{\partial u}{\partial t} = u(1-u) - \frac{bv(u-a)}{u+a} + \epsilon D \nabla^2 u \tag{1.22a}$$

$$\partial v / \partial t = u - v + D' \nabla^2 v \,. \tag{1.22b}$$

The reaction–diffusion system corresponding to the FN model (in a single spatial dimension) in the case in which the diffusion constant corresponding to the variable v vanishes is known as the *nerve conduction equation*, since it has been found to model the propagation of the action potential along a nerve fiber.

We can think of systems like (1.22) as being composed of a set of spatially distributed local elements displaying limit cycles and excitable behavior (often called *active functional elements*) which are connected by diffusion. In fact, this manner of thinking can be applied to a much wider class of systems than that represented by reaction–diffusion equations. For example, the same reasoning applies to the reduced equations describing Bénard convection. Note that (1.12) describes individual planar, rotor-like active functional elements, while the NW equation (1.16) describes a system in which such individual elements, distributed in two dimensions, are coupled via diffusion (although not simple diffusion in this case).

We have described in this chapter the starting point for the study of dissipative structures. With the fundamental concept of eliminating all nonessential complexity and complicated effects, we regard a spatially extended system as consisting of an idealized homogeneous field of active functional elements extending to infinity. As we will see, this manner of thinking is extremely effective.

2. Amplitude Equations and Their Applications

In the preceding chapter, we saw how in the neighborhood of a bifurcation point at which a new spatial pattern arises, the equation describing the system in question can be reduced to a relatively simple form we refer to as an amplitude equation. Then, as a representative example of this reduction procedure, we derived the Newell–Whitehead (NW) equation using phenomenological considerations. In this chapter, we investigate how different types of amplitude equations are derived to describe a variety of physical conditions. We then study the properties of these equations.

We first consider the NW equation and show how, through a certain mechanism, its fundamental stationary solution, consisting of a one-dimensional periodic pattern, becomes unstable. Next we consider the Ginzburg–Pitaevskii equation, a simple equation which applies to certain types of systems consisting of anisotropic media (for example, liquid crystal systems exhibiting convective behavior), unlike the NW equation, which arises in situations involving isotropic media. After sketching the derivation of this equation, we discuss the structure and motion of topological defects that appear in its solutions. Then, on the basis of phenomenological arguments, we derive the complex Ginzburg–Landau equation as one example of an amplitude equation describing the appearance of limit cycle oscillations and discuss its properties. In particular, we investigate uniform oscillating solutions and planar traveling waves and also touch upon the topics of rotating spiral waves, hole solutions, and spatio-temporal intermittency.

2.1 The Newell–Whitehead Equation and the Stability of Periodic Solutions

The NW equation was originally derived in the context of Bénard convection. However, this equation possesses a universality which transcends the realm of phenomena related to fluid systems. In fact it can be used to describe many types of systems in which one-dimensional structure appears. For example, as we will see in Chap. 5, the same kind of equation arises in the study of reaction–diffusion systems. Following a similar line of reasoning, we would like to stress that the scope of the discussion given below concerning the destabilization of periodic structure is not limited to convection phenomena. However, experimentally at least, most research with regard to the dynamics of periodic patterns has been performed in the context of thermal convection structures. For this reason, the following discussion will be cast in terms of this type of system.

Appropriately scaling the amplitude W and the spatial coordinates in the NW equation (1.16), we obtain

$$\frac{\partial W}{\partial t} = \mu W - |W|^2 W + \left(\frac{\partial}{\partial x} - \frac{\mathrm{i}}{2k_\mathrm{c}} \frac{\partial^2}{\partial y^2} \right)^2 W \,. \tag{2.1}$$

Note that the quantity k_c appearing here is the critical wave number in this new, scaled system. It is possible, through further scaling, to put the equation in an even simpler form in which μW is replaced by $\pm W$, but since we wish to trace the effect of this term, we leave the parameter μ in the equation. The NW equation can be derived from the potential

$$\Psi = \iint \mathrm{d}x\mathrm{d}y \left(-\mu |W|^2 + \tfrac{1}{2}|W|^4 + \left| \frac{\partial W}{\partial x} - \frac{\mathrm{i}}{2k_\mathrm{c}} \frac{\partial^2 W}{\partial y^2} \right|^2 \right) . \tag{2.2}$$

by use of the functional derivative

$$\frac{\partial W}{\partial t} = -\frac{\delta \Psi}{\delta \bar{W}} \,. \tag{2.3}$$

The fact that the value of Ψ decreases monotonically as W evolves in time is demonstrated by the relations

$$\begin{aligned} \frac{\mathrm{d}\Psi}{\mathrm{d}t} &= \iint \mathrm{d}x\mathrm{d}y \left(\frac{\delta \Psi}{\delta W} \frac{\partial W}{\partial t} + \frac{\delta \Psi}{\delta \bar{W}} \frac{\partial \bar{W}}{\partial t} \right) \\ &= -2 \iint \mathrm{d}x\mathrm{d}y \left| \frac{\partial \Psi}{\partial W} \right|^2 \leq 0 \,. \end{aligned} \tag{2.4}$$

The system evolves toward the minimum of Ψ, and if at some time it were to reach this minimum, it would remain there forever. Systems such as this which possess potentials are simple in the sense that they do not display oscillatory or chaotic behavior. However, in many cases these systems have a very large number of metastable states and, in such cases, their behavior is indeed not simple. Interest in pure relaxational systems often stems from the multiple forms of stability that they display.

In the $\mu > 0$ supercritical region, (2.1) possesses a family of stationary solutions, denoted by W_k, in the form

$$W_k = A_k \mathrm{e}^{\mathrm{i}(\delta k x + \psi_0)} \,, \tag{2.5a}$$

$$A_k = \sqrt{\mu - (\delta k)^2} \,, \tag{2.5b}$$

where ψ_0 is the arbitrary phase parameter arising from the translational symmetry of the NW equation. The second arbitrary parameter, δk, is very important in the subsequent discussion. In the case of Bénard convection, the above solution represents a stationary roll pattern with wave number shifted slightly from the the critical value k_c. From (2.4b), we have $k = k_c + \delta k$. It is not the case that all members of this family of periodic solutions are stable. In order to investigate the linear stability of W_k, let us write $W = (A_k + \rho)\exp[\mathrm{i}(\delta k \cdot x + \psi)]$ and derive the linearized equation of motion for the amplitude perturbation ρ and the phase perturbation ψ. The equations are given as

$$\begin{aligned}\frac{\partial \rho}{\partial t} &= \left\{-2\left[\mu - (\delta k)^2\right] + \frac{\partial^2}{\partial x^2} + \frac{\delta k}{k_c}\frac{\partial^2}{\partial y^2} - \frac{1}{(2k_c)^2}\frac{\partial^4}{\partial y^4}\right\}\rho \\ &\quad + A_k\left(-2\delta k\frac{\partial}{\partial x} + \frac{1}{k_c}\frac{\partial^3}{\partial x \partial y^2}\right)\psi\,, \end{aligned} \tag{2.6a}$$

$$\begin{aligned}\frac{\partial \psi}{\partial t} &= A_k^{-1}\left(2\delta k\frac{\partial}{\partial x} - \frac{1}{k_c}\frac{\partial^3}{\partial x \partial y^2}\right)\rho \\ &\quad + \left(\frac{\partial^2}{\partial x^2} + \frac{\delta k}{k_c}\frac{\partial^2}{\partial y^2} - \frac{1}{(2k_c)^2}\frac{\partial^4}{\partial y^4}\right)\psi. \end{aligned} \tag{2.6b}$$

In order to determine the stability of a disturbance with wave vector $\boldsymbol{q} = (q_x, q_y)$, it is sufficient to assume the form $\exp[\mathrm{i}(q_x x + q_y y) + \lambda t]$ for both ρ and ψ, substitute this form into the above equation, solve the quadratic equation for the growth rate of λ, and check the sign of the real part of the value so obtained. Following this procedure, we obtain the two roots λ_+ and λ_- given by

$$\begin{aligned}\lambda_\pm &= -\left\{\mu - (\delta k)^2 + q_x^2 + \frac{\delta k}{k_c}q_y^2 + \frac{q_y^4}{(2k_c)^2}\right\} \\ &\quad \pm \left[\{\mu - (\delta k)^2\}^2 + 4q_x^2\left(\delta k + \frac{2}{k_c}q_y^2\right)^2\right]^{1/2}. \end{aligned} \tag{2.7}$$

When $q_x = q_y = 0$, we have $\lambda_+ = 0$ and $\lambda_- < 0$, and therefore we see that the branch $\lambda_+(q)$ represents the spectral eigenvalue corresponding to the phase perturbation, and $\lambda_-(q)$ is that corresponding to the amplitude perturbation. For sufficiently small q_x and q_y, λ_+ can become positive; that is, there exists the possibility of instability with respect to sufficiently long-wavelength phase perturbations. To see this, let us expand λ_+ in q_x and q_y:

$$\begin{aligned}\lambda_+ &= \frac{\mu - 3(\delta k)^2}{\mu - (\delta k)^2}q_x^2 - \frac{\delta k}{k_c}q_y^2 - \frac{2(\delta k)^4}{\{\mu - (\delta k)^2\}^3}q_x^4 - \frac{1}{(2k_c)^2}q_y^4 \\ &\quad + \frac{8\delta k}{\{\mu - (\delta k)^2\}k_c}q_x^2 q_y^2 + \cdots\,. \end{aligned} \tag{2.8}$$

Long-wavelength phase instability occurs when the coefficient of either the q_x^2 or the q_y^2 term is positive.

The instability arising in the case in which the coefficient of the q_x^2 term is positive is known as the *Eckhaus instability*, and that in which the q_y^2 term is positive is known as the *zig-zag instability*. The former occurs when the inequality

$$|\delta k| > \sqrt{\frac{\mu}{3}} \tag{2.9}$$

is satisfied, while the latter occurs when

$$\delta k < 0 \, . \tag{2.10}$$

In the context of Bénard convection, for the Eckhaus instability, the uniform rolls aligned along the y-axis remain essentially straight and parallel, while the system becomes unstable with respect to the growth of certain perturbations in the roll spacing. In the case of the zig-zag instability, perturbations disturbing the uniformity of the spacing between rolls do not arise, but instead the system becomes unstable with respect to the growth of perturbations which bend the rolls, destroying their uniform alignment. Figure 2.1 displays the region of existence of the solution W_k in the δk–μ plane, divided into regions in which this solution is either stable or unstable. For a system that is located in the region of the Eckhaus instability or the zig-zag instability, it is in general difficult to determine the way in which a disturbance of the pattern develops and into what state the system eventually evolves. This point will be discussed in the following section and in Sect. 4.1.

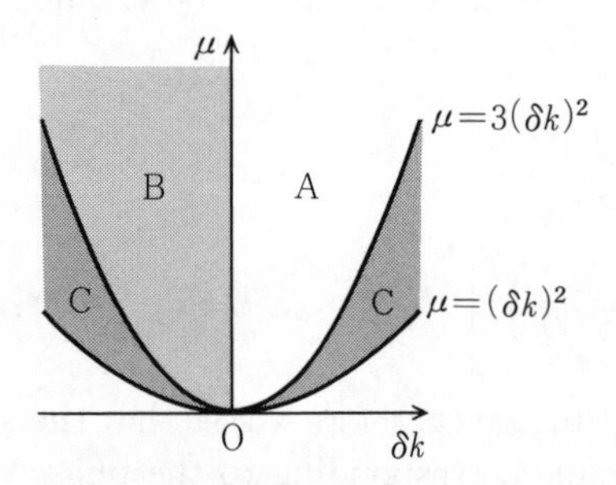

Fig. 2.1. The region of existence of the plane wave solution to the Newell–Whitehead equation, W_k. This region is divided into subregions determined by the stability of this solution. In region A the solution is stable, while in regions B and C it is unstable with respect to the zig-zag and Eckhaus instabilities, respectively. Here, $\mu = (R - R_c)/R_c$ is the bifurcation parameter, and δk represents the deviation from the critical wave number k_c. The plane wave solution exists when $\mu > (\delta k)^2$

In actual Bénard convection, in addition to the above-mentioned instabilities, the so-called *oscillatory instability* and *skewed varicose instability* have been observed. Their existence has also been demonstrated in an exact linear stability analysis performed on the Boussinesq equation. [Busse and Clever (1979)]. The oscillatory instability, as the zig-zag instability, represents an instability of the system with respect to the growth of long-wavelength phase disturbances which bend the roll axis, but unlike the zig-zag instability, in this case the eigenvalue λ is a complex number whose real part is positive. This type of instability leads to a wave motion of the rolls which propagates along the roll axis. This is easily observed in systems with small Prandtl number. The skewed varicose instability is similar to both the Eckhaus and zig-zag instabilities in the sense that it corresponds to a dependence of the real part of λ causing it to change sign in response to long-wavelength phase perturbations, but it differs from these two instabilities in that it is an instability arising in the system when $|\boldsymbol{q}| \to 0$ while q_x/q_y remains finite.

The fact that the oscillation and skewed varicose instabilities can be observed for values of R very close to R_c suggests that, in the neighborhood of R_c, the NW equation is not necessarily a good model of Bénard convection. In the derivation of this equation, it was assumed that since the dynamics of modes with wave numbers in the neighborhood of K_c have extremely long time scales, all other degrees of freedom adiabatically follow these modes. However, we ignored some important modes in this derivation: those corresponding to long-wavelength fluid flow in the horizontal plane. The dynamics of these modes, like those of the critical mode, are characterized by a very long time scale. This is due to the fact that the Navier–Stokes equation is Galilean invariant, and therefore if the aspect ratio is sufficiently large and the effect of the tangential stress at the upper and lower horizontal walls is sufficiently small, long-wavelength components of the fluid velocity in the horizontal plane are approximately conserved quantities. In fact, we cannot ignore these long time scale degrees of freedom, nor can we assume that they are adiabatically eliminated. Like the critical mode, these modes must be treated with more careful consideration.

Based on the above form of reasoning, E.D. Siggia and A. Zipperius (1981a) derived a more generally valid version of the NW equation. From the analysis of this equation, many aspects concerning the destabilization of the stationary roll structures and the spontaneous appearance of various disturbances have come to be more fully understood. For example, the oscillatory instability can be observed in the new amplitude equation, reflecting the fact that there is no potential Ψ for the amended form of the NW equation. Amplitude equations of this type, for which there exists no potential, are thought to be necessary to properly describe spatio-temporal chaos and turbulence. However, in attempting to precisely describe actual convection phenomena, the model in question, which in its earlier form transcended the context of fluid systems, comes to lose its universal nature. We often encounter a dilemma of this type in the study of nonequilibrium dissipative structures. In such a situation, it is important to keep in mind these reciprocal tendencies arising in the modeling of physical phenomena, being aware of what is being kept and what is being discarded in any given case.

2.2 Anisotropic Fluids and the Ginzburg–Pitaevskii Equation

Bénard convection is a type of thermal convection that occurs in isotropic fluid systems. As an example of a type of anisotropic fluid, we can consider liquid crystals, which under nonequilibrium conditions can exhibit an electrohydrodynamic instability. The study of these systems has been a very active field of research in recent years.

Liquid crystals are fluids composed of rod-like organic molecules. They can be classified into several groups (nematic, smectic, cholesteric, etc.) which differ with respect to the thermodynamically stable molecular arrangements by which they are characterized. Nematic liquid crystals, which do not exhibit a reduced symmetry state other than that in which all molecules are aligned in the same direction, are used in convection experiments. An often used liquid crystal is MMBA [N-(p-methoxybenzyliden)-p-butylaniline]. In a typical experiment, the liquid crystal material is situated in the space between two parallel plates, across which an electric potential is created. The molecules are then aligned so that each is oriented along some given direction parallel to the plates. (The orientation of a molecule in a liquid crystal system is referred to as its *director*.) Assuming that these plates define the x–y plane, we take the x direction to be that defined by the molecules' directors. Then, we consider the case in which an alternating electric field is applied along the z-axis. Under these conditions, for an electric field with a fixed frequency of oscillation f, and for an amplitude V greater than some critical value V_c (which depends on f), convective behavior will be displayed.

In order for convection to appear as a result of an electro-hydrodynamic instability, there must exist a mechanism able to increase the amplitude of fluctuations in the molecular orientation. Such a mechanism does in fact exist. Fluctuations in the molecular orientation induce the flow of electric current in the x–y plane and in this way create an inhomogeneity in the distribution of electric charge. As a result of this inhomogeneity, an electric force acts on the electric charge, taking the form of a rotational force which tends to promote fluctuations in the molecular orientation. Furthermore, owing to the anisotropic dielectric properties of the system, the molecules attempt to align themselves along a direction perpendicular to the electric field but, because of the presence of the additional electric field created by the inhomogeneous distribution of electric charge, the direction of the total electric field is altered, resulting in an additional rotational force acting to further change the distribution of molecular orientations. In this way, fluctuations in the molecular orientation tend to become enhanced. However, this mechanism must compete with the tendency for such fluctuations to be damped by the material's elasticity with respect to bending.

The description given in the previous paragraph is a valid interpretation of the destabilizing mechanism acting in a system subject to a D.C. field. In actual practice, however, for technical reasons, an A.C. field is used. Despite this fact, if the time scale f^{-1} of the polarity inversion of the electric field is much larger than the relaxation time of the charge distribution but much shorter than the relaxation time of the molecular orientations, during such a duration the charge distribution will relax fully, while the molecular orientations will change very little. In this situation, the destabilizing mechanism acting in the A.C. case is essentially the same as that acting in the D.C. case. In the following discussion, we assume that the frequency of the A.C. field satisfies these conditions.

The convection pattern appearing in this system consists of periodic structure, just as in the case of Bénard convection. For a liquid crystal convection system, this pattern is known as a *Williams domain*. Here, there are two possible situations with regard to the alignment of the rolls, one in which they are parallel to the y-axis ('normal rolls') and one in which they are rotated by some finite angle from this axis ('oblique rolls'). The alignment that actually appears depends on the value of f. For values of f less than some characteristic value f_L, oblique rolls are realized, while for values greater than f_L, normal rolls are realized (see Fig. 1.1 for definition of axes).

In the study of nonequilibrium pattern formation, the electro-hydrodynamic convection appearing in liquid crystals has several advantages over Bénard convection. First, the precision afforded by the use of electricity as the controlling physical quantity is much higher than that afforded by the use of heat. For this reason, in the former case the study of systems with large aspect ratios can be easily carried out. In addition, both the strength of the electric field, which corresponds to the Rayleigh number, and the frequency of the A.C. field, which corresponds to the Prandtl number, can be changed continuously, and therefore a variety of patterns can be observed in a single system. However, while liquid crystal systems have these clear experimental advantages, there is a complication inherent in their theoretical investigation: the equation governing the anisotropic fluid is itself quite complicated, and, in addition, this equation is coupled to Maxwell's equations. For this reason, the linear stability analysis of the quiescent state involves an extremely cumbersome calculation. However, as discussed in Chap. 1, for the study of nonequilibrium open systems, this is not a fatal problem: the new and novel nature of the theoretical framework we employ, as a field of physics, is expressed by the nature of its methodology, as set forth in Chap. 1, in which a given system is considered from a point of view which transcends the complexity encountered in the consideration of its material composition.

Given that convection appears in liquid crystal systems as a result of an electro-hydrodynamic destabilization, it is expected that a NW-type equation can provide an approximate description of such a system's dynamics in the region near the emergence of convection. The only apparent fundamental difference between the liquid crystal system and the type of system for which the NW equation was derived is the anisotropic nature of the fluid in the former. With this in mind, we will phenomenologically derive an amplitude equation with a form appropriate for the description of such a fluid. In contrast to the isotropic case, here there is no rotational arbitrariness associated with the direction of the critical wave vector k_c, and (1.13) should thus be replaced by an expression of the form

$$\lambda(\boldsymbol{k}) = \mu - D_1(k_x - k_{cx})^2 - D_2(k_y - k_{cy})^2 \ . \tag{2.11}$$

Here, as in the situation considered above, μ represents an appropriately defined bifurcation parameter. In the present context, $\mu \propto (V - V_c)/V_c$. Then, for the case of normal rolls, $k_{cx} \neq 0$ and $k_{cy} = 0$, while for oblique rolls, $k_{cx}, k_{cy} \neq 0$.

In the neighborhood of the critical point, we should be able to express various physical quantities and the deviations of these quantities from their stationary state values using a linear combination of the complex amplitude $Z = W \exp[\mathrm{i}(k_{cx}x + k_{cy})]$ and its complex conjugate $\overline{Z}$. The linearized equation of motion for Z corresponding to (2.10) is

$$\frac{\partial Z}{\partial t} = \mu Z + D_1 \left(\frac{\partial}{\partial x} - \mathrm{i}k_{cx}\right)^2 Z + D_2 \left(\frac{\partial}{\partial y} - \mathrm{i}k_{cy}\right)^2 Z \ , \tag{2.12}$$

and thus for W we have

$$\frac{\partial W}{\partial t} = \mu W + D_1 \frac{\partial^2 W}{\partial x^2} + D_2 \frac{\partial^2 W}{\partial y^2} \ . \tag{2.13}$$

Precisely as in the case of Bénard convection, considering the symmetry of the system along with the smallness of W, the most important nonlinear term with which this equation should be supplemented is $\sim |W|^2 W$. Adding the term $-g|W|^2 W$ to (2.12) and scaling W, x, y and t as $W \to \sqrt{\mu/g}W, x \to \sqrt{D_1/\mu}x, y \to \sqrt{D_2/\mu}$, and $t \to \mu^{-1}t$, considering the supercritical region, $\mu > 0$, we obtain

$$\partial W/\partial t = W - |W|^2 W + \nabla_\perp^2 W \ . \tag{2.14}$$

Here, $\nabla_\perp^2$ is the two-dimensional Laplacian. Note that we have assumed a normal bifurcation ($g > 0$) here. Equation (2.13) is called the *Ginzburg–Pitaevskii (GP) equation* and is essentially the same equation as that used in the study of superfluid behavior. It is interesting that the equation found here to describe an anisotropic fluid has been reduced to a form that is more simple than the NW equation, used to describe an isotropic fluid. As with the NW equation, there exists a potential Ψ corresponding to (2.13):

$$\Psi = \iint \mathrm{d}x\mathrm{d}y \left(-|W|^2 + \tfrac{1}{2}|W|^4 + \left|\tfrac{\partial W}{\partial x}\right|^2 + \left|\tfrac{\partial W}{\partial y}\right|^2\right) \ . \tag{2.15}$$

2.3 Topological Defects and Their Motion

In the NW equation, or alternatively the GP equation, let us write the complex amplitude W as $W = A\exp(i\psi)$ with the real amplitude A and phase ψ. For the NW equation, (2.4b) applies to A for the case of steady rolls. In this case, A assumes a constant value throughout space. When such a periodic pattern is disturbed, A becomes spatially nonuniform. However, if the disturbance is weak, the value of A will everywhere be close to that for the steady-state case, as defined by (2.4b). Thus, in this case we can assume that at no point will A vanish and thereby cause ψ to become undefined.

However, we expect that in response to sufficiently strong disturbances phase singular points will be generated. The cause of such a disturbance may be something inherent in the system arising as a result of the destabilization of the periodic pattern, or it may be due to some externally applied perturbation. An isolated phase singular point has the property that in traversing a path around this point the phase changes by 2π (or, more generally, by an integer multiple of 2π). It then follows that such a point cannot be removed by only a local deformation of the pattern. In order for a phase singular point to be destroyed, it must either be absorbed by a side wall or collide with another phase singular point. For this reason, structural defects containing phase singular points are physically very stable. This type of defect is termed a *topological defect* and represents a very common type of defect appearing at the microscopic level in many systems, including crystals, spin systems, and liquid helium. These defects have a very important effect on the material properties of such systems.

Phase singular points are also often observed in Bénard and liquid crystal convection. As displayed in Fig. 2.2, their appearance results in the local warping of the pattern (or of the lines representing equal phase values – 'isophase lines'). As seen in Fig. 2.2b, where the center of the figure corresponds to the position of the defect, the lower half of the figure possesses one more roll than the upper half, and so the wave number in the lower half is slightly larger. Thus if the defect were to move upward, the average wave number of the system would increase slightly, and if it were to move downward, the value would decrease slightly. In this way, the existence and motion of defects serve to adjust the wave number of the periodic structure. While for a large range of wave numbers the corresponding periodic structure will be locally stable, the global stability of each should differ. It is through the motion of defects that systems evolve toward an increasingly stable structure.

The motion mentioned above, in which the defect moves along the roll axis, is called a 'climb'. Defects can also move across rolls at a variety of angles. This kind of motion is referred to as a 'glide'. As shown in Fig. 2.3, glide motion causes the connection between isophase lines to change and, for this reason, the slopes of these lines become slightly altered. In this way, glide motion adjusts the slope of the periodic pattern, serving to alter the alignment of rolls along increasingly stable directions.

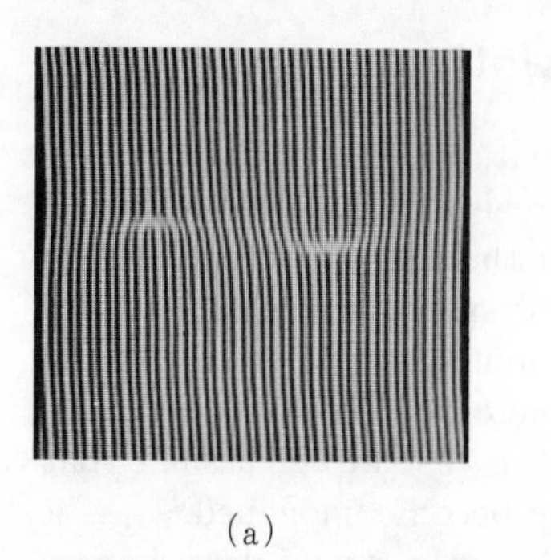

(a)

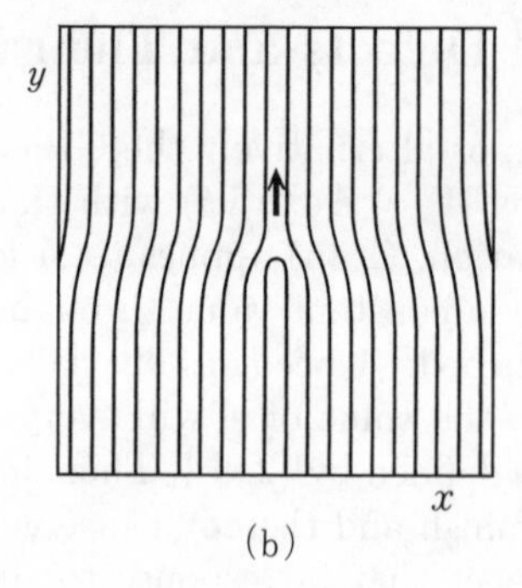

(b)

Fig. 2.2. (**a**) A defect structure embedded in the roll pattern of liquid crystal convection (figure used by permission of Shoichi Kai). (**b**) The climb of a defect. As the defect moves upward (downward), the wave number of the roll pattern increases (decreases) slightly

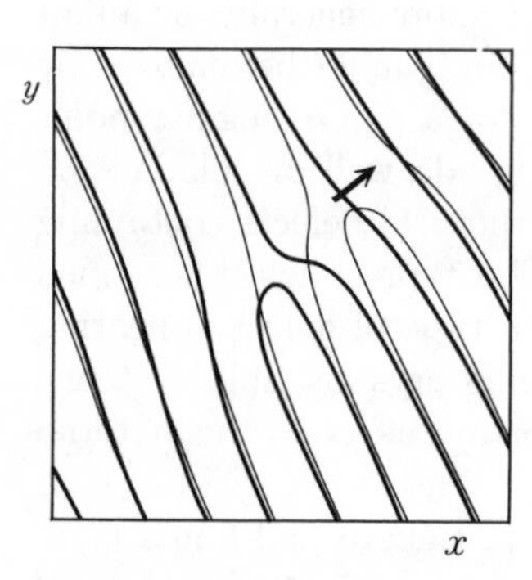

Fig. 2.3. As a consequence of the glide motion of the defect, the connection of isophase lines is changed, and the orientation of the roll pattern is slightly altered

We now discuss the motion of an isolated defect in the context of the GP equation. By reinterpreting the defect solutions of this equation (for which the anisotropic nature of the physical system has become somewhat hidden) in terms of the underlying anisotropic fluid, the manner in which the climb and glide motions discussed above actually appear becomes clear. We could also discuss the defect dynamics exhibited by isotropic fluid systems and the corresponding NW equation. However, the mathematical treatment for this case is more difficult than for the GP equation, and in addition, in this isotropic case, glide motion does not appear. The context of the discussion below is thus limited to the GP equation. To treat defect motion, we recast defect solutions of the GP equation in terms of the original system represented by the amplitude Z. Then we consider the case in which the defect is contained in a system consisting of a normal roll pattern; that is, the case $\boldsymbol{k}_{\mathrm{c}} = (k_{\mathrm{c}}, 0)$. In addition, we assume from this point that k_{c} is positive.

For the GP equation (2.13), there exists a steady-state periodic pattern solution with wave vector $\boldsymbol{q} = (q_x, q_y)$ of the form

$$W = \sqrt{1 - q_x^2 - q_y^2}\,\mathrm{e}^{\mathrm{i}(q_x x + q_y y)} \,. \tag{2.16}$$

In the previous section, we gave a linear stability analysis of a similar solution for the NW equation. If we apply the same type of analysis to (2.15), we easily find that this solution is linearly stable for $q_x^2 + q_y^2 < 1/3$, and for $q_x^2 + q_y^2 > 1/3$ it is subject to an Eckhaus instability. The wave vector of the periodic structure of the original system (i.e., with the field Z) is

$$\boldsymbol{k} = \left(k_{\mathrm{c}} + \sqrt{\frac{\mu}{D_1}} q_x, \sqrt{\frac{\mu}{D_2}} q_y \right) . \tag{2.17}$$

The GP equation also possesses a stationary defect solution corresponding to the inclusion of a defect in the uniform W state. Let us position this defect at the origin of a polar coordinate system (r, θ). We can then write the defect solution as

$$W = A(r)\mathrm{e}^{\mathrm{i}\theta} . \tag{2.18}$$

Substituting this form into the GP equation, we find the equation that A must satisfy:

$$\frac{\mathrm{d}^2 A}{\mathrm{d}r^2} + \frac{1}{r}\frac{\mathrm{d}A}{\mathrm{d}r} - \frac{A}{r^2} + A - A^3 = 0 . \tag{2.19}$$

From this equation, it is apparent that at the origin $A = 0$, and as r approaches ∞, A approaches its supremum, 1. This equation cannot be solved exactly, but it is easy to show that near the origin $A(r) \propto r$, and for large r the value of A increases monotonically, approaching the form $1 - (2r^2)^{-1}$. For the defect solution (2.17), if we traverse a path around the defect at the origin in the counter-clockwise direction, the phase is increased by 2π. Therefore, returning to the original system and the field Z, this situation corresponds to that of Fig. 2.2b, in which there is one extra roll in the lower half of the system. If we make the transformation $\theta \to -\theta$ in (2.13), we obtain a defect of the opposite sign. The defect solution so obtained corresponds to that for the case in which Fig. 2.2b is rotated by 180°.

The type of defect discussed above represents that for which, considering the description in terms of Z, there is a defect in a periodic structure characterized by the critical wave vector $\boldsymbol{k}_{\mathrm{c}}$. More generally, there can of course also exist defects in periodic patterns characterized by the wave vector appearing in (2.16). However, in this case, the defect is not stationary, possessing some nonzero velocity $\boldsymbol{c}(\boldsymbol{q})$ determined by $\boldsymbol{q}$. The goal of the subsequent discussion is to obtain an equation for $\boldsymbol{c}(\boldsymbol{q})$.

First, from symmetry considerations, we can assume that $\boldsymbol{c}$ is perpendicular to $\boldsymbol{q}$. This is due to the fact that the GP equation is isotropic, and therefore periodic structure for this equation with wave vector $\boldsymbol{q}$ is invariant with respect to the transformation $\boldsymbol{q} \to -\boldsymbol{q}$. We expect a defect contained in such a system to possess the same symmetry. Let us now determine a new orthogonal coordinate system ξ–η, obtained from the x–y system through a rotation φ $(-\pi/2 < \varphi \le \pi/2)$ such that the η component of $\boldsymbol{q}$ vanishes. Of course, the ξ component of $\boldsymbol{q}$ is thus given by $q = \pm|\boldsymbol{q}|$ (+ and − corresponding to the cases $q_x > 0$ and $q_x < 0$, respectively).

According to our assumption, the defect moves in the η direction at constant speed. We thus look for a solution of the form

$$W = w(\xi, \eta - ct)\mathrm{e}^{\mathrm{i}q\xi} \,. \tag{2.20}$$

Sufficiently far from the defect, we should have $|w| \to \sqrt{1-q^2}$, and here W should approach a periodic function. However, traversing a path around the defect, the phase of W changes by either 2π or -2π, and thus far from the defect we can assume the following asymptotic form for w:

$$w_\pm \approx \sqrt{1-q^2}\mathrm{e}^{\pm \mathrm{i}\psi(\xi,\eta)} \,. \tag{2.21}$$

Here, following a path once around the defect in the counter-clockwise direction, the phase ψ increases by 2π. Note that if $q = 0$, $\psi = \theta$. When $q > 0$, w_+ and w_- correspond to the pattern in Fig. 2.2b and its image under spatial inversion with respect to the y dimension, respectively. When $q < 0$, this correspondence is reversed.

If we insert (2.19) into the GP equation, we obtain

$$-c\frac{\partial w}{\partial \eta} = \left\{\left(\frac{\partial}{\partial \xi} + \mathrm{i}q\right)^2 + \frac{\partial^2}{\partial \eta^2} + 1\right\} w - |w|^2 w \,. \tag{2.22}$$

In principle, c and w are simultaneously determined by this equation, but it is not feasible for us to solve for these functions directly. However, with the simple considerations described below, we can understand the qualitative behavior of $c(q)$. First, we multiply both sides of (2.21) by $\partial\overline{w}/\partial\eta$ and then integrate each over the entire ξ–η space. Assuming that the surface integral vanishes, we obtain

$$\begin{aligned} c\iint_{-\infty}^{\infty} \mathrm{d}\xi\mathrm{d}\eta \left|\frac{\partial w}{\partial \eta}\right|^2 &= \mathrm{i}q\left\{\iint_{-\infty}^{\infty} D\xi\mathrm{d}\eta\frac{\partial \overline{w}}{\partial \xi}\frac{\partial w}{\partial \eta} - \mathrm{c.c.}\right\} \\ &= \mathrm{i}q\left[\int_{-\infty}^{\infty} \mathrm{d}\xi w\frac{\partial \overline{w}}{\partial \xi}\right]_{\eta=-\infty}^{\infty} \\ &\quad -\mathrm{i}q\left[\int_{-\infty}^{\infty} \mathrm{d}\eta w\frac{\partial \overline{w}}{\partial \eta}\right]_{\xi=-\infty}^{\infty} \,, \end{aligned} \tag{2.23}$$

where we have used integration by parts. The sum of the two terms on the right-hand side of the second equation reduces to a single integral along the closed path $(\infty,\infty) \to (\infty,-\infty) \to (-\infty,-\infty) \to (-\infty,\infty)$, $\oint \mathrm{d}s w\partial\overline{w}/\partial s$ (ignoring the prefactor iq here). Thinking of w as w_+, from the asymptotic form of this function, we find that the value of this integral is $2\pi\mathrm{i}(1-q^2)$. Thus we obtain the following relation for c:

$$\begin{aligned} c(q) &= -\mu_{\mathrm{d}} q(1-q^2) \,, \\ \mu_{\mathrm{d}} &= 2\pi\left\{\iint \mathrm{d}\xi\mathrm{d}\eta\left|\frac{\partial w}{\partial \eta}\right|^2\right\}^{-1} \,. \end{aligned} \tag{2.24}$$

The quantity $\mu_{\rm d}$ can be thought of as the defect's mobility. The direction of the motion of the defect relative to the roll axis depends on $\boldsymbol{q}$. A pure climb results when $q_y = 0$. In this case, the ξ–η and x–y axes coincide, and $q = q_x$. For $q > 0$ – in other words, when the distance between rolls is smaller than the critical wavelength $\lambda_{\rm c} = 2\pi/k_{\rm c}$ – the defect falls along a line parallel to the y-axis. Thus the defect moves in such a way as to remove a single roll from the system, thereby slightly increasing the spacing between rolls. Contrastingly, when $q < 0$, the defect rises, and in so doing acts to increase the number of rolls in the system by 1. As a result, the periodic structure of the system becomes slightly finer. In this manner, the motion of defects acts to push the system toward the wavelength $\lambda_{\rm c}$.

A pure glide occurs in the case $q_x = 0$. Here, $\xi = y$, $\eta = -x$, and $q = q_y$. If $q > 0$, the roll axis is somewhat tilted with respect to the y-axis. Here, the defect moves in the negative η (positive x) direction. In this situation, the defect moves almost normally across the rolls. As is seen in Fig. 2.3, the detachment and reattachment mechanism associated with the glide motion causes the slope of the tilted rolls to increase slightly. Similarly, in the $q < 0$ case, the glide motion causes the slope to decrease. In the case in which both q_x and q_y are nonzero, the defect moves diagonally, causing the wave vector to become slightly closer to the critical wave vector $\boldsymbol{k}_{\rm c}$. A change in the wave vector necessarily implies a large-scale change in the pattern, and this change is naturally facilitated by the motion of topological defects.

In the above discussion, we did not consider the mobility $\mu_{\rm d}$. If the defect speed c is sufficiently small, it may be thought possible to use the form of w for the case of a stationary defect, $w = A(x)\exp(\mathrm{i}\theta)$, in the expression for $\mu_{\rm d}$. However, this is not valid. If this were done, the integrand in the region far from the defect would behave as $r^{-2}\cos 2\theta$, and the integral would thus diverge. In fact, analysis reveals that if we use the asymptotic form (2.20) for a defect solution with finite velocity, the integral is proportional to $\ln|c|$.

Physically, the above discussion implies the following. Defects are not composed of localized, rigid formations, but rather, they and their motion arise out of some soft, deformable structures. If c is small, the corresponding deformation is also small, but even in this case we cannot ignore the deformation. It is not possible for the Fourier modes from which some localized structure is constructed to adiabatically follow the slow motion of a defect, because it is always the case that even slower long-wavelength modes exist. For this reason, structural deformation appears even in regions far from the defect. Here, the r^{-2} dependence of the integrand changes to a decaying exponential, and thus the integral has a finite value. The reason long-wavelength Fourier modes of indefinitely long time scale exist is that the defect is embedded in a neutrally stable 'phase field'. As a result, we cannot treat the defect as a particle or rigid body with no internal degrees of freedom, nor can we treat it as a set of degrees of freedom isolated from the surrounding field. To put this into the context of phase dynamics (discussed in Chap. 4), this is to say that, at least with respect to the present problem, defect dynamics and phase dynamics are inseparable.

2.4 The Amplitude Equation of an Oscillating Field

To this point the discussion in this chapter has been in regard to the amplitude equations arising in connection with the development of spatially ordered structures. However, nonequilibrium dissipative fields are often accompanied by temporal order (that is, limit cycles) as well. This type of behavior results from what is known as a *Hopf bifurcation*. In simple situations, a spatially uniform stationary state becomes unstable, and as a result, a spatially uniform oscillating state is realized. For an oscillatory chemical reaction such as the BZ reaction, if it is possible to continuously alter some appropriate control parameter, the appearance of this reaction can be understood as an example of such bifurcation phenomena. The corresponding amplitude equation is called the *complex Ginzburg–Landau (GL) equation.*

Oscillatory motion often arises in convection as well as other types of fluid motion, but, in general, such systems include some spatial nonuniformity, and their description requires amplitude equations in the form of modified versions of the complex GL equation containing additional terms. While a study of such phenomena is indeed interesting, in this section it is our goal to make the nature of oscillating fields as clear as possible, and we thus consider only the simplest case, in which a spatially uniform oscillating field appears out of a spatially uniform stationary field, ignoring the more complicated case including spatial non-uniformity. Thus reaction–diffusion systems become the archetypal subject of our study.[1] However, while the present investigation involving the complex GL equation is set in this limited physical context, this equation, like the NW and GP equations, has a universality which implies its validity in application to a wide range of physical systems. In fact, through the analysis of the complex GL equation, many general properties exhibited universally by self-excited oscillating fields become clear.

In reaction–diffusion systems, the fundamental field functions represent the concentrations of the various materials present. Here we use $X_1, X_2, X_3, \dots, X_n$ for a system with n such fields. Let us represent the deviation of each of these concentrations from their values in the uniform state by $x_1, x_2, x_3, \dots, x_n$. The discussion given below parallels that given in Chap. 1 for the derivation of the NW equation. The only difference here is that in place of a single mode oscillating as a function of space, we now have a single

[1] We are considering the situation in which a spatially uniform state becomes oscillatorily unstable against spatially uniform perturbation. Obviously, diffusion plays no role in this instability, but we can include a diffusion term in the equation obtained to describe this system. Then, due to the presence of this term, the resulting equation (the complex GL equation) can produce nonuniform patterns. Such patterns are the main focus of our study.

critical mode oscillating as a function of time. The neutrally stable solution in this case, corresponding to (1.8), (1.9), etc., is given by

$$\begin{aligned} x_\nu &= \alpha_\nu Z + \text{c.c.}\,, \\ Z &= W\mathrm{e}^{\mathrm{i}\omega_0 t}, \end{aligned} \tag{2.25}$$

where α_ν is a complex constant. In this manner, at the critical point all concentration functions become interrelated through the complex parameter W. Then, let us assume that in the case in which the system is slightly removed from the critical point this form remains approximately valid. However, in this case we must reinterpret W as a slowly varying function of time. We can assume that in the neighborhood of the critical point the critical mode eigenvalue can be written as $\lambda = \mathrm{i}\omega_0 + \mu\lambda_1$, where λ_1 is a complex number. From this we obtain, in the linear approximation, $\dot{Z} = (\mathrm{i}\omega_0 + \mu\lambda_1)Z$ or, equivalently, $\dot{W} = \mu\lambda_1 W$.

As in the case of equations studied in previous sections, symmetry considerations are used in determining the nonlinear terms that should be added to (2.24). First, from the form of the neutral solution, we immediately see that the transformation $W \to W\exp(\mathrm{i}\psi)$ is equivalent to the time translation $t \to t + \omega_0^1\psi$. That is, the state of the system approximately represented by (2.24) is invariant with respect to the transformation under which simultaneously $W \to W\exp(\mathrm{i}\psi)$ and $t \to t + \omega_0^1\psi$. The equation we seek must be invariant under this transformation. Thus, using considerations similar to those applied in the derivation of (1.12), we obtain an amplitude equation of the form

$$\mathrm{d}W/\mathrm{d}t = \mu\lambda_1 W - g|W|^2 W\,. \tag{2.26}$$

In contrast to the NW equation, here g is in general a complex number. This follows from the fact that in the present case there is no reason to require invariance under the transformation $W \to \overline{W}$.

Now, if we consider W to be a slowly varying function of space, we can express $\lambda(\boldsymbol{k})$ for relevant values of k as a long-wavelength expansion:

$$\lambda(\boldsymbol{k}) = \mathrm{i}\omega_0 + \mu\lambda_1 - Dk^2\,. \tag{2.27}$$

Here, D is in general a complex number, and since we assume that $k = 0$ is the first mode to become unstable upon increasing μ, the real part of D must be positive. Equation (2.26) implies that in considering W as a function of space, we must add the linear term $D\nabla^2 W$ to the amplitude equation. Thus we obtain

$$\partial W/\partial t = \mu\lambda_1 W - g|W|^2 W + D\nabla^2 W \tag{2.28}$$

as the form of the amplitude equation in the present context, generalized to allow for spatial variation. This is the complex Ginzburg–Landau equation.

A solution to the complex GL equation oscillating uniformly in space is realized for $\mathrm{Re}\,g > 0$ (< 0) with $\mu > 0$ (< 0), and is given by

$$W_0 = \sqrt{\frac{\mu \operatorname{Re} \lambda_1}{\operatorname{Re} g}} e^{i\omega t} , \qquad (2.29)$$

$$\omega = \mu \left(\operatorname{Im} \lambda_1 - \frac{\operatorname{Im} g \operatorname{Re} \lambda_1}{\operatorname{Re} g} \right) .$$

These equations represent an oscillatory solution to (2.25) which is stable if $\operatorname{Re} g > 0$ (normal bifurcation) and unstable if $\operatorname{Re} g < 0$ (inverted bifurcation). However, it is not known whether (2.28) in the former case is stable when considered as a solution of (2.27). This is because it may be unstable with respect to spatially inhomogeneous disturbances. We discuss this point further in the next chapter. In the following, we assume a normal bifurcation and consider the case $\mu > 0$.

Under an appropriate scale transformation, the complex GL equation can be written

$$\partial W / \partial t = (1 + i c_0) W + (1 + i c_1) \nabla^2 W - (1 + i c_2) |W|^2 W . \qquad (2.30)$$

Here, $c_0 = \operatorname{Im} \lambda_1 / \operatorname{Re} \lambda_1$, $c_1 = \operatorname{Im} D / \operatorname{Re} D$, and $c_2 = \operatorname{Im} g / \operatorname{Re} g$. The parameter c_0 can be made to vanish if we make the transformation $W \to W \exp(i c_0 t)$, and thus the fundamentally important parameters for this equation are c_1 and c_2 only. The complex GL equation possesses no potential, and it is therefore possible for this equation to display complicated behavior which cannot be seen in the NW or GP equation. In the next chapter we consider some of this behavior.

The complex GL equation is one idealized model of self-excited oscillatory fields, but it is not appropriate for modeling actual oscillatory chemical reactions, such as the BZ reaction. This is due to the fact that, while the complex GL equation describes the motion of a very smoothly oscillating field, the BZ reaction consists of extremely sudden bursts which are perhaps better termed 'excitations' than 'oscillations'. While it can be said that the types of wave motion appearing in these oscillating fields have some common aspects, it is not particularly meaningful to make a detailed comparison of the two. Thus in this chapter we refrain from giving a detailed analysis of the complex GL equation in terms of its relation to actual physical systems. The analysis of wave motion patterns pertaining to more realistic reaction-diffusion systems is reconsidered in Chap. 3.

2.5 The Properties of the Complex Ginzburg–Landau Equation

2.5.1 Propagating Planar Wave Solutions and Their Stability

As is the case for the NW and GP equations, the complex GL equation possesses a family of planar wave solutions and solutions containing phase singular points. In a space of infinite extent, (2.24) obviously has the uniformly oscillating solution $W_0 = \exp[\mathrm{i}(c_0 - c_2)t]$. More generally, there exists a continuous family of planar oscillating solutions parameterized by the corresponding wave number possessing the following form:

$$W_k = A_k \mathrm{e}^{\mathrm{i}(\boldsymbol{k}\cdot\boldsymbol{r}+\omega_k t)} , \tag{2.31}$$

$$A_k = \sqrt{1-k^2} ,$$

$$\omega_k = c_0 - c_2 + (c_2 - c_1)k^2 ,$$

$$k = |\boldsymbol{k}| .$$

Only wave numbers satisfying the condition $k < 1$ are allowed. A change in the sign of $\boldsymbol{k}$ results in a reverse in the direction of propagation of the plane wave. Note that the above propagating plane waves are very different from those of linear wave equations. In the present case, the amplitude is uniquely determined by k, and the principle of superposition does not apply. Also, the oscillation frequency of the medium can suffer a change as a result of the propagation through the system of planar waves interpolating between regions of differing wave numbers. The propagation of such waves is due to the flexibility of the medium. For example, if for some reason one portion of the system comes to possess a higher wave number than that of the bulk, wave motion will appear spontaneously in the medium, and the frequency of the entire system will become synchronized to the high-frequency region. We develop this idea in more concrete terms from the point of view of phase dynamics in Chap. 4.

It is possible to conduct a general investigation of the linear stability of W_k, but this is a very difficult task, and therefore our primary concern here is to treat the stability of the uniformly oscillating solution W_0, limiting our discussion of the linear stability of W_k to a few important results.

We first set $W = W_0(1+\rho)\exp(\mathrm{i}\psi)$, where ρ and ψ represent the shift of the amplitude and phase from their values for the uniformly oscillating solution. The resulting linearized equation of motion is as follows:

$$\partial\rho/\partial t = (-2+\nabla^2)\rho - c_1\nabla^2\psi , \tag{2.32a}$$

$$\partial\psi/\partial t = (-2c_2 + c_1\nabla^2)\rho + \nabla^2\psi . \tag{2.32b}$$

In order to investigate the stability of the uniformly oscillating solutions with respect to a disturbance of wave vector $\boldsymbol{q}$, we assume both ρ and ψ behave as $\sim \exp(\mathrm{i}\boldsymbol{q}\cdot\boldsymbol{r}+\lambda t)$ and derive the quadratic equation determining the eigenvalue λ. The two roots of λ are given by

$$\lambda_\pm = -(1+q^2) \pm \{(1+q^2)^2 - 2(1+c_1c_2)q^2 - (1+c_1^2)q^4\}^{1/2} , \tag{2.33}$$

$$q = |\boldsymbol{q}| .$$

The λ_+ branch here corresponds to the phase and the λ_- branch to the amplitude. Destabilization is caused by long-wavelength planar disturbances. Expanding λ_+ in q, we have

$$\lambda_+(q) = -(1 + c_1c_2)q^2 - \tfrac{1}{2}c_1^2(1 + c_2^2)q^4 + \cdots , \tag{2.34}$$

and thus in the case that the relation

$$1 + c_1c_2 \equiv \alpha < 0 \tag{2.35}$$

is satisfied, phase instability is realized. The kind of instability seen in the complex GL equation is known as the *Benjamin–Feir (BF) instability.* The behavior of the system in response to this instability will be discussed in the final part of this section and also in Chap. 4. According to stability analysis, all propagating solutions W_k possessing finite wave numbers are unstable when $\alpha < 0$. When $\alpha > 0$, for values of k satisfying

$$k > \sqrt{\frac{\alpha}{3 + c_1c_2 + 2c_2^2}} , \tag{2.36}$$

W_k is unstable. This instability, like the Eckhaus instability of stationary patterns, is one with respect to long-wavelength disturbances whose wave vectors are parallel to $\boldsymbol{k}$.

2.5.2 Rotating Spiral Waves

The two-dimensional complex GL equation is known to possess *rotating spiral wave* solutions containing phase singular point topological defects (although such solutions cannot be easily obtained analytically). Such a phenomenon is also observed in the BZ reaction. In fact, because it is fairly easily produced in such systems, detailed experiments exist. However, as discussed above, the behavior of an oscillating medium described by the complex GL equation is very different from that for a BZ system, and thus it is not meaningful to compare such wave phenomena in the two systems. For the BZ system, this is more easily understood by using an approach based on an excitable model. This is considered in Chap. 3.

We expect rotating spiral waves to possess forms similar to that of the defect solution (2.17), but, more generally, we assume the form

$$W(r, \theta, t) = A(r)\mathrm{e}^{\mathrm{i}(\theta + S(r) + \omega t)} . \tag{2.37}$$

This solution represents a pattern rotating steadily with frequency ω. If such a solution exists, the corresponding oppositely rotating solution (that obtained by taking $\theta \to -\theta$) also exists, as can be seen by considering the symmetry of the equation. It is possible to investigate the nature of each of the unknown quantities $A(r)$, $S(r)$, and ω to a certain extent analytically, and we give a some results of this analysis below.

The amplitude A is a monotonically increasing function of r, proportional to r near the origin, and asymptotically approaching some value $A_\infty < 1$ for large r. The function S behaves smoothly in the neighborhood of the origin, assuming the approximate form $S \approx S_0 + S_1r^2$. Far from the origin, S becomes proportional to r, behaving as $S \approx kr$. In this way, in the distant region, the isophase lines approach the form of Archimedes spirals, converging to plane waves with wave number k. Thus, according to (2.30), we have $A_\infty = \sqrt{1-k^2}$ and $\omega = c_0 - c_2 + (c_2 - c_1)k^2$. In sharp contrast to the situation for the GP equation, in which the defect structure does not produce a global effect on the roll structure forming the background, for the complex GL equation, wave motion is induced in such a manner as to cause the global synchronization of the medium to the oscillation (of frequency ω) exhibited by the defect. Figure 2.4 displays the spiral wave pattern obtained through the direct numerical simulation of the complex GL equation.

Fig. 2.4. The spiral wave pattern obtained from a numerical simulation of the two-dimensional complex GL equation with $c_1 = -2.0$ and $c_2 = 0.6$ (figure used by permission of L. Kramer)

2.5.3 Hole Solutions and Disordered Patterns

The one-dimensional complex GL equation possesses a family of solutions for which the amplitude displays regions of local depression. These are called *hole solutions.* Their analytic form was determined by Nozaki and Bekki (1984), and they are therefore also referred to as *Nozaki–Bekki solutions.* Generically, hole solutions propagate with finite speeds, but the zero velocity hole solution W_{H} is particularly important. This solution possesses the simple form

$$W_{\mathrm{H}}(x,t) = \sqrt{1-Q^2}\tanh(kx)\exp[\mathrm{i}(\theta(x)+\omega t)]\,, \tag{2.38}$$

as can be easily confirmed by direct substitution into the complex GL equation. Here, θ is a function of x satisfying the equation $\mathrm{d}\theta/\mathrm{d}x = -Q\tanh(kx)$, and ω, Q, and k are related according to

$$\omega = c_0 - c_2 + (c_2 - c_1)Q^2\,, \tag{2.39a}$$

$$Q = \frac{2k^2-1}{3kc_1}\,. \tag{2.39b}$$

The quantity k^{-1} represents the width of the locally depressed region. Its value is determined by

$$\begin{aligned}&\left\{4\left(c_2-c_1\right)+18c_1\left(1+c_1^2\right)\right\}k^4\\&\quad-\left\{4\left(c_2-c_1\right)+9c_1\left(1+c_1c_2\right)\right\}k^2+c_2-c_1=0\,.\end{aligned} \tag{2.40}$$

Note that if a real root k of this equation exists, the root $-k$ also exists. The form of $W_{\rm H}$ is made clear in Fig. 2.5. The amplitude of this solution is 0 at the center of the 'hole' and approaches the value $\sqrt{1-Q^2}$ far from this point. The phase of $W_{\rm H}$ jumps by π at the hole, but its real and imaginary parts behave smoothly here. In the limits as $x \to \pm\infty$, $W_{\rm H}$ approaches the propagating plane wave $\pm\sqrt{1-Q^2}\exp(\mp {\rm i}Qx + {\rm i}\omega t)$.

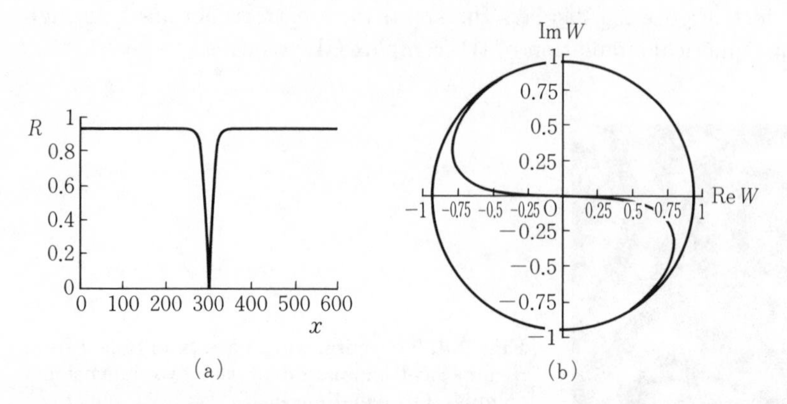

Fig. 2.5. The symmetric hole solution (2.37) of the one-dimensional complex GL equation. (**a**) Stationary pattern of the amplitude $|W_{\rm H}|$. (**b**) For a given time, the complex amplitude $W_{\rm H}(x,t)$ describes an S-shaped curve parameterized by x. As $x \to \infty$, this curve wraps around a circle of radius $(1-Q^2)^{1/2}$. As t is changed, the curve maintains its shape while rotating about the circle

A hole solution is qualitatively equivalent to the function that would be obtained by taking a radial slice of the two-dimensional rotating spiral solution discussed above. When Q and ω have the same sign, the defect becomes a source of phase waves traveling from left to right. In the case of opposite signs, it becomes a sink for such waves. The existence of an Eckhaus instability for the asymptotic plane wave forms of these phase waves depends on the values of c_1 and c_2. However, the lack of an Eckhaus instability for the asymptotic solution does not necessarily imply the stability or instability of the core region and, conversely, the stability of the core region does not necessarily imply the lack or presence of a BF instability of the asymptotic plane wave. In addition, the instability of the core structure and/or the asymptotic plane wave does not necessarily imply the immediate destruction of the hole's basic structure. The elucidation of these points included, there are several matters with regard to the stability of holes which have not yet been settled. However, from the results of numerical simulations performed using a variety of initial conditions, we know that at least the local hole-like structure is relatively stable for a wide range of parameter values. Also, this has been verified experimentally by the study of roll structures exhibiting destabilization with respect to oscillatory motion in Bénard convection induced in toroidal containers. [Lega et al. (1992)].

The BF instability of uniformly oscillating solutions of the complex GL equation which we have been discussing causes a variety of turbulent states in which both spatial and temporal disturbances are displayed. If the instability is sufficiently weak, as in the case considered in Chap. 4, the behavior of the system is characterized by phase turbulence, but when the instability becomes strong, the turbulent behavior is not confined exclusively to the phase. In such a situation, in one spatial dimension, the spontaneous production and destruction of hole-like structure arises. Similarly, in two dimensions, a shift from phase turbulence to defect production can be seen. To complicate matters further, even in the case of 'BF stability', depending on the initial conditions, extremely turbulent states may exist in which holes and defects are produced and destroyed randomly. The strong fluctuations displayed by such states often become spatially and temporally localized. This behavior, characteristic of hole-defect turbulence, is called *spatio-temporal intermittency*. Figure 2.6 presents an example of the spatio-temporal disorder exhibited by the one-dimensional complex GL equation. Although research concerning spontaneously arising disorder in extended nonequilibrium media is still at an early stage, we believe that considerable progress can be made toward the understanding of such phenomena by using the very important and complementary concepts of phase and defects. We hope that the discussion presented in this chapter can help point the way to the development of this understanding.

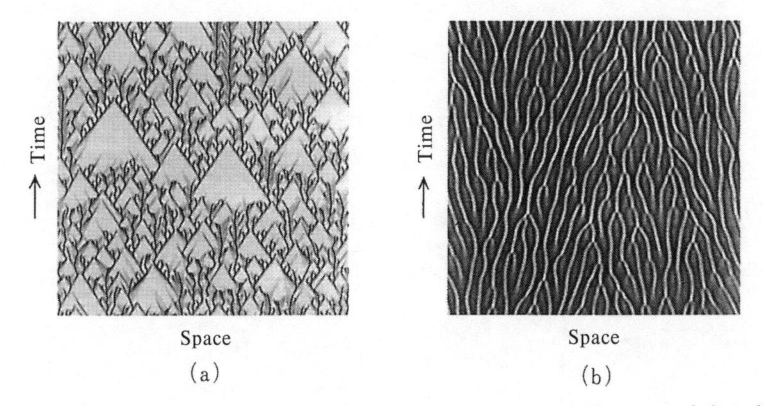

Fig. 2.6. An example of the disordered space–time behavior exhibited by the one-dimensional complex GL equation. The degree of shading reflects the size of $|W|$. (**a**) The shattered glass-like space–time pattern seen in the Benjamin–Feir stable region ($c_1 = 0$, $c_2 = 2.0$). (**b**) The disordered pattern seen in the Benjamin–Feir unstable region ($c_1 = 1.0$, $c_2 = -1.25$) (figure used by permission of H. Chaté)

3. Reaction–Diffusion Systems and Interface Dynamics

Existing studies on reaction–diffusion systems have been carried out with the Belousov–Zhabotinskii reaction acting as the standard example. Approaches based on amplitude equations are in general not suited to describe patterns peculiar to such 'excitable' systems because, while excitability originates in a particular property of global flow in phase space, amplitude equations are obtained by considering only local flow. In fact, BZ reaction systems display wave patterns that lack both the temporal and spatial smoothness displayed by solutions to amplitude equations, and thus it is much more natural to treat excitable systems using local *interface* structure representing the sudden change in state of a system over a small distance. Patterns containing interfaces arise out of the cooperative dynamics of degrees of freedom undergoing rapid and slow temporal change. Rapidly changing degrees of freedom form a bistable partial subsystem consisting of two states, an active (or excitable) state and an inactive (or nonexcitable) state. The speed with which the transition between these two states is carried out is controlled by the slowly changing degrees of freedom.

The main aim of this chapter is to elucidate the method used in describing patterns associated with interface motion based on two-component reaction–diffusion models, where the components in question correspond to the fast and slow degrees of freedom. The types of phenomena we consider include isolated waves, periodic traveling waves, rotating spiral waves in two-dimensional systems, the Turing instability, and interface instabilities.

3.1 Interfaces in Single-Component Bistable Systems

As preparation for the following sections, we consider here a single-component reaction–diffusion system of the following form:

$$\frac{\partial u}{\partial t} = \epsilon^{-1} f(u; v) + D\frac{\partial^2 u}{\partial x^2} \,. \tag{3.1}$$

The spatial dimension of the system is taken to be 1. The parameter ϵ represents the characteristic time scale of u, and D is the diffusion constant. By appropriately choosing spatial and temporal units, both of these can be set to 1, but in order to make clear the origin of the physical effects in which we are interested, we leave these parameters explicitly expressed. In the above equation, f is a nonmonotonic function of u that also depends on v. For simplicity, we assume the form $f = f_0(u) - v$. As shown in Fig. 3.1, we stipulate that $f_0(u)$ is an increasing function over a single continuous range of values of u and that everywhere else it is a decreasing function. Thus the value of u at which the the equality $f = 0$ is realized, i.e., the value of u corresponding to the stationary solution of the elemental equation $\dot{u} = \epsilon^{-1} f(u)$, depends on v. More specifically, three stationary points exist in the region defined by the inequality

$$v_{\min} < v < v_{\max} , \tag{3.2}$$

while outside this region there is just a single stationary point. In the former case, the outermost such points, u_- and u_+ ($u_- < u_+$), represent stable states, while the central point represents an unstable state. Thus, under the conditions of (3.2) the system is bistable, while outside the region satisfying these conditions it is monostable. The representative form

$$f = u - u^3 - v \tag{3.3}$$

is often used for f. For later discussion, we also cite the following piecewise linear model:

$$f = -u + \theta(u) - \tfrac{1}{2} - v . \tag{3.4}$$

Here, $\theta(u)$ is the unit step function ($\theta = 1$ for $u \geq 0$ and 0 otherwise).

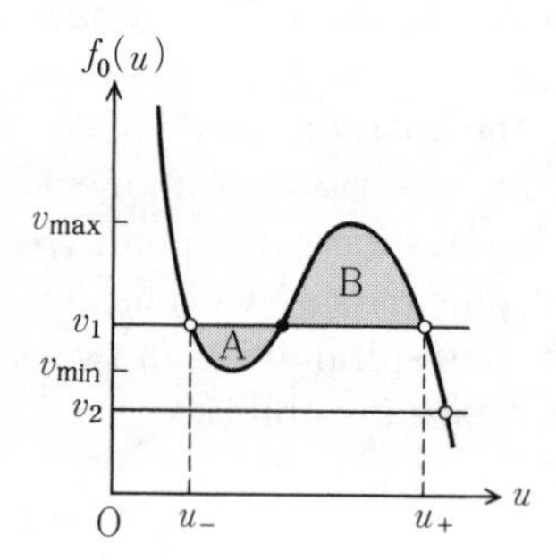

Fig. 3.1. The form of the nonlinear function $f_0(u)$ for the single-component reaction–diffusion system (3.1). Depending on the value of the parameter v, the number of values of u satisfying the equality $f_0(u) = v$, i.e., the number of uniform stationary solutions of (3.1), is either 1 or 3. The latter case is depicted in the figure. Here, the solution represented by the centrally located *black dot* is unstable, while the solutions u_+ and u_- represented by the *white dots* on either side of this are both stable

From this point we assume that v satisfies the bistability condition (3.2). After being prepared with some arbitrary initial pattern, the system will quickly become partitioned into regions consisting of two types of domains, those in which $u \approx u_+$ and those in which $u \approx u_-$. The switch between domains of these two types takes place over a sharp transition layer. Such a boundary (ignoring its thickness) consists of an $(n-1)$-dimensional object embedded in an n-dimensional system (a point, curve, and surface for $n = 1$, 2, and 3). However, without regard to the spatial dimension of the system, we will refer to the type of local structure representing a sudden state transition of this kind as an *interface*. In bistable systems, we also refer to interfaces as *kinks*.

For the moment, let us consider the case in which only a single interface exists and in which $u \to u_\mp$ as $x \to \pm\infty$. In this case, as $t \to \infty$, the interface assumes a fixed shape and propagates at a constant speed, c. If we write such a steadily propagating solution as $u_0(z)$, with $z = x - ct$, (3.1) yields the following eigenvalue problem parameterized by c:

$$-c\frac{\mathrm{d}u_0}{\mathrm{d}z} = \epsilon^{-1} f(u_0) + D\frac{\partial^2 u_0}{\partial z^2} . \tag{3.5}$$

If we multiply both sides of this equation by $\mathrm{d}u_0/\mathrm{d}z$ and integrate over all space, we obtain

$$c = \frac{\epsilon^{-1}\displaystyle\int_{u_-}^{u_+} f(u)\mathrm{d}u}{\displaystyle\int_{-\infty}^{\infty}\left(\frac{\mathrm{d}u_0}{\mathrm{d}z}\right)^2 \mathrm{d}z} . \tag{3.6}$$

Now, let us define v^* to be the value of v for which the areas of regions A and B in Fig. 3.1 are equal. Then from (3.6), it is seen that $v > v^*$ implies $c < 0$, and the interface will thus propagate in such a manner as to enlarge the u_- region. If $v < v^*$, the opposite behavior results. When $v = v^*$, the balance between the two regions is maintained, and the interface is stationary. This is similar to the situation of phase equilibrium described by the Maxwell principle of equal area. In the present case, the quantity corresponding to the chemical potential is the kinetic potential,

$$\Psi(u) = -\epsilon^{-1}\int_0^u f(u')\mathrm{d}u' . \tag{3.7}$$

Note that $\Psi(u)$ has minima at u_+ and u_-. The relation between the depths of these two minima depends on the value of v. When $v = v_+$, these depths are equal. Writing (3.6) as

$$c = [\Psi(u_-) - \Psi(u_+)] \Big/ \int_{-\infty}^{\infty}\left(\frac{\mathrm{d}u_0}{\mathrm{d}z}\right)^2 \mathrm{d}z , \tag{3.8}$$

it is clear that the interface moves in such a way as to expand the domain corresponding to the smaller potential. In the following, we will consider c to be a function of v, writing $c = c_0(v)$.

The steady-state propagating solutions corresponding to both (3.3) and (3.4) are known. Under the previously stated boundary conditions, for (3.3) we easily obtain

$$u_0(z) = \frac{1}{2}\left\{u_+ + u_- - (u_+ - u_-)\tanh\left(\frac{(u_+ - u_-)z}{\sqrt{8\epsilon D}}\right)\right\} \tag{3.9a}$$

$$c_0(v) = s\sqrt{\frac{D}{2\epsilon}}(u_+ + u_-) . \tag{3.9b}$$

Note that we have used $v^* = 0$ here.

The steady-state propagating solution of the piecewise-linear model (3.4) can also be easily found. In this case, in order to satisfy the surface conditions at $\pm\infty$, we solve (3.5) for the two cases $z \geq 0$ and $z < 0$ separately. In so doing, we set $u_0(0) = 0$ and require $u_0(z)$ and its first derivative to be continuous at $z = 0$. Following this procedure, for the solution in question we obtain the form

$$u_0(z) = \begin{cases} u_-(1 - e^{\alpha_- z}) & (z \geq 0)\,, \\ u_+(1 - e^{\alpha_+ z}) & (z < 0)\,, \end{cases} \tag{3.10a}$$

$$\alpha_\pm = (2D)^{-1}\left\{-c_0 \pm \sqrt{c_0^2 + \frac{4D}{\epsilon}}\right\}, \tag{3.10b}$$

$$c_0(v) = -4v\sqrt{\frac{D}{\epsilon(1-4v^2)}}\,. \tag{3.10c}$$

In general, regardless of the form of f, the thickness of the interface is $\sim O(\epsilon^{1/2})$, while the propagation speed is $\sim O(\epsilon^{-1/2})$. Thus when ϵ is small, if we consider the parameter v (assumed above to be a constant) to be a sufficiently slowly changing function of time and space, the essential points of our discussion remain unchanged. More specifically, if the time scale on which v fluctuates is much longer than ϵ, and if the spatial scale of its modulation is much longer than $\epsilon^{1/2}$ (the scale of the interface thickness), we can simply replace v by $v(x,t)$ in all of the above expressions. In this case, we expect that at all points in space u will be slaved to v. Also, we can set v equal to a constant in the neighborhood of the interface, and therefore the propagation speed of the interface at each instant can be written as a function of the value of v assumed at the interface, v_i:

$$c_0(t) = c_0(v_i(t))\,. \tag{3.11}$$

In this case the detailed structure of the interface will be closely approximated by the steady-state propagating solution $u_0(z)$. (Note that the form of $u_0(z)$ above assumes that the interface approaches $u_+(v_i)$ to the left of the interface and $u_-(v_i)$ to the right. To realize the opposite case, we need only change the signs of z and c_0 here.) The sign of c_0 depends on the relative sizes of v_i and v^*. For convenience, let us refer to interfaces for which $v_i < v^*$ as 'fronts' and those for which $v_i > v^*$ as 'backs'. (The reason for this choice of names is made clear in the next section.) Hence if we designate a value v_i corresponding to a front as v_f and that for a back as v_b, $c_0(v_f)$ and $c_0(v_b)$ have opposite signs.

In the sections that follow, we treat $v(x,t)$ not as some predetermined quantity, but as a function that develops in time while interacting with $u(x,t)$. In this way, v comes to represent the system's second degree of freedom, and we thus treat the interface as an autonomously developing structure.

3.2 Solitary Wave Pulses and Periodic Wave Pulse Trains in Excitable Systems

In the last section we considered the bistable field $u(x,t)$. If we reinterpret the parameter v used there as a state function obeying some appropriate evolution equation, we obtain an *excitable system.* The Keener–Tyson (KT) and FitzHugh–Nagumo (FN) equations introduced in Chap. 1 in connection with the BZ reaction are typical examples of excitable systems. Note that if we consider the functions v appearing in these models as parameters, they too become bistable.

In this section we consider v as a state function, and we extend the considerations of the previous section to the two-component reaction–diffusion system

$$\frac{\partial u}{\partial t} = \epsilon^{-1} f(u,v) + D\frac{\partial^2 u}{\partial x^2} , \tag{3.12a}$$

$$\frac{\partial v}{\partial t} = g(u,v) + D'\frac{\partial^2 v}{\partial x^2} , \tag{3.12b}$$

where $f(u,v)$ has the same general nature as the corresponding function in the previous section. Thus the nullcline representing $f(u,v)=0$ is given by an S-shaped curve. Let us call this curve Σ. Then, as in the KT and FN equations, we assume that the nullcline defined by $g(u,v)=0$ is monotonic. We refer to this curve as Γ. It is assumed that Σ and Γ have shapes and relative positionings similar to those displayed in Fig. 1.3(b), characteristic of excitable systems. In the following discussion we assume that ϵ is a small parameter and that D and D' are both $O(1)$ constants. These assumptions are appropriate in the context of modeling the BZ equation.

Considering (3.12a), if we think of $v(x,t)$ as a slowly varying function of both time and space, as pointed out in the preceding section, all of the discussion given there applies almost unchanged to the present situation. Let us now take the $\epsilon \to 0$ limit and thus think of the interface as possessing zero width. If we do this, u can be written as $u_\pm(v)$ at every point except at that single point constituting the interface. Thus at all points except those on the interface we can write

$$\frac{\partial v}{\partial t} = g(u_\pm(v), v) \equiv G_\pm(v) , \tag{3.13}$$

where we have assumed that the scale of the spatial modulation of v is sufficiently large to allow omission of the diffusion term. The validity of this assumption depends on the nature of the pattern actually realized from the dynamics, a point which must be checked after the fact. We will consider this problem again later. Now, given a particular model, using (3.11) and (3.13) as our basis, a more detailed discussion of interface dynamics can be entered into. In that case, in general, it is necessary to require the function v to be continuous across the moving interface.

Let us now consider the simple models defined by the following relations corresponding to (3.3) and (3.4):

$$f = u - u^3 - v, \qquad g = u + \frac{1}{\sqrt{3}} + a \tag{3.14}$$

$$f = -u + \theta(u) - \tfrac{1}{2} - v, \qquad g = u + a\,. \tag{3.15}$$

We note that the model corresponding to (3.14) can be put into a form equivalent to the FN equation under appropriate linear transformations of u and v. The pair of evolution equations obtained from (3.15) is known as the McKean model. Its nullclines are shown in Fig. 3.2. It is clear from the discussion in Chap. 1 that in the case $a \gtrsim 0$, the two models corresponding to (3.14) and (3.15) represent excitable systems. In what follows, we develop general ideas concerning excitable systems by considering the McKean model as their representative.

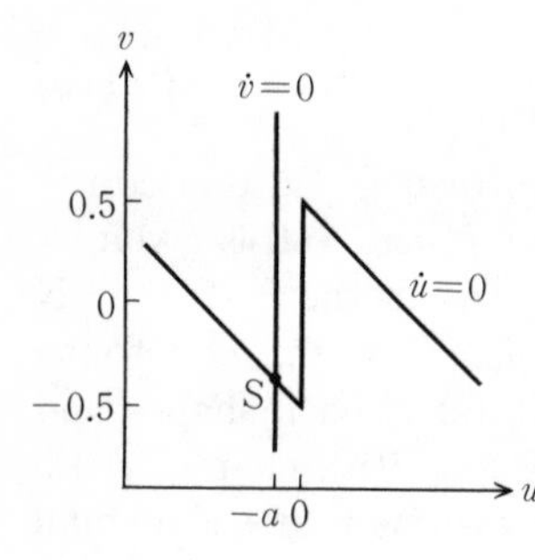

Fig. 3.2. The nullclines of the McKean model (3.15). When a is a small positive parameter, the system displays excitability

The excitable medium in question possesses a stable uniform stationary state (rest state) represented by (u_s, v_s). For the above piecewise linear model, this state is given by

$$u_s = -a, \qquad v_s = -\tfrac{1}{2} + a\,. \tag{3.16}$$

If a finite strength perturbation is applied to a part of such an excitable system existing in a uniform rest state, a pair of isolated propagating waves (e.g., *pulses*, *trigger waves*, *excitation waves*), one propagating in either direction, can be created. Initially, the region of the system subjected to the perturbation exhibits transient excitation in accordance with the dynamics of excitable systems. This excitation then acts to stimulate the regions neighboring the affected location. As a result, these regions, after a delay of some interval, begin to display transient excitation according to the same principle as that responsible for the original excitation. In this way, a transient excitation in the form of a localized traveling wave begins to propagate in each direction. If two excitations created in this way collide, they mutually annihilate.

The form of a single isolated pulse traveling in such a system is determined uniquely, and in order to preserve this shape the pulse must travel at a fixed speed. Let us find such a solution. As depicted in Fig. 3.3, we find a stationary solution $u(z)$, $v(z)$ satisfying the boundary conditions

$$u(z) \to u_s, \qquad v(z) \to v_s \qquad (z \to \pm\infty) , \tag{3.17}$$

with a front at $z = 0$ and a back at $z = d$. For such a solution, it is appropriate to require the following conditions. First, for $z \geq 0$ we require

$$u(z) = u_s, \qquad v(z) = v_s , \tag{3.18}$$

and thus

$$v_f = v_s . \tag{3.19}$$

For $-d \leq z < 0$ we require

$$u(z) = u_+(v(z)) , \tag{3.20a}$$

$$-c\frac{dv}{dz} = G_+(v) , \tag{3.20b}$$

and for $z < -d$,

$$u(z) = u_-(v(z)) , \tag{3.21a}$$

$$-c\frac{dv}{dz} = G_-(v) . \tag{3.21b}$$

Furthermore, at $z = 0$ and $-d$ we enforce the boundary conditions

$$v(0^+) = v(0^-) = v_f , \tag{3.22}$$

$$v(-d^+) = v(-d^-) = v_b . \tag{3.23}$$

In order for the front and back to propagate at equal speeds, we have

$$c = c_0(v_f) = -c_0(v_b) . \tag{3.24}$$

From the above equations, the solution in question is uniquely determined.

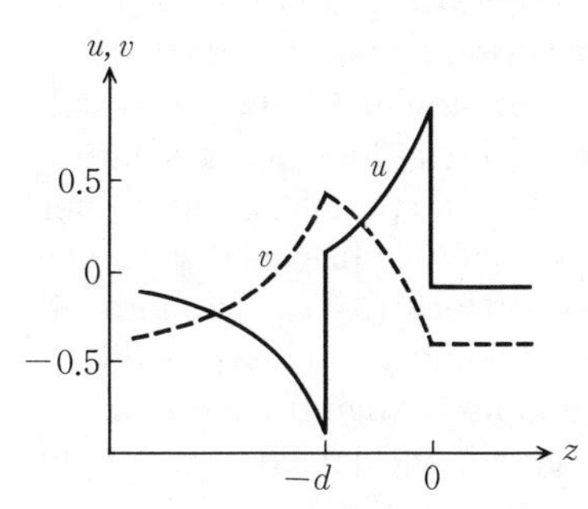

Fig. 3.3. Shape of the isolated pulse obtained in the McKean model in the $\epsilon \to 0$ limit

It is not difficult to obtain an explicit formula for the solution of the McKean model (3.15). Beginning with (3.10c), we see that the propagation speed of this solution c is defined by the relation

$$c = 4v_{\mathrm{f}}\sqrt{\frac{D}{\epsilon(1-4v_{\mathrm{f}}^2)}}\ . \tag{3.25}$$

Next, from the second equality in (3.24), we have

$$v_{\mathrm{b}} = -v_{\mathrm{f}} = \tfrac{1}{2} - a\ . \tag{3.26}$$

Then, under the boundary conditions (3.17), (3.22), and (3.23), we solve the linear equations (3.20b) and (3.21b). Finally, combining these results, we obtain the wave form

$$u = -a, \quad v = -\tfrac{1}{2} + a \qquad (z \geq 0)\ , \tag{3.27a}$$

$$\left.\begin{aligned} u &= -a + \mathrm{e}^{z/c} \\ v &= \tfrac{1}{2} + a - \mathrm{e}^{z/c} \end{aligned}\right\} \qquad (-d \leq z < 0)\ , \tag{3.27b}$$

$$\left.\begin{aligned} u &= -a + (2a-1)\mathrm{e}^{(z+d)/c} \\ v &= \left(a - \tfrac{1}{2}\right)\left(1 - 2\mathrm{e}^{(z+d)/c}\right) \end{aligned}\right\} \qquad (z < -d)\ . \tag{3.27c}$$

In addition, from the first equality in (3.23) we obtain $d = -c \ln 2a$ for the pulse width d. The quantities c and d are both $O(\epsilon^{-1/2})$. This implies that $v(z)$ is a slowly varying function of z, undergoing modulation with a length scale $\sim O(\epsilon^{-1/2})$. Thus the diffusion term $[\sim O(\epsilon)]$ is much smaller than the reaction term g and can be ignored. This confirms the validity of the assumption made above.

The solution found above represents the approximate pulse shape found by ignoring the internal structure of the interface. This is known as the 'outer solution'. The fine structure of the interface is given by the 'inner solution'. The inner solution can be found independently by treating this region as a single-component system containing the fixed parameter v, following the procedure presented in Sect. 3.1.

As can be inferred from the manner in which a pair of pulses is created from a single, locally applied external stimulation, if such a stimulus is applied repeatedly to a given local region, pulse trains propagating in both directions will be created. A solution representing such a periodic pulse train traveling in a single direction can be obtained in a manner similar to that in which the isolated pulse solution was obtained above. Let us look for a solution possessing a front at $z = 0$, a back at $z = -d$, and a spatial period λ over which this structure is repeated, as depicted in Fig. 3.4. In contrast to the case of a single pulse, where the boundary conditions (3.17) are enforced at $z = \pm\infty$, in the present case we have the periodic boundary condition $v(0) = v(-\lambda)$. Aside from this difference, there is no change from the above treatment; (3.20), (3.21), (3.23), and (3.24) all remain unchanged. If we apply these to the McKean model, we obtain the following result. First, c is given by (3.25), substituting for v_{f} according to

$$v_{\mathrm{f}} = -\left(a + \tfrac{1}{2}\right)\frac{1 - \mathrm{e}^{-d/c}}{1 + \mathrm{e}^{-d/c}} \,. \tag{3.28}$$

The pulse width d appearing in this expression can be expressed as a function of λ:

$$\begin{aligned} d &= -c\,ln\left[a(1-y) + \{a^2 + (1-2a^2)y + a^2y^2\}^{1/2}\right] , \\ y &= \mathrm{e}^{-\lambda/c}. \end{aligned} \tag{3.29}$$

Equations (3.25), (3.28), and (3.29) implicitly determine the dispersion function $c(\lambda)$. In the limit as $\lambda \to \infty$, this set of equations becomes equivalent to an expression for the velocity in the case of a single isolated pulse, as should be the case. Considering the opposite extreme of small λ, c is given by

$$\begin{aligned} c &\approx \epsilon^{-1/2}\sqrt{\tfrac{1}{2} - 2a^2}D^{1/4}\tilde{\lambda}^{1/2} , \\ \tilde{\lambda} &= \sqrt{\epsilon}\lambda. \end{aligned} \tag{3.30}$$

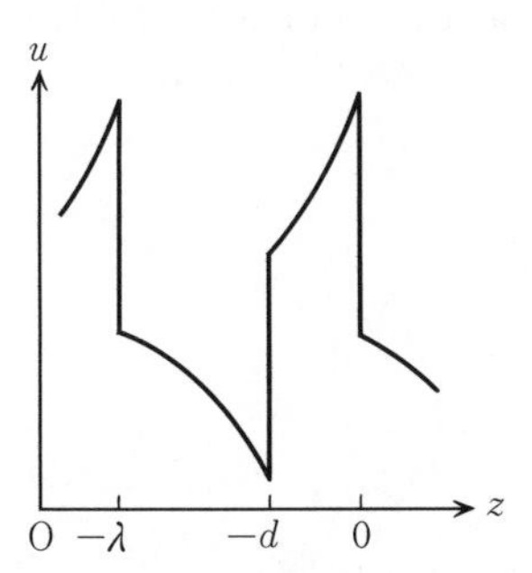

Fig. 3.4. The shape of the periodic pulse train (wavelength λ) for the McKean model obtained in the limit as $\epsilon \to 0$

It is thus seen that, in the $\epsilon \to 0$ limit we have considered here, pulse trains of arbitrarily small wavelength λ exist. However, for a finite ϵ, more realistic model, there exists a lower bound on allowed values of λ. This reflects the fact that at frequencies of the applied impulse exceeding some maximum value, the system cannot respond with excitations of the same frequency. Once a transient excitation has occurred, in order for the following application of the external stimulus to be effective in causing subsequent excitation, the state of the system in the region in question must first return to a 'relaxed' state relatively close to the stationary state. For extremely large frequencies of the applied perturbation, this condition cannot be satisfied, and the application of an impulse succeeding excitation will not induce further excitation. The time during which a system is unable to respond to stimulation in this way is referred to as a *refractory period.*

3.3 Spiral Waves in Excitable Systems

To this point we have limited our investigation to systems consisting of a single spatial dimension. The actual systems in which the BZ reaction is usually studied experimentally are quasi-two-dimensional. In this section we discuss a representative pattern appearing in such systems, the rotating spiral wave. In Sect. 2.5 we discussed spiral waves from the point of view of amplitude equations describing oscillatory media. However, spiral waves occurring in excitable media are, as we will see, of a very different nature.

A pair of rotating spiral waves appearing in a system undergoing a BZ reaction is shown in Fig. 3.5. The nature of the wave front constituted by such a pattern is essentially the same as the excitation waves discussed in the previous section. If the wave front here were stretched into a straight line and extended to infinity, it would be precisely the same as the one-dimensional pulse studied above. In fact, however, a spiral wave differs in two important ways from a pulse. The first difference, of course, is that the wave front of a spiral wave consists of a spiraling curve. The second and more important difference is that a spiral wave has endpoints. Determining the structure in the neighborhood of the endpoints is perhaps the most important problem involved in understanding rotating spiral waves.

Fig. 3.5. Rotating spiral waves seen in the Belousov–Zhabotinskii reaction (from Winfree (1974))

There are several different methods to induce a spiral wave excitation in the BZ reaction. We describe one typical such method here. First, if one part of the medium is subjected to an appropriate stimulation, a single circular wave can be produced. This wave will then expand at a nearly constant speed. (This propagation corresponds to the propagation of a pair of pulses, one in either direction, in a one-dimensional system.) Then, using some appropriate method of perturbation (e.g., application of a disturbance to the reagent, irradiation with light, etc.), it is possible to destroy a portion of the circular wave front. As a result, the wave comes to possess a pair of endpoints, and the wave begins to curl in upon itself, with these endpoints acting as the centers of the curling motion. In this way, from the endpoints a pair of oppositely spiraling patterns develops. Let us think of the situation in which one of the spiral waves of a given pair exists independently. Such a pattern, after developing for a sufficiently long time and under some appropriate set of conditions, will spiral in such a manner as to maintain some fixed shape. We are interested in determining the form of the equation representing such a single-curve spiral wave. With polar coordinates (r, θ), let us assume that the curve representing the spiral wave $\theta(r)$ spirals steadily with some constant angular frequency ω. In this case each point of the curve propagates in a direction normal to the curve at that point with some speed c. This, together with the assumption that the shape of the spiral remains constant, implies that the change produced by a rotation of the curve by an infinitesimal angle $\mathrm{d}\theta = \omega \mathrm{d}t$ is equivalent to the change suffered when each point moves ahead in the normal direction by a distance $c\,\mathrm{d}t$.

As can be understood from the construction shown in Fig. 3.6, this situation is represented by the equation

$$\tan\phi = \frac{\mathrm{d}r}{r\mathrm{d}\theta}, \tag{3.31a}$$

$$\sin\phi = \frac{c\mathrm{d}t}{r\mathrm{d}\theta} = \frac{c}{r\omega}. \tag{3.31b}$$

Eliminating ϕ from these equations leaves

$$c = \frac{r\omega}{\sqrt{1+\psi^2}}, \tag{3.32}$$

$$\psi = r\frac{\mathrm{d}\theta}{\mathrm{d}r}.$$

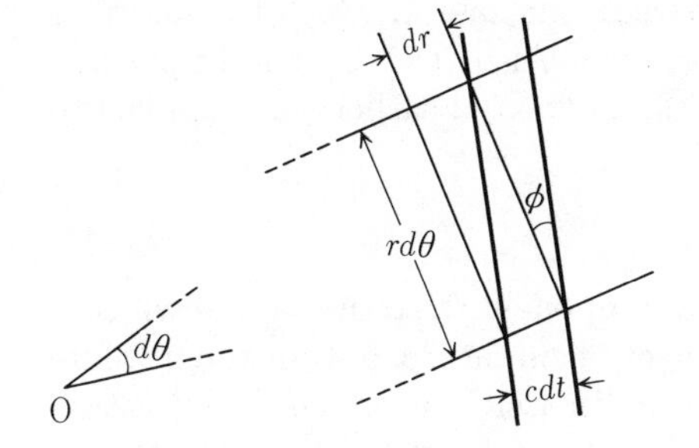

Fig. 3.6. As a result of an infinitesimal rotation $\mathrm{d}\theta$ about the spiral wave's center located at the origin, an arbitrary small segment of the wave front (heavy line) moves ahead by a distance $c\,\mathrm{d}t$

When c is a constant, the above becomes a first-order differential equation in θ with the solution

$$\pm\theta(r) = \left(\frac{r^2}{r_0^2} - 1\right)^{1/2} - \cos^{-1}\frac{r_0}{r} + \omega t \tag{3.33}$$

$$r_0 = c/\omega,$$

but this solution exists only for $r \geq r_0$. This fact is made clear by Fig. 3.7. Here, an involute with inner radius r_0 appears. The endpoint of the involute

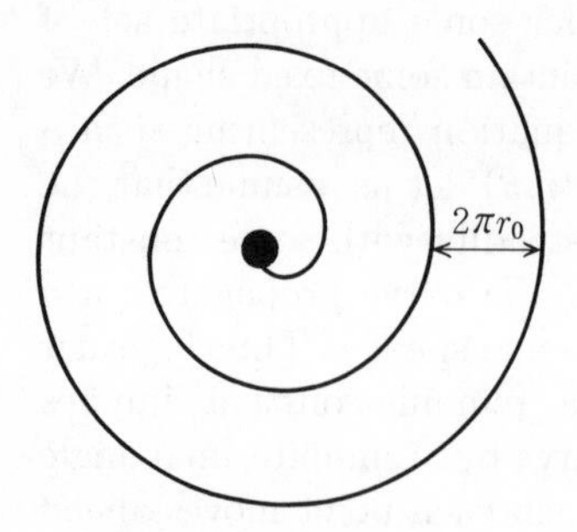

Fig. 3.7. The involute (3.33) defined by a circle (central *black dot*) of radius r_0

travels around this circle. For $r \gg r_0$, this involute is approximated by an Archimedes spiral of wavelength $2\pi r_0$, defined by $\pm\theta = r/r_0$. While we have obtained these results using only simple considerations, they capture several of the characteristics of actual spiral waves quite well. For example, it has been confirmed experimentally that the core of the spiral consists of an endpoint that travels around a circularly shaped region, and the shapes of spiral waves are closely described by involutes and Archimedes spirals. However, despite these apparent successes of our naive treatment, it has several problems. First, the assumption that the propagation speed of excitation waves is given everywhere by the constant c, whose value was determined above, is incorrect. As we discuss below, the propagation speed c of the interface deviates from the value c_0 of a flat interface (appearing in Sect. 3.1) by a term which is proportional to the curvature κ:

$$c = c_0 - D\kappa \,. \tag{3.34}$$

Here, we take the value of κ to be positive in the case in which the interface is convex in the direction of the propagation. Also, D is a strictly positive constant. In the two-component model (3.12) treated earlier (now applied to two spatial dimensions), we have

$$c = c_0(v_\mathrm{i}) - D\kappa \,, \tag{3.35}$$

where D is the diffusion constant for the component u. In the case in which κ is sufficiently small, we can ignore the effect of curvature, but for the involute obtained from the above phenomenological discussion, in which the curvature effect was ignored, the curvature at the endpoint is infinite, and the theory is self-contradictory. The fact that ω is left as an undetermined parameter is also a problem in the above theory. In order to have a closed theory with no undetermined quantities, it is necessary to make clear the detailed structure in the core region. Below, we describe an attempt to construct a theory that represents an improvement with respect to these various points.

First, let us discuss the reason why the speed behaves as expressed by (3.35) in the context of the two-dimensional version of the two-component excitable system (3.12). We assume that ϵ is sufficiently small. In this case, the function $v(x, y, t)$ in (3.12a) can be considered as a parameter. (This holds in the case of a two-dimensional medium just as in the one-dimensional case.) In addition, the most important problem becomes that involving the effect of the curvature, so we set $v(x, y, t)$ to a constant value. Thus the equation that we should consider is

$$\frac{\partial u}{\partial t} = \epsilon^{-1} f(u, v) + D\left(\frac{\partial^2}{\partial x^2} + \frac{\partial^2}{\partial y^2}\right) u . \tag{3.36}$$

For some time t_0, we consider the neighborhood of a point P on the interface at which the curvature is κ. Then, as depicted in Fig. 3.8, we define a local orthogonal coordinate system x–y such that the x-axis intersects the interface perpendicularly at the point P. In the neighborhood of P, u_0 (given in Sect. 3.1) should be an approximate solution of (3.36), where $z = x - c(t - t_0) + (\kappa y^2/2)$. If we substitute this form into (3.36), at P we obtain

$$-(c + D\kappa)\frac{\mathrm{d}u_0}{\mathrm{d}z} = \epsilon^{-1} f(u_0) + D\frac{\mathrm{d}^2 u_0}{\mathrm{d}z^2} . \tag{3.37}$$

Note here that if $\kappa = 0$, we have $c = c_0$, and thus it is clear that (3.35) (with $v_{\mathrm{i}} = v$) follows from this approximate treatment. We give a more general treatment of the dynamics of curved interfaces based on phase dynamics methods in Chap. 4.

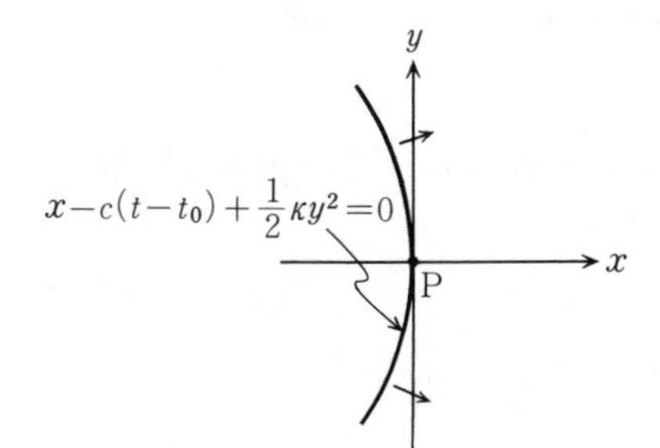

Fig. 3.8. Local shape of an interface of curvature κ at the point P

Now, let us return to our improved treatment of rotating spiral waves. In (3.32), c is now a nonconstant quantity, and (3.35) should be employed. However, κ can itself be expressed as a function of ψ according to

$$\kappa = \pm\left\{\frac{\mathrm{d}\psi/\mathrm{d}r}{(1+\psi^2)^{3/2}} + \frac{\psi}{r(1+\psi^2)^{1/2}}\right\} \quad (\psi \gtrless 0) . \tag{3.38}$$

This follows from simple geometric considerations. Using this equation for κ, and rewriting (3.32) as

$$\frac{r\omega}{(1+\psi^2)^{1/2}} = c_0 \mp D\left\{\frac{\mathrm{d}\psi/\mathrm{d}r}{(1+\psi^2)^{3/2}} - \frac{\psi}{r(1+\psi^2)^{1/2}}\right\} \quad (\psi \gtrless 0) , \tag{3.39}$$

we should be able to obtain a somewhat improved version of the theory given earlier. However, it is not yet possible to solve the problem in closed form with no undetermined quantities.

In the discussion appearing to this point, we have considered the spiral wave as a single interface $\theta(r)$, not distinguishing between the front and back. Let us now revise our thinking and treat the spiral as two distinct interfaces, $\theta_{\mathrm{f}}(r)$ and $\theta_{\mathrm{b}}(r)$. Then, we require each of the quantities ψ_{f} and ψ_{b} corresponding to these interfaces to satisfy (3.39). As displayed explicitly in Fig. 3.9, we are thinking of the following situation. The front and back are joined smoothly, and the point at which they are joined contacts the core of radius r_0 as it orbits about this core in a clockwise direction with constant angular frequency ω. Furthermore, we define the quantity c_0 at the front and back as $c_0(v_{\mathrm{f}}) \equiv c_{\mathrm{f}}(v_{\mathrm{f}})$ and $c_0(v_{\mathrm{b}}) \equiv -c_{\mathrm{b}}(v_{\mathrm{b}})$, where v_{f} and v_{b} are *not* constant along the interface but, rather, are quantities that must be determined directly from the evolution equation for v, (3.13). Note that we again assume that the effect of diffusion for v is small enough that it can be ignored. Thus the equation that we consider becomes

$$\frac{r\omega}{(1+\psi_j^2)^{1/2}} = c_j(v_j) \mp D\left\{\frac{\mathrm{d}\psi_j/\mathrm{d}r}{(1+\psi_j^2)^{3/2}} + \frac{\psi_j}{r(1+\psi_j^2)^{1/2}}\right\} \quad (\psi_j \gtrless 0,\ j=\mathrm{f},\mathrm{b}) \,. \tag{3.40}$$

The boundary conditions that must be satisfied by ψ_j include, first, from the stipulation that θ_j constantly remains in contact with the core,

$$\psi_{\mathrm{f}}(r_0) = -\infty, \qquad \psi_{\mathrm{b}}(r_0) = \infty \,. \tag{3.41}$$

Next, requiring that the front and back both asymptotically approach Archimedes spirals at large distances from the core,

$$\psi_{\mathrm{f}}(r), \psi_{\mathrm{b}}(r) \to kr \qquad (r \to \infty) \tag{3.42}$$

must hold. Here, $k = \omega/c_\infty$ and $c_\infty = c_{\mathrm{f}}(r=\infty) = c_{\mathrm{b}}(r=\infty)$.

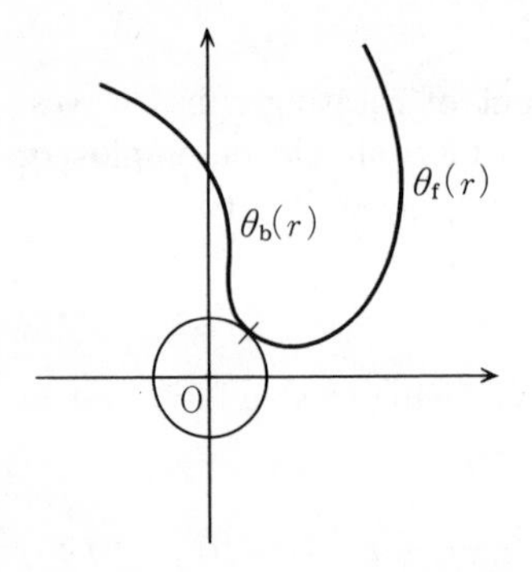

Fig. 3.9. The shape of the assumed spiral wave interface in the neighborhood of the core (*central circle*). The front and back consist of smoothly joined curves. The joining point remains in contact with the core, and the curve maintains its shape as it traverses a path around the core with angular frequency ω

Equation (3.40) contains $v_\mathrm{f}(r)$ and $v_\mathrm{b}(r)$, and if we can express these in terms of $\psi_\mathrm{f}(r)$ and $\psi_\mathrm{b}(r)$, the equation becomes closed. For this purpose, we first note the fact that since the pattern is rotating in a fixed manner, we can write $v(r,\theta,t) = v(r,\theta+\omega t)$, and thus (3.13) becomes

$$\omega\frac{\partial v}{\partial \theta} = G_\pm(v) \, , \tag{3.43}$$

where for values of θ satisfying $\theta_\mathrm{f} < \theta < \theta_\mathrm{b}$ we take G_+, and for all other values of θ we take G_-.

Let us consider a point P separated from the center by a distance r. Then, let us define the time interval $2\pi/\omega_+(r)$ to be that between the times at which the front and back cross P. In other words, this is the time interval during which the system is in the excited state at P. From the definition of $\omega_+(r)$, we have

$$\frac{\omega}{\omega_+(r)} = \frac{\theta_\mathrm{b}(r) - \theta_\mathrm{f}(r)}{2\pi} \, , \tag{3.44}$$

or, taking the derivative of this with respect to r, we obtain

$$\omega r \frac{\mathrm{d}\omega_+^{-1}}{\mathrm{d}r} = \frac{\psi_\mathrm{b} - \psi_\mathrm{f}}{2\pi} \, . \tag{3.45}$$

Integrating (3.43), ω_+ can also be expressed by

$$\frac{2\pi}{\omega_+(r)} = \int_{v_\mathrm{f}}^{v_\mathrm{b}} \frac{\mathrm{d}v}{G_+(v)} \, . \tag{3.46}$$

Similarly, for a single period, the interval during which P is in the resting state is clearly given by

$$2\pi\left(\frac{1}{\omega} - \frac{1}{\omega_+}\right) = -\int_{v_\mathrm{f}}^{v_\mathrm{b}} \frac{\mathrm{d}v}{G_-(v)} \, . \tag{3.47}$$

The above two equations together constitute an expression for $v_\mathrm{f}(r)$ and $v_\mathrm{b}(r)$ in terms of $\omega_+(r)$ and ω and thus imply a function $c_j(\omega_+,\omega)$ through the function $c_j(v_j)$. Thus, considering (3.40) and (3.45) as simultaneous equations and enforcing the boundary conditions (3.41) and (3.42), the problem becomes closed. This is a nonlinear eigenvalue problem, and a solution for $\psi_j(r)$ exists only for certain values of ω and r_0.

As previously mentioned, for the McKean model, the explicit form of $c_j(v_j)$ is known, and, in addition, here we can obtain the following expressions for v_f and v_b through a simple calculation:

$$v_\mathrm{f} = a + \tfrac{1}{2} - \frac{1-\exp[2\pi(\omega_+^{-1} - \omega^{-1})]}{1-\exp(-2\pi/\omega)} \, , \tag{3.48a}$$

$$v_\mathrm{b} = a + \tfrac{1}{2} + \frac{\exp(-2\pi/\omega) - \exp(-2\pi/\omega_+)}{1-\exp(-2\pi/\omega)} \, . \tag{3.48b}$$

Using these, it is possible to solve the above nonlinear eigenvalue problem numerically [Pelcé and Sun (1991)].

3.4 Multiple Spiral Waves and the Turing Pattern

In order to constantly maintain the nonequilibrium state of a chemical reaction, it is necessary to continuously add fresh reagents to the system. Until recently, because it is experimentally quite difficult to carry out such a process without disturbing the spatial pattern in the system, the observation of such patterns has taken place in systems evolving naturally toward equilibrium, in which these patterns represent transient structure. However, with such an observational method, it is not possible to perform qualitative analysis based on a long time measurement. For example, if we attempt to clearly describe the transition from one pattern to another in terms of bifurcation phenomena, it is absolutely necessary to maintain the consistency of the medium.

An experimental method developed in recent years for systems such as those involving the BZ reaction is groundbreaking with regard to the maintenance of nonequilibrium states and has opened up new possibilities in the study of dissipative structures in reaction–diffusion systems. This is a method involving the observation of patterns of reagents absorbed into thin layers of gel, and it is very well suited for the study of two-dimensional spiral patterns. In such a system, a sufficiently well stirred reagent in a reaction chamber is steadily supplied to a thin layer of gel (situated above this chamber). This reagent is diffusively infused into the gel through a two-dimensionally aligned array of filaments, one end of each filament being situated in the reaction chamber, and the opposite end in the gel. In this way, the isotropic nature of the infusion process is maintained, and it is possible to suppress the outbreak of flow.

3.4.1 Compound Spiral Rotation

One of the concrete results obtained using the experimental method described above is the qualitative analysis of the compound rotation of spirals. As described in the previous section, under appropriate conditions, spirals exhibit steady rotational motion. However, by changing the conditions of the system (for example, by changing the excitability of the medium), the steady rotational motion of the spiral's central region becomes unstable, and more complicated motion appears. This motion is referred to as *compound rotation* or *meandering*. Experimentally, the concentration of $KBrO_3$ is used as a control parameter for such systems. The form exhibited by compound rotation is shown in Fig. 3.10. This phenomenon appears supercritically from a Hopf bifurcation out of simple rotational motion, as found by experiment. Thus, typical compound rotation is quasi-periodic. Here, even when changing the system parameters, the two fundamental wave numbers (those corresponding to the distance around the inner circle and the distance between arms) do not lock into some rational relationship. Similar compound rotational motion has been observed in numerical simulations of the FN equation, but at present no theoretical explanation exists for this type of behavior.

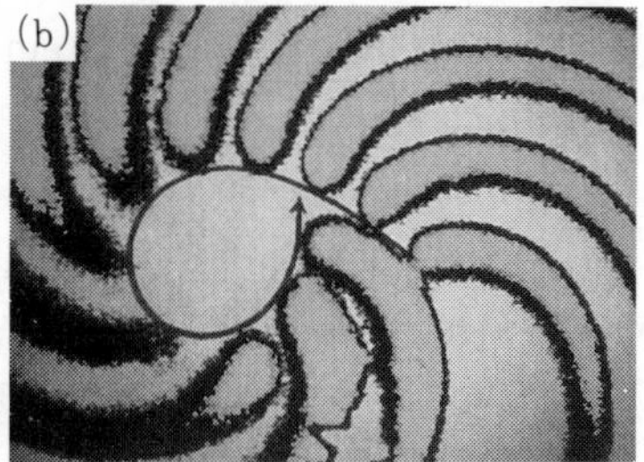

Fig. 3.10. Simple and complex rotation of a spiral wave. (**a**) The endpoints of the spiral arms move in such a way as to describe a stationary circle. (**b**) When the concentration of $KBrO_3$ is changed, this circular motion destabilizes and complex rotation (biperiodic motion) results, [from Skinner and Swinney (1991)]

3.4.2 The Turing Pattern

An additional noteworthy result obtained using an experimental method similar to that described above is the discovery of the *Turing pattern* [Ouyang and Swinney (1991)]. We will briefly discuss the experiment itself below, but first let us discuss the Turing instability. In this subsection we stray somewhat from interface dynamics, which is the main theme of this chapter.

There exist cases of reaction–diffusion systems in which a spatially uniform stationary state is unstable with respect to inhomogeneous fluctuations and in which nonuniform structure appears spontaneously. Such phenomena result from the *Turing instability*, named after A. Turing, who theoretically demonstrated its existence in 1952. With respect to questions concerning morphological development in biological systems and, in particular, the naive question, "Why from a uniform and spatially isotropic egg cell does a form arise which breaks this original symmetry?", which touches on the very root of biological development, Turing's discovery provides a metaphorical answer, and for this and other reasons, it came to generate a great deal of interest. It is thought that I. Prigogine's formulation of the concept of *dissipative structure* was based strongly on consideration of the Turing pattern, and the mathematical analysis of many hypothetical reaction–diffusion models was actively carried out by a number of researchers who were to originate the 'Prigogine school'. However, despite the great activity that it generated, evidence confirming Turing's prediction was not found in a real reaction–diffusion system for nearly 40 years.

The interesting feature of the Turing instability is that, because of the existence of diffusion, the spatial inhomogeneity of the system becomes more pronounced. The mathematical explanation and physical interpretation of the cause behind this behavior are given below.

Let us consider the two-variable system $\dot{u} = f(u,v)$, $\dot{v} = g(u,v)$, with nullclines intersecting as shown in Fig. 3.11. To study the stability of the stationary point S, we linearize the system around this point. This yields

$$du/dt = au - bv \,, \tag{3.49a}$$

$$dv/dt = cu - dv \,, \tag{3.49b}$$

where we have redefined u and v to represent the distance of each variable from S. From the signs of $\dot{u}$ and $\dot{v}$, which can be read off Fig. 3.11, we are able to assume that $a, b, c, d > 0$. Also, from the manner in which the nullclines intersect, we obtain the inequality

$$bc > ad \,. \tag{3.50}$$

From the quadratic equation for the linear growth rate λ of u and v, we know that if the inequality

$$d > a \tag{3.51}$$

is satisfied, S is stable, while otherwise it becomes unstable with respect to the growth of oscillatory motion. In each section of this chapter, we have considered the case in which one of the variables has a characteristic timescale ϵ which is extremely short. In the present case this implies that $a \gg d$, and thus that S is unstable. However, below we study the case in which (3.50) and the stability condition (3.51) are both satisfied.

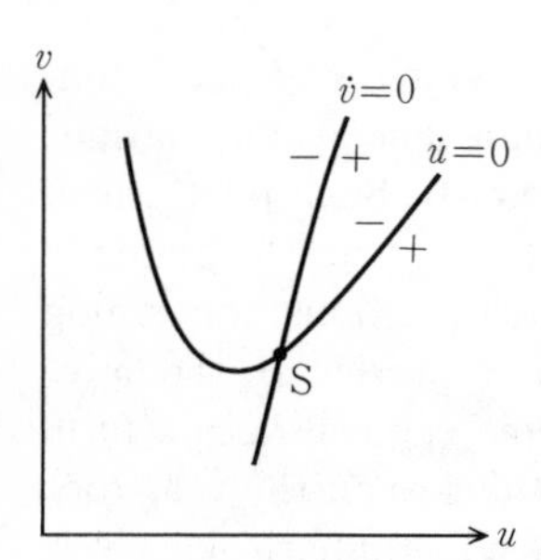

Fig. 3.11. The crossing of the nullclines in an activator–inhibitor system

In some cases, it is possible for the elements in a system such as this, each individually possessing some stable stationary state S, to become mutually coupled by the addition of diffusion terms, and for the state S to become unstable as a result. For the sake of simplicity, let us consider a system in one spatial dimension and add diffusion terms $D\partial^2 u/\partial x^2$ and $D'\partial^2 v/\partial x^2$ to (3.49a) and (3.49b). Then, for the Fourier modes u_k and v_k of the resulting linear reaction–diffusion system of equations, we have

$$du_k/dt = (a - Dk^2)u_k - bv_k \,, \tag{3.52a}$$

$$dv_k/dt = cu_k - (d + D'k^2)v_k \,. \tag{3.52b}$$

This is of the same form as (3.49), with $a - Dk^2$ and $d + D'k^2$ replacing a and d. Thus the necessary and sufficient conditions for the uniform stationary state to be stable for a given value of k become

$$bc > (a - Dk^2)(d + D'k^2) \tag{3.53}$$

$$d + D'k^2 > a - Dk^2 \,. \tag{3.54}$$

Under condition (3.51), the inequality (3.54) is automatically satisfied, and thus the oscillatory instability cannot arise. However, there are situations in which the inequality (3.53) is violated. In particular, for any given value of k, by choosing a sufficiently small value of D and a sufficiently large value of D' we can cause this inequality to become reversed. The Turing point is that at which the inequality (3.53) becomes an equality. This occurs at a particular value of k, which we denote by k_c. Thus this instability is essentially of the same type as that resulting in the appearance of Bénard convection, and thus, in a system of sufficiently large spatial extent, it should be possible to observe a periodic pattern.

How can we intuitively understand the cause of this type of instability, brought on by diffusion? For this purpose, it is first necessary to consider the mechanism responsible for maintaining the stability of the stationary state in an elemental mechanical system such as that considered here. As seen from (3.49a), in the case in which v does not exist, the $u = 0$ state is unstable and u will grow. On the other hand, (3.49b) expresses the situation in which the growth of the inhibitor v is caused solely by the presence of component u. Constituents u and v with this type of reciprocal relationship are referred to, respectively, as an *activator* and an *inhibitor*. Here, maintenance of the stability of the stationary state depends on the predominance of the inhibiting effect of component v. If this effect is not sufficiently strong, the characteristic instability of u is realized. Then, introducing diffusion gives rise to a change in the restraining effect experienced by the system. In particular, when the diffusion of the v component is relatively fast, the strength of the restraining effect diminishes. To see why this is the case, consider the situation in which, for whatever reason, the local concentration of the activator becomes high in some given region. In this situation, inhibitor will be produced, but, despite this production, before the inhibitor is able to restrain the growth of the activator, inhibitor will flow out of the region in which it was produced into the surrounding region. For this reason, not only will the inhibitor be ineffective in restraining the growth of the activator in the region in question, but its outflow into the surrounding region will over-suppress the growth of u (and even cause u to become negative) there. In this way, the inhomogeneity of the u component can be amplified by diffusion.

In Fig. 3.12, several Turing patterns observed by Ouyang and Swinney are shown. Depending on the conditions present, hexagonal or striped periodic structure appears. The system used here consists of the so-called CIMA (chloride–iodide–malonic acid) reaction, which can be used as an alternative to the BZ reaction. An interesting point regarding this experiment is the fact that by continuously changing the temperature of the medium, Ouyang and Swinney succeeded in demonstrating that the appearance of such patterns can be understood as bifurcation phenomena.

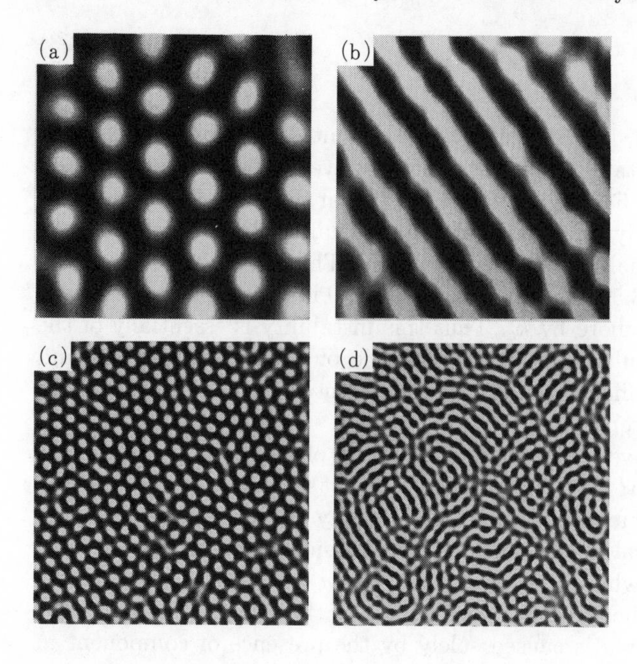

Fig. 3.12. Several examples of the Turing pattern in a system undergoing the CIMA reaction (figure used by permission of H.L. Swinney)

3.5 The Instability of Interfaces and Formation of Structure

The change in shape of a physical object can be thought of as the change in shape of the boundary separating the body's internal region from the external world. In the realm of nonequilibrium open systems, we can observe a large number of phenomena in which a flat boundary of this type spontaneously changes shape to take on various different morphologies. Morphological development in biological systems represents the most conspicuous example of this nature, but there are examples among nonliving systems as well, the growth of snowflakes and dendritic crystals being typical examples. Reaction–diffusion models are often useful in the study of these types of phenomena, while treatments focusing on interface dynamics have also been found to be quite effective. In what follows, we consider crystal growth in a pure supercooled melt and discuss a simple macroscopic model, the phase field model, that deals with the stability of the interface along with the morphological development following its destabilization.

The phase field model, proposed by R. Kobayashi in 1993, is not limited in application to crystal growth, but captures a certain universal aspect of the dynamics of interfaces separating two different phases. This model is similar to the two-component excitable system model described earlier in this chapter in the sense that it can be thought of as a model obtained from a one-component bistable evolution equation by the reinterpretation of a parameter appearing in this equation as a variable. The first equation in this system is as follows:

$$\tau \frac{\partial u}{\partial t} = \epsilon^{-1} u(1-u)(u-a) + \nabla^2 u \,. \tag{3.55}$$

Here, τ is a positive parameter, and for the time being, a is also considered to be a parameter satisfying the condition $0 < a < 1$. Changing the time scale and applying a linear transformation to u, it is clear that (3.55) can be made to assume the form of the bistable reaction–diffusion system defined by (3.1) and (3.3).

The bistable model (3.55) possesses a potential Ψ corresponding to the potential (3.7) for the model (3.1). The points $u = 0$ and $u = 1$ represent minima of this potential. We will interpret the former as a liquid phase and the latter as a solid phase for our system. Thus we can interpret Ψ as a free energy. If $a < 1/2$, the solid phase has the lower energy of the two states and is therefore more stable. For $a > 1/2$, the reverse holds. Now, thinking in physical terms, the relative stability of the two phases should depend strongly on the temperature T, and therefore it is appropriate to consider a as a function of the temperature. Furthermore, since the solid phase should be more stable at low temperature, we can assume that a is a monotonically increasing function of T. Let us refer to the temperature at which $a(T) = 1/2$ as T_0. This is the melting temperature of the system. Then, choosing the temperature of some supercooled liquid state of the system as the reference temperature, $T = 0$, we have $T_0 > 0$.

In the situation in which a solid–liquid interface exists and its position evolves over time, it is not feasible for the temperature to be uniform throughout the system. On very general grounds, we expect the temperature to obey a diffusion equation, but in the present case, the latent heat effect accompanying the phase change must also be accounted for. Then, assuming the rate of diffusion of the temperature to be equal in the two phases, we can write the evolution equation for the temperature in the form

$$\frac{\partial T}{\partial t} = \nabla^2 T + K \frac{\partial u}{\partial t} \,, \tag{3.56}$$

where K is a positive parameter that expresses the strength of the latent heat effect. The phase field model is defined by the two-component reaction–diffusion system constituted by (3.55) and (3.56). However, in order to use these as a model of crystal growth, it is necessary to introduce an anisotropy effect. This is done by assuming the parameters τ, ϵ, and a to be direction-dependent quantities. However, for the sake of simplicity, we will ignore any anisotropic effect in the following discussion.

The parameter τ is included in (3.55) in order to allow us the freedom to choose the time scale such that the diffusion coefficient in the temperature equation of motion is 1. If we now think of a (and thus τ) as a fixed parameter, it is clear that (3.55) possesses a one-dimensional steady-state traveling wave solution corresponding to (3.9a), given by

$$u(x,t) = \tfrac{1}{2}\left(1 - \tanh \tfrac{x-\phi}{2\sqrt{2\epsilon}}\right), \tag{3.57}$$

$$\phi = c(T)t\,.$$

In the two-dimensional system of interest to be considered below, this solution represents a traveling wave with an interface possessing only one-dimensional structure (i.e., a perfectly straight interface). In correspondence with (3.9b), c is given by

$$c(T) = \tau^{-1}\sqrt{\frac{2}{\epsilon}}\left(\tfrac{1}{2} - a\right)\,. \tag{3.58}$$

In an actual system, $a(T)$ is a variable, changing in both time and space, and hence it is not easy to find a steady-state traveling wave solution $u(x - ct), T(x - ct)$ and to investigate its stability in two dimensions in a general manner. Thus, to make the analysis more simple, we consider the $\epsilon \to 0$ limit. In this case, (3.57) reduces to

$$u(x,y,t) = \begin{cases} 1 & (x < \phi)\,, \\ 0 & (x > \phi)\,. \end{cases} \tag{3.59}$$

The fact that in the limit as $\epsilon \to 0$ the expression for u reduces to this form for the simple case considered above leads us to believe that the same expression will be valid in the more realistic case in which $a(T)$ changes with x, y and t and in which the interface coordinate ϕ possesses a y dependence. For this reason, we wrote $u(x,y,t)$ in place of $u(x,t)$ in the above equation. However, in this more complicated case, as in Sect. 1.3, we must rederive the propagation speed. We now turn to this derivation.

We first fix T appearing in a at the value $T(x = \phi)$ corresponding to the interface $x = \phi(y,t)$. Then, the effect of the local curvature of the interface, κ, will appear. With this newly defined system, c represents the propagation speed in the direction normal to the interface, and its relationship with $\partial\phi/\partial t$ is given by

$$\frac{\partial \phi}{\partial t} = c\sqrt{1 + \left(\frac{\partial \phi}{\partial y}\right)^2}\,. \tag{3.60}$$

Thus, from (3.58) we obtain the expression

$$\frac{\partial \phi}{\partial t} = \tau^{-1}\sqrt{1 + \left(\frac{\partial \phi}{\partial y}\right)^2}\left[\sqrt{\frac{2}{\epsilon}}\left\{\frac{1}{2} - a(T(x = \phi))\right\} - \kappa\right]\,. \tag{3.61}$$

If we can find $T(x = \phi)$, this will become an equation describing the evolution of the curved interface $\phi(y,t)$ in a closed form.

In order to determine $T(x = \phi)$, we first note that the term representing the latent heat production becomes $Kc\delta(x - \phi)$ in the case that we consider. From this, we see that the temperature gradient in the direction normal to the interface, $\partial T/\partial n$, satisfies the jump condition

$$(\partial T/\partial n)_{x=\phi(y)^+} - (\partial T/\partial n)_{x=\phi(y)^-} = -Kc\,, \tag{3.62}$$

while the continuity of T itself must of course be maintained:

$$T(x=\phi(y)^+) = T(x=\phi(y)^-)\,. \tag{3.63}$$

The behavior of T is thus determined by the diffusion equation

$$\partial T/\partial t = \nabla^2 T\,. \tag{3.64}$$

Equations (3.61)–(3.64) form a closed system. They constitute the fundamental relations of our description of interface dynamics.

In the following analysis, we first seek a flat, steady-state traveling wave solution as a special solution for the two-dimensional system, and then we consider the stability of this solution. For this kind of steady propagating interface, the propagation speed c assumes a constant value c_0, and $\phi(y,t) = \phi_0 \equiv c_0 t$. If we set the temperature at $x=\infty$ of the supercooled liquid to 0, we can obtain the steady-state traveling wave solution

$$T = \begin{cases} K & (x < \phi_0)\,, \quad (3.65a) \\ K\mathrm{e}^{-c_0(x-\phi_0)} & (x > \phi_0) \quad (3.65b) \end{cases}$$

from (3.61)–(3.64). Then, $T(x=\phi_0) = K$, and thus the propagation speed is given by

$$c_0 = \tau^{-1}\sqrt{\frac{2}{\epsilon}}\left(\frac{1}{2} - a(K)\right)\,. \tag{3.66}$$

Let us investigate the stability of the above flat propagating interface using (3.61)–(3.64). For this purpose, we apply an infinitesimal deformation to the straight interface. As an eigenmode of such a deformation, let us consider a sine wave with wave number k. Representing the growth rate of this wave by λ, we have

$$\phi = c_0 t + \delta\phi\mathrm{e}^{\lambda t}\sin ky\,, \tag{3.67}$$

where $\delta\phi$ is a sufficiently small quantity. As a result of the application of this deformation, the temperature on both sides of the interface will in general also suffer a small change, $\delta T(x,y,t)$. Then, since T obeys a diffusion equation, we expect the spatial variation of an eigenmode of δT to take the form $\exp[-\gamma(x-\phi_0)]\sin ky$. In other words, we assume that T takes the form

$$T = \begin{cases} K + \delta T_s \mathrm{e}^{-\gamma_s(x-c_0t)+\lambda t}\sin ky & (x < \phi)\,, \quad (3.68a) \\ K\mathrm{e}^{-c_0(x-c_0t)} + \delta T_l \mathrm{e}^{-\gamma_l(x-c_0t)+\lambda t}\sin ky & (x > \phi)\,. \quad (3.68b) \end{cases}$$

In fact, this form does satisfy the diffusion equation (3.64), where γ_s and γ_l are the two roots of the quadratic equation $\lambda + c_0\gamma = \gamma^2 - k^2$. Since the inequalities $\gamma_s < 0$ and $\gamma_l > 0$ must be satisfied, we have $\gamma_{s,l} = \frac{1}{2}\{c_0 \mp \sqrt{c_0^2 + 4(k^2+\lambda)}\}$. Substituting (3.67), (3.68a), and (3.68b) into the fundamental equations (3.61)–(3.63) and linearizing in $\delta\phi$ and $\delta T_{s,l}$ gives

$$K(c_0^2 + \lambda)\delta\phi + \gamma_s\delta T_s - \gamma_l\delta T_l = 0 \tag{3.69a}$$

$$-Kc_0\delta\phi - \delta T_s + \delta T_l = 0 \tag{3.69b}$$

$$(\tau\lambda + k^2)\delta\phi + \sqrt{\frac{2}{\epsilon}}a'(K)\delta T_s = 0 \,. \tag{3.69c}$$

Then, by setting the coefficient matrix equation equal to 0, we can obtain the eigenvalue λ. For a long-wavelength mode we can expand λ in powers of k^2, obtaining

$$\lambda = -b_2k^2 - b_4k^4 + \cdots, \tag{3.70}$$

where

$$b_2 = \tau^{-1}\left(1 - \sqrt{\frac{2}{\epsilon}}\frac{Ka'(K)}{c_0}\right) \tag{3.71a}$$

$$b_4 = \sqrt{\frac{2}{\epsilon}}\frac{Ka'(K)}{4\tau c_0^3}(1-b_2)(7+b_2) \,. \tag{3.71b}$$

Under the condition $c_0 < \sqrt{2/\epsilon}Ka'(K)$, b_2 becomes negative. In this situation the interface is unstable with respect to deformations of sufficiently long wavelength. This phase instability is similar to the zig-zag instability discussed in Chap. 2, but in the present context of interface instability, it is referred to as the *Mullins–Sekerka* instability. In the neighborhood of the instability point, the inequality $b_4 > 0$ holds, and thus the interface is stable with respect to short-wavelength deformations.

The reason for the appearance of the Mullins–Sekerka instability can be understood qualitatively in the following way. Consider a propagating front which, by virtue of some cause, develops a local uneven region. The curvature effect will tend to flatten this uneven region, but there is an additional effect which acts to work against this flattening tendency. To see this, note that, as depicted schematically in Fig. 3.13 in the neighborhood of the interface, the isotherms do not necessarily run parallel to the interface. In this situation, along the interface, an inhomogeneity in the temperature appears. In particular, when the rate of heat diffusion is fast, the isotherms will become straight. Stronger diffusion implies straighter curves. Thus the temperature will be lowered for the leading portion P, while for the lagging portions Q and Q′, the temperature will be elevated. Now, keeping in mind that the propagation velocity of the interface is a decreasing function of temperature, we see that the propagation speed of a leading region will increase as a result, while that of a lagging region will decrease. This implies that the deformation of the front will tend to grow larger. From the relative strengths of this tendency toward the destabilization of a flat front and the stabilizing tendency introduced by the curvature effect, the ultimate stability of the front is determined. We can already see these two opposing effects in the interface evolution equation, (3.61); namely, the term proportional to the curvature κ represents the tendency toward stabilization. Working against this tendency is the destabilizing effect proportional to $-a'(K)\delta T$, which tends to produce a velocity difference along the interface.

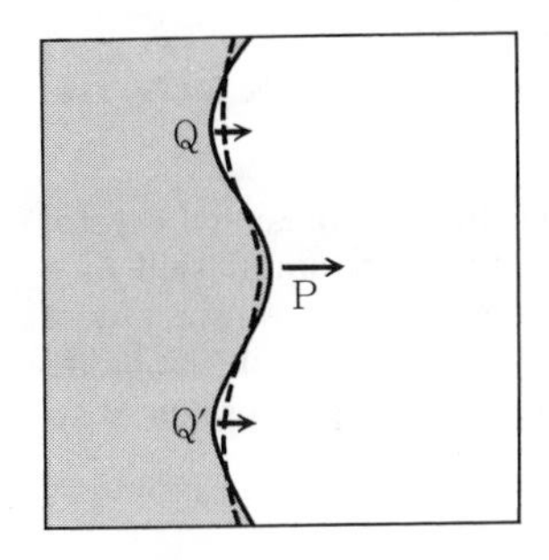

Fig. 3.13. A curved interface (*solid curve*) and the isotherm in its neighborhood (*broken curve*). In comparison with the case of a flat interface, the temperature at the point P is slightly decreased, while those at Q and Q′ are slightly increased

We can draw an analogy between the interface destabilizing mechanism described above and the mechanism of the Turing instability in the activator–inhibitor model discussed in the previous section. In this case, the order parameter u plays the role of the activator, while the temperature T plays the role of the inhibitor. In the case in which the diffusion rate of heat is sufficiently fast, the inhibitor T moves quickly out of the protruding region, and because of the resulting decrease in the strength of the restraining effect in this region, its speed increases further. Contrastingly, in the depressed region, an excess of inhibitor will build up, causing the propagation of this region to become increasingly retarded. However, this interface instability begins at infinitely long wavelengths, in contrast to the Turing instability. This arises from the spatial translational invariance of the system and the fact that the spatial translation of the interface represents a marginally stable mode.

The linear stability theory discussed above can be extended to the realm of a weakly nonlinear theory, and the evolution equation of the interface coordinates can be derived approximately from the point of view of the phase dynamics discussed in the next chapter. This point will be reconsidered in Sect. 4.3. However, we should note that at present there exists no general theory capable of dealing with very strongly nonlinear systems, such as dendritic crystal growth. For this reason, in most cases when dealing with such problems, the only method of analysis is the direct numerical simulation of model equations. For the case of the phase field model, there is a detailed numerical analysis [Kobayashi (1993)]. Figure 3.14 displays one result of this analysis. With these computations, by inserting a parameter to account for the anisotropy of the system, patterns can be obtained which are surprisingly similar to those produced in real systems.

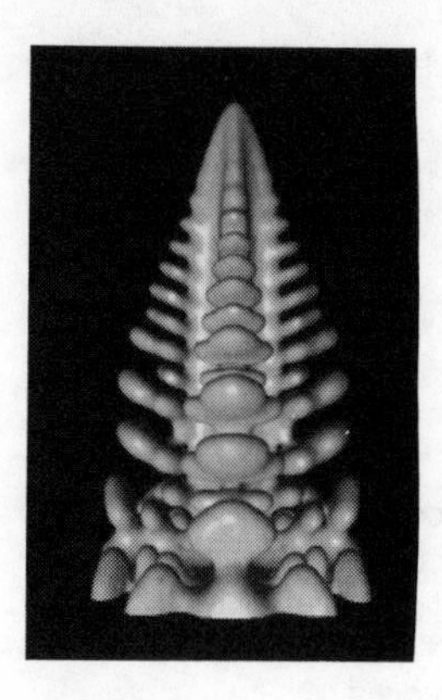

Fig. 3.14. An example of the three-dimensional crystal growth pattern obtained by numerical simulation of a phase field model, taking into account the anisotropy of the function $a(T)$. The surface of the object in the figure corresponds to $u = 1/2$ (figure used by permission of R. Kobayashi)

4. Phase Dynamics

In the previous two chapters, we considered several different types of nonequilibrium dissipative fields. There, the physical point of view defined by the statement, "Slow degrees of freedom govern the dynamics of the system," was extremely effective. In Chap. 2, the weakly unstable mode in the neighborhood of the bifurcation point served as the slow degree of freedom, while in Chap. 3 this was the concentration of the inhibiting substance present. An important slow mode that emerges in conjunction with pattern dynamics is the phase degree of freedom. In this chapter, utilizing this fact, we outline a phenomenological theory consisting of a reduction method, known as *phase dynamics*. We then apply this theory to several concrete examples.

When a nonequilibrium pattern exists in an extended field, it is often the case that the continuous symmetry of the system becomes spontaneously broken. In general, in this situation a marginally stable quantity called a *phase* exists. Because of its marginally stable nature, the phase mode possesses an unlimitedly long time scale. This mode is similar to the critical mode arising in bifurcation phenomena, and just as it is possible to obtain a closed description of the critical mode, it is possible to obtain a closed description of the phase mode.

Phase dynamics is an approach that is widely applicable to fluid systems, to reaction–diffusion systems, and in general to dissipative systems in which a phase can be defined. In this chapter we derive phase field equations of motion in a phenomenological manner, much as we did for the amplitude equations in Chaps. 1 and 2. We leave the theoretical foundation for this treatment to Chap. 5.

4.1 Weak Turbulence of Periodic Structures and the Phase Equation

Let us consider an extended, uniform, isotropic field in two dimensions in which a periodic one-dimensional structure exists at the initial time. A Bénard convection system with large aspect ratio is one example in which such a situation can exist. In the case of thermal convection the field variables in such a system represent each component of the flow velocity in addition to the temperature and pressure, and in the case of a reaction–diffusion system, they represent the concentration of each chemical species. In a more abstract system, for example, a system of the type we reduced to amplitude equations in Chap. 2, the amplitude W is considered as the field variable.

In general, let us express one of the field variables by $X(\boldsymbol{r}, t)$. For a stationary periodic pattern possessing one-dimensional structure with wave number k, in general X can be represented as

$$X(\boldsymbol{r}, t) = X_0(\phi), \qquad \phi = kx + \psi \,. \tag{4.1}$$

Here, $X_0(\phi)$ is a 2π-periodic function of ϕ, and ψ is an arbitrary constant. In general, a pattern whose appearance destroys spatial translational invariance carries with it such an arbitrary phase constant. In addition to periodicity, we assume that X_0 possesses spatial inversion symmetry:

$$X_0(-\phi) = X_0(\phi) \,. \tag{4.2}$$

Now, let us assume that such a pattern is subject to a long-scale spatial modulation. For example, in the case of thermal convection we suppose that the roll axes suffer a gentle but large-amplitude bending, and that as a result the distance between rolls, when observed on a large spatial scale, displays an inhomogeneity. Locally, however, such a system appears to possess a well-preserved periodic structure. In general, on a global scale, such a pattern cannot be represented even approximately by the addition of a small correctional term to the left-hand side of the first equation in (4.1). This becomes clear if one considers the special case in which the pattern is not deformed but simply translated uniformly by some finite distance.

A situation quite similar to this is encountered in systems subject to secular perturbations, such as an oscillator whose period has been thus modulated. It has been found that such systems can be successfully treated by applying the idea of 'absorbing' the secular effect. Analogously, in the case that we are presently considering, it is possible to think in terms of a method of absorbing the spatial modulation of the pattern into a spatial dependence of the phase ψ. In this way, we expect that the form (4.1) can be made a sufficiently good approximation of the true solution.

When a stationary pattern suffers a spatial deformation, in general this deformation will possess some nontrivial time dependence. In our treatment, a time dependence of this type thus also comes to be absorbed into ψ. We expect that the time evolution of such a spatial deformation will grow slower as the length scale of the deformation increases, and that in the limit of spatial uniformity, ψ will become some arbitrary constant; that is, $\dot{\psi} = 0$. Recall that we used similar reasoning when deriving amplitude equations near bifurcation points. In that case, at the critical point we reinterpreted the arbitrary constant W as a quantity depending on time and space variables, and we absorbed the gentle temporal and spatial fluctuations of the pattern into this quantity. Thus (4.1) corresponds to the neutral solution represented by (1.8).

What kind of equation of motion does $\psi(x, y, t)$ obey? In Chap. 5 we attempt to answer this question in an orthodox fashion, beginning from the condition that (4.1) is an approximate solution to the original evolution equation. Here, however, let us find the form of the phase equation from phenomenological considerations.

First, we assume that $\dot{\psi}$ can be expressed as a function of ψ and its spatial derivatives. We set

$$\frac{\partial \psi}{\partial t} = H\left(\psi, \frac{\partial \psi}{\partial x}, \frac{\partial \psi}{\partial y}, \frac{\partial^2 \psi}{\partial x^2}, \frac{\partial^2 \psi}{\partial y^2}, \cdots\right) . \tag{4.3}$$

Now, the system state $X_0(\psi)$ is invariant with respect to transformations of the following three types:

(i) $\psi \to \psi + \psi_0$, $x \to x - k^{-1}\psi_0$ (ψ_0 arbitrary),

(ii) $\psi \to -\psi$, $x \to -x$,

(iii) $y \to -y$.

Therefore $\dot{\psi}$ too must satisfy these invariance conditions. First, from (i) we see that H cannot contain a bare ψ dependence (i.e., only its spatial derivatives can appear). Then, if we expand H in derivatives of ψ, invariance with respect to (ii) and (iii) determines the types of terms that can appear. Let us write out the lowest-order allowed terms. To second order in ψ, we have

$$\begin{aligned} \frac{\partial \psi}{\partial t} &= a_2\frac{\partial^2 \psi}{\partial x^2} + b_2\frac{\partial^2 \psi}{\partial y^2} - a_4\frac{\partial^4 \psi}{\partial x^4} - b_4\frac{\partial^4 \psi}{\partial y^4} + \left(g_1\frac{\partial^2 \psi}{\partial x^2} + g_2\frac{\partial^2 \psi}{\partial y^2}\right)\frac{\partial \psi}{\partial x} \\ &\quad + \left\{h_1\frac{\partial \psi}{\partial x} + h_2\left(\frac{\partial \psi}{\partial y}\right)^2\right\}\frac{\partial^2 \psi}{\partial y^2} + h_3\frac{\partial \psi}{\partial x}\left(\frac{\partial \psi}{\partial y}\right)^2 + \cdots . \end{aligned} \tag{4.4}$$

Considering a pattern deformation of sufficiently gentle spatial modulation, it is necessary to consider only the lowest-order terms in the above expansion. Hence we obtain the anisotropic phase diffusion equation

$$\frac{\partial \psi}{\partial t} = a_2\frac{\partial^2 \psi}{\partial x^2} + b_2\frac{\partial^2 \psi}{\partial y^2} . \tag{4.5}$$

If both a_2 and b_2 are positive, the deformation eventually disappears due to diffusion. Applying (4.5) to a stationary roll of the Newell–Whitehead equation with wave number δk (2.4a,b), by comparison with the eigenvalue of the phase branch of (2.7), we find that the coefficients of the phase diffusion are given by

$$a_2(k) = \frac{\mu - 3(\delta k)^2}{\mu - (\delta k)^2} , \tag{4.6a}$$

$$b_2(k) = \frac{\delta k}{k_c} , \tag{4.6b}$$

where $k = k_c + \delta k$.

In the region of phase instability, at least one of the diffusion coefficients becomes negative, and the phase diffusion equation loses its validity. In this case, higher-order derivatives from the above expansion or nonlinear terms must be included. In the case in which the instability is weak, from scaling considerations that we discuss below, we can determine the leading terms of these higher-order derivatives. Let us first study the system in the neighborhood of the Eckhaus instability point, corresponding to $a_2 = 0$. For this purpose, we set $a_2 = -\epsilon$. (We assume that $b_2 > 0$.) The small parameter ϵ introduces the long space and time scales into ψ. We write the characteristic scales of t, x, and y as $|\epsilon|^{-\nu}$, $|\epsilon|^{-\mu_1}$, and $|\epsilon|^{-\mu_2}$. Then, let us further assume that the characteristic size of the envelope of ψ varies as $|\epsilon|^{\delta}$ ($\delta \geq 0$). As a result, the scaling form of ψ becomes

$$\psi(x, y, t) = |\epsilon|^{\delta} f(|\epsilon|^{\mu_1} x, |\epsilon|^{\mu_2} y, |\epsilon|^{\nu} t) . \tag{4.7}$$

If we substitute this form into (4.4), the size of each term on the right-hand side becomes expressed as a power (appearing as some combination of μ_1, μ_2 and δ) of ϵ: for example, $-\epsilon\partial^2\psi/\partial x^2 \sim |\epsilon|^{1+2\mu_1+\delta}$, $g_2(\partial^2\psi/\partial y^2)(\partial\psi/\partial x) \sim |\epsilon|^{2\mu_2+\mu_1+2\delta}$, etc.

If the sizes of the exponents μ_1, μ_2, and δ were known a priori, the selection of the leading terms would be clear at this point. However, in actual practice, it is not possible to determine these exponents in the absence of physical considerations. We thus proceed by making some simple, physically reasonable assumptions. First, we cannot drop the term responsible for the phase instability, $-\epsilon\partial^2\psi/\partial x^2$. Then, the the lowest-order term preserving the stability along the y direction, $\partial^2\psi/\partial y^2$, and the term insuring attenuation in the small-wavelength region, $\partial^4\psi/\partial x^4$, are clearly necessary. From the condition that all three of these are of roughly 'equal importance' (more precisely, that they are of the same order in ϵ) we obtain the relations $\mu_1 = 1/2$ and $\mu_2 = 1$. Then, we need at least one nonlinear term. This choice too is quite simple, because we know immediately that the largest such term (independent of the size of δ, provided that $\delta \geq 0$) is $(\partial^2\psi/\partial x^2)(\partial\psi/\partial x)$, which is of order $|\epsilon|^{3\mu_1+2\delta}$. Then, the order of this term must be equal to that of the above linear terms, and thus we have the condition $\delta = 1/2$. Similar considerations for $\dot{\psi}$ give $\nu = 2$. We thus obtain the nonlinear phase equation

$$\frac{\partial\psi}{\partial t} = a_2\frac{\partial^2\psi}{\partial x^2} + b_2\frac{\partial^2\psi}{\partial y^2} - a_4\frac{\partial^4\psi}{\partial x^4} + g\frac{\partial\psi}{\partial x^2}\frac{\partial\psi}{\partial x} . \tag{4.8}$$

The same kinds of arguments can also be used in the neighborhood of the zig-zag instability. In this case, we set $b_2 = -\epsilon$ (with $a_2 > 0$), and once again we assume the scaling form (4.7). In this case, we obtain $\mu_1 = 1$, $\mu_2 = 1/2$, $\delta = 0$, and $\nu = 2$, and the phase equation takes on the following form:

$$\begin{aligned}\frac{\partial\psi}{\partial t} &= b_2\frac{\partial^2\psi}{\partial y^2} - b_4\frac{\partial^4\psi}{\partial y^4} + a_2\frac{\partial^2\psi}{\partial x^2} + h_1\frac{\partial\psi}{\partial x}\frac{\partial^2\psi}{\partial y^2} \\ &\quad + h_2\left(\frac{\partial\psi}{\partial y}\right)^2\frac{\partial^2\psi}{\partial y^2} + h_3\frac{\partial^2\psi}{\partial x\partial y}\frac{\partial\psi}{\partial y} .\end{aligned} \tag{4.9}$$

The nonlinear terms in (4.8) and (4.9) have simple physical interpretations, and, using this, we can find a general equation for their coefficients. First, in order to discuss (4.8), let us assume that the following is a solution of this equation:

$$\psi(x, y, t) = \kappa x + \psi'(x, y, t) \; . \tag{4.10}$$

Here, we assume that ψ' is sufficiently small and that the spatial modulation is gradual enough that we can ignore fourth-order spatial derivatives. With these assumptions, the equation of motion for ψ' becomes

$$\frac{\partial \psi'}{\partial t} = (a_2(k) + g\kappa)\frac{\partial^2 \psi'}{\partial x^2} + b_2(k)\frac{\partial^2 \psi'}{\partial y^2} \; . \tag{4.11}$$

The derivation of (4.8) is based on the premise that the spatial modulation of ψ is very gradual, and thus to avoid a contradiction, we must assume here that $|k|$ is small. We point out that, as can be seen from the form of (4.10), ψ' represents a small phase disturbance of the wave vector $(k+\kappa, 0)$ periodic pattern. Therefore the conditions $a_2(k) + g\kappa = a_2(k+\kappa)$ and $|\kappa/k| \ll 1$, concerning diffusion in the x-direction, must be satisfied. This implies that

$$g = \frac{\mathrm{d}a_2(k)}{\mathrm{d}k} \; . \tag{4.12}$$

In order to determine the coefficient of the nonlinear term in (4.9), let us consider

$$\psi = \kappa_1 x + \kappa_2 y + \psi'(x, y, t) \tag{4.13}$$

as a solution to this equation.

Making assumptions with regard to ψ' that are similar to those made above, the relation

$$\frac{\partial \psi'}{\partial t} = a_2(k)\frac{\partial^2 \psi'}{\partial x^2} + (b_2(k) + h_1\kappa_1 + h_2\kappa_2^2)\frac{\partial^2 \psi'}{\partial y^2} + h_3\kappa_2\frac{\partial^2 \psi'}{\partial x \partial y} \tag{4.14}$$

holds. Now, ψ' represents a small phase disturbance of the periodic wave pattern characterized by the wave vector $(k+\kappa_1, \kappa_2)$. Let us next rotate the coordinate system by a small angle θ and introduce the orthogonal coordinate system ξ–η (Fig. 4.1) for which the wave vector in question lies entirely along the ξ-axis. In this new coordinate system, the phase diffusion equation becomes

$$\frac{\partial \psi'}{\partial t} = a_2(k')\frac{\partial^2 \psi'}{\partial \xi^2} + b_2(k')\frac{\partial^2 \psi'}{\partial \eta^2} \; , \tag{4.15}$$

where $k' = \{(k+\kappa_1)^2 + \kappa_2^2\}^{1/2}$ is the wave number of the new periodic pattern. Using the identities $x = \xi\cos\phi - \eta\sin\theta$ and $y = \xi\sin\theta + \eta\cos\theta$, (4.15) can be transformed into the form of (4.14). Then, comparing coefficients of these two expressions, we find that

$$b_2(k) + h_1\kappa_1 + h_2\kappa_2^2 = a_2(k')\sin^2\theta + b_2(k')\cos^2\theta \tag{4.16a}$$

$$h_3\kappa_2 = 2\{a_2(k') - b_2(k')\}\sin\theta\cos\theta\ . \tag{4.16b}$$

Next, noting the relations $\cos\theta = (k+\kappa_1)/k'$ and $\sin\theta = \kappa_2/k'$, and requiring the equalities (4.16a) and (4.16b) to hold to first order in κ_1 and second order in κ_2, we obtain the following formulas for the coefficients of the nonlinear terms:

$$h_1 = \frac{\mathrm{d}b_2(k)}{\mathrm{d}k} \tag{4.17a}$$

$$h_2 = \frac{a_2(k) - b_2(k)}{k^2} + \frac{1}{2k}\frac{\mathrm{d}b_2(k)}{\mathrm{d}k} \tag{4.17b}$$

$$h_3 = \frac{2}{k}\{a_2(k) - b_2(k)\}\ . \tag{4.17c}$$

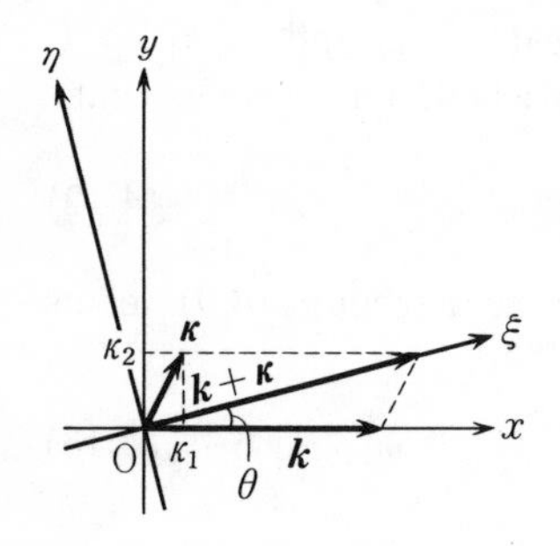

Fig. 4.1. The orthogonal coordinate system ξ–η obtained from x–y by a small-angle rotation

Now, applying (4.12) and (4.17a)–(4.17c) to the phase equation corresponding to a stationary periodic pattern of the NW equation, we can determine explicit expressions for the various coefficients in terms of the fundamental parameters of the system. The coefficients a_2 and b_2 are already determined by (4.6a) and (4.6b). Then, a_4 and b_4 can be determined from the coefficients of the q_x^4 and q_y^4 terms in the eigenvalue equation (2.7):

$$a_4 = \frac{2(\delta k)^4}{\{\mu - (\delta k)^2\}^3}\ , \tag{4.18a}$$

$$b_4 = \frac{1}{4k_c^2}\ . \tag{4.18b}$$

In determining the coefficients of the nonlinear terms, there is one point we need to keep in mind which has not yet been mentioned. Solutions to the NW equation of the form (2.4) exist only in the case $\delta k < \mu^{1/2}$, as can clearly be inferred from the amplitude, (2.4b). Thus, in the equations appearing above for g ,h_1 ,h_2 , and h_3 we can consider $|\delta k|$ to be $O(\mu^{1/2})$. Then, retaining only terms of lowest order in μ, we have

$$g = -\frac{4\mu\delta k}{\{\mu - (\delta k)^2\}^2}, \tag{4.19a}$$

$$h_1 = 1/k_c, \tag{4.19b}$$

$$h_2 = \frac{1}{2k_c^2}\frac{3\mu - 7(\delta k)^2}{\mu - (\delta k)^2}, \tag{4.19c}$$

$$h_3 = \frac{2\{\mu - 3(\delta k)^2\}}{k_c\{\mu - (\delta k)^2\}}. \tag{4.19d}$$

Using the nonlinear phase equations (4.8) and (4.9), we can, to a certain extent, predict the state to which the pattern will ultimately converge as a result of the phase instability. Let us first assume that $a_4, b_4 > 0$. Now, considering the Eckhaus instability, an inhomogeneous perturbation depending only on y will be damped. We can thus assume that the y dependence of ψ in this case is not particularly important; we ignore it. So, instead of (4.8), we consider the relation

$$\frac{\partial \psi}{\partial t} = a_2 \frac{\partial^2 \psi}{\partial x^2} - a_4 \frac{\partial^4 \psi}{\partial x^4} + g \frac{\partial^2 \psi}{\partial x^2}\frac{\partial \psi}{\partial x}. \tag{4.20}$$

Then, subsuming the nonlinear term, the effective coefficient for the local phase diffusion term becomes $a_2 + g(\partial\psi/\partial x)$, and thus in this case, acceleration phenomena associated with destabilization will appear. The reason for this is clear: local regions will emerge in which the phase gradient $\partial\psi/\partial x$ has a sign opposite to that of g. Then, if $a_2 < |g(\partial\psi/\partial x)|$, the effective diffusion coefficient will come to assume an increasingly large negative value, causing the magnitude of the phase deformation to become increasingly large. In fact, according to numerical analysis of (4.8), its solution diverges as a result of the spatial concentration of inhomogeneity. Allowing for a y dependence of ψ and including terms containing $\partial\psi/\partial y$ would not change this result.

It may be thought that the higher-order terms not considered in (4.8) would suppress this instability. In particular, the presence of the term $(\partial\psi/\partial x)^2\partial^2\psi/\partial x^2$ with a positive coefficient would have such an effect. However, from a slightly extended version of the treatment given above concerning the determination of the coefficients, it is known that for the NW equation, the coefficient for this term is $-2\mu\{\mu + 3(\delta k)^2\}/\{\mu - (\delta k)^2\}^3 < 0$. It is expected that the situation is similar for other higher-order terms, and thus it can be thought that the growth of the phase disturbance caused by the Eckhaus instability leads to the formation of topological defects (or in the one-dimensional case, to a change in the average wave number), whose appearance destroys the validity of the phase description itself.

In the case of the zig-zag instability, an anisotropic disturbance possessing only an x dependence will become attenuated. This dependence is thus assumed to be unimportant and is ignored. Then, setting $\partial\psi/\partial y = v$, (4.9) becomes

$$\frac{\partial v}{\partial t} = \frac{\partial^2}{\partial y^2}\left(b_2 v - b_4 \frac{\partial^2 v}{\partial y^2} + \frac{h_2}{3} v^3\right). \tag{4.21}$$

This expresses the motion of the conserved order parameter v possessing the Ginzburg–Landau type potential

$$\Psi = \int \mathrm{d}y \left\{ f(v) + \frac{b_4}{2} \left(\frac{\partial v}{\partial y} \right)^2 \right\} , \tag{4.22}$$

$$f(v) = \frac{b_2}{2} v^2 + \frac{h_2}{12} v^4 .$$

Thus if a phase instability occurs, in the case in which $h_2 > 0$, the nonlinear effect serves to stabilize the system, and a pattern possessing a weak phase deformation tends to come to rest. Conversely, if $h_2 < 0$, the nonlinear effect increasingly adds to the instability, and the phase description again fails. As above, we can interpret this behavior as signifying that the deformation has reached the point of producing defects. From (4.19c) we can see that, for the NW equation, either sign for h_2 is possible. In the $h_2 > 0$ case, if $b_2 < 0$, the potential f possesses two minima, given by $v = \pm\{3|b_2|/h_2\}^{1/2}$. In this case there appear two coexisting domains (one corresponding to each of these minima) whose respective areas are in such a proportion as to preserve the total integral of v. In the context of Bénard convection, this coexistence represents the coexistence of two types of order whose wave vectors point in two slightly different directions. In a real system we expect this to be observed as a zig-zag state stationary roll.

4.2 Phase Waves and Phase Turbulence of Oscillating Fields

In Chap. 2 we dealt with the generation of uniform oscillation in an isotropic reaction–diffusion medium. In this section we consider the situation in which this kind of oscillation possesses some pre-existing finite amplitude, and paralleling the discussion given in the previous chapter we derive a phase equation. We then analyze several special characteristic types of behavior displayed by the solutions of these equations.

4.2.1 The Phase Equation of an Oscillating Field and Its Applications

An arbitrary field X that oscillates with angular frequency ω while maintaining its spatial uniformity can be represented as

$$X(\boldsymbol{r}, t) = X_0(\phi), \qquad \phi = \omega_0 t + \psi , \tag{4.23}$$

where $X_0(\phi)$ possesses 2π periodicity. However, in contrast to the solution (4.1), this quantity in general does not possess symmetry with respect to time reversal, represented by (4.2). Let us write the equation of motion for ψ as

$$\partial\psi/\partial t = H(\psi, \nabla\psi, \nabla^2\psi, \cdots) . \tag{4.24}$$

We will proceed by expanding H in spatial derivatives of ψ. The function $X_0(\phi)$ expressed by (4.23) is invariant with respect to the following simultaneous transformations:

(i) $\psi \to \psi + \psi_0,\ t \to \omega_0^{-1}\psi_0$ (ψ_0 arbitrary).

Also, $X_0(\phi)$, like the original reaction–diffusion system itself, must be symmetric with respect to spatial inversion. That is, both are symmetric with respect to the transformation

(ii) $\nabla \to -\nabla$.

These invariance properties must be reflected by the evolution equation for ψ.

From (i), H can depend only on spatial derivatives of ψ. Then, from (ii) we find that in the expansion for H, each term must contain the operator ∇ an even number of times. We thus have

$$\partial\psi/\partial t = a_2\nabla^2\psi + g(\nabla\psi)^2 - a_4\nabla^4\psi + \cdots . \tag{4.25}$$

Formally, if we consider ∇ as a small parameter, to lowest order in ∇, we obtain

$$\partial\psi/\partial t = a\nabla^2\psi + g(\nabla\psi)^2 , \tag{4.26}$$

where we have written the coefficient a_2 simply as a. If we define $\nabla\psi = \boldsymbol{v}$, the above equation corresponds to the following equation for $\boldsymbol{v}$:

$$\partial\boldsymbol{v}/\partial t = a\nabla^2\boldsymbol{v} + 2g\boldsymbol{v}\cdot\nabla\boldsymbol{v} . \tag{4.27}$$

In a single spatial dimension, this equation is known as the *Burgers equation*. This equation can be reduced to a diffusion equation under a nonlinear transformation known as the Hopf–Cole transformation. Almost equivalently, it can be shown that, writing $\psi = g^{-1}a\ln Q$, (4.26) also reduces to a diffusion equation:

$$\partial Q/\partial t = a\nabla^2 Q . \tag{4.28}$$

Let us now return to (4.25). In the neighborhood of the phase instability point, setting $a_2 = -\epsilon$ and following the scaling idea presented in the previous section, we can select the leading terms in an expansion in derivatives of ψ. Let us assume the scaling form of ψ as

$$\psi(\boldsymbol{r}, t) = |\epsilon|^\delta f(|\epsilon|^\mu \boldsymbol{r}, |\epsilon|^\nu t) , \tag{4.29}$$

where μ, ν and δ are non–negative exponents. If we insert this form into (4.25), each term in the expansion can be evaluated in terms of its power in ϵ. First, we stipulate that the term responsible for the linear instability, $a_2\nabla^2\psi$, and the dissipative term, $a_4\nabla^4\psi$, must be present. From the required balance of these terms (i.e., from the condition that they are of the same order in ϵ) we obtain the relation $1 + \delta + 2\mu = 4\mu + \delta$, and thus we have $\mu = 1/2$. The size of each remaining term depends on the size of the only unknown parameter, δ. However, provided that $\delta \geq 0$, the largest of these remaining terms is $(\nabla\psi)^2$, independent of the size of δ, as can easily be seen. It follows, then, from the balance of this term and those mentioned above, that $\delta = 1$. Then, requiring that these three terms be of the same order in ϵ as $\dot{\psi}$, we find that $\nu = 2$. Following this reasoning, we obtain the phase equation

$$\partial\psi/\partial t = a_2\nabla^2\psi - a_4\nabla^4\psi + g(\nabla\psi)^2 \,. \tag{4.30}$$

The above equation can also be derived from the complex Ginzburg–Landau equation (2.29). In this case, the relations for the coefficients a_2 and a_4 become

$$a_2 = 1 + c_1c_2 \,, \tag{4.31a}$$

$$a_4 = \tfrac{1}{2}c_1^2(1 + c_2^2) \,, \tag{4.31b}$$

as can be understood by comparing (4.30) with the equation corresponding to the eigenvalue spectrum for the phase branch, defined by (2.33). Then, in order to determine the form of the coefficient g of the nonlinear term, we consider the particular solution

$$\psi = kx + gk^2t \tag{4.32}$$

of (4.26). This represents the long-wavelength approximation of the plane wave solution (2.30). Thus, by comparison with the equation for ω_k corresponding to this same equation, we obtain

$$g = c_2 - c_1 \,. \tag{4.33}$$

In the case in which $a_2 < 0$ and $a_4 > 0$, with the scale transformations $\boldsymbol{r} \to (a_4/|a_2|)^{1/2}\boldsymbol{r}$, $t \to |a_2|^{-2}a_4 t$, and $\psi \to g^{-1}|a_2|\psi$, (4.30) reduces to the parameter-free equation

$$\partial\psi/\partial t = -\nabla^2\psi - \nabla^4\psi + (\nabla\psi)^2 \,. \tag{4.34}$$

This is known as the Kuramoto–Sivashinsky equation. As we will discuss below, this equation exhibits spatio-temporal chaos.

4.2.2 Phase Waves and the Target Pattern

It is worthwhile pausing for a time to investigate the phase equation (4.26). For the particular solution (4.32) of this equation, we have $X = X_0((\omega_0 + gk^2)t + kx)$, which represents a plane wave of wave number k. Arising from the phase non–uniformity, this so-called *phase wave* has essentially no effect on the amplitude of the pattern represented by X. As a result of the excitation of this wave number k phase wave, the frequency of oscillation of the medium changes by gk^2. From experience, we know that, in general, the presence of waves causes the frequency to increase, and we thus assume that $g > 0$. The propagation speed of the phase wave c is given by

$$c = k^{-1}(\omega_0 + gk^2) . \tag{4.35}$$

The fact that $c \to \infty$ as $k \to 0$ is characteristic of phase waves.

It is easy to verify the existence of the solution

$$\psi = \frac{a}{g} \ln \left(\cosh \left[\frac{qg}{a}(x - x_s(t)) \right] \right) + px + g(p^2 + q^2)t \equiv \psi_s , \tag{4.36a}$$

$$x_s = -2pgt \tag{4.36b}$$

to (4.26). This can be thought of as the generalization of the particular solution (4.32). The parameters p and q here are arbitrary. The behavior of the function $\phi_s \equiv \omega_0 + \psi_s$ for large X is found from (4.36) to be

$$\phi_s \approx \omega_\pm t + k_\pm x \qquad (x \gtrless x_s) , \tag{4.37a}$$

$$k_\pm = p \pm |q| , \tag{4.37b}$$

$$\omega_\pm = \omega_0 + gk_\pm^2 . \tag{4.37c}$$

Thus the solution ϕ_s represents something of a joining of two phase waves of different wave numbers in the region near $x = x_s$. As expressed by (4.36b), x_s moves at a fixed speed, and therefore, with the passage of time, one of the two phase wave regions comes to control the system. In the following discussion, we assume that $\omega_0 > 0$. As can easily be seen from (4.36b), (4.36b), and (4.36c), the region with large wave number eventually dominates; that is, $\omega_+ \gtrless \omega_-$ implies $\dot{x}_s \lessgtr 0$.

When two nonlinear oscillators possessing two distinct frequency eigenvalues are coupled, behavior in which their frequencies become identical is known as *entrainment* or *synchronization.* The phenomenon discussed above is regarded as one type of entrainment. With the motion of x_s, the 'entrained' region becomes spatially extended. In the context of entrainment phenomena, usually the frequency tends to be entrained to the higher-frequency side. Thus let us assume that $g > 0$. The solution represented by (4.36) is known as a *shock wave solution* and corresponds to a solution of the Burgers equation.

Phenomena in which the entrainment region expands can be observed in a system in which, for whatever reason, one region of an originally uniformly oscillating medium comes to possess an increased frequency. In many cases, this results in a so-called *target pattern.* In order to investigate this problem concretely, let us consider the model equation

$$\partial\phi/\partial t = \omega_0 + a\nabla^2\phi + g(\nabla\phi)^2 + \sigma(\boldsymbol{r}) , \tag{4.38}$$

obtained from the phase equation (4.26) by adding the term $\sigma(\boldsymbol{r})$ expressing the inhomogeneity of the characteristic frequency. We define σ as a nonnegative function which is nonzero only in the neighborhood of the origin. As in the derivation of (4.28), with the nonlinear transformation $\phi(\boldsymbol{r}, t) = g^{-1}a \ln Q(\boldsymbol{r}, t)$, (4.38) is reduced to the linear equation

$$\frac{\partial Q}{\partial t} = a[\omega_0 g a^{-2} + \nabla^2 - U(\boldsymbol{r})]Q \tag{4.39}$$

$$U = -g a^{-2}\sigma(\boldsymbol{r}) .$$

Writing the fundamental solution to this equation as

$$Q(\boldsymbol{r},t) = q(\boldsymbol{r})\exp[(\omega_0 g a^{-1} + a\lambda)t] , \tag{4.40}$$

we obtain the eigenvalue problem

$$\lambda q(\boldsymbol{r}) = (\nabla^2 - U(\boldsymbol{r}))q(\boldsymbol{r}) . \tag{4.41}$$

The time development of the phase pattern can be understood by solving this equation.

Equation (4.41) is formally equivalent to the quantum-mechanical problem in which a single particle is subject to the attractive potential $U(\boldsymbol{r})$. Here, for a bound state we have $\lambda > 0$, and the corresponding wave function Q grows exponentially with time. Thus the long-time behavior is controlled by the ground state. In that case, sufficiently far from the local potential, the spatial behavior of q is given by $\exp(-\sqrt{\lambda}|\boldsymbol{r}|)$. Accordingly, if the system is initially in the uniform state $Q = Q_0 (> 0)$, for large t and $r = |\boldsymbol{r}|$, Q takes on the form

$$Q \approx \mathrm{e}^{\omega_0 g a^{-1} t}(Q_0 + Q_1 \mathrm{e}^{a\lambda t - \sqrt{\lambda} r}) . \tag{4.42}$$

The corresponding phase pattern then becomes

$$\begin{aligned} \phi &= \omega_0(t - t_0) + g^{-1} a \ln[1 + \exp(a\lambda(t - t_0) - \sqrt{\lambda} r)] + \text{const.} , \\ t_0 &= -(a\lambda)^{-1} \ln \frac{Q_1}{Q_0} . \end{aligned} \tag{4.43}$$

The relation between the relative sizes of the two terms inside the logarithmic term in the above equation for ϕ reverses sharply at $r = r_\mathrm{s}(t)$, where we have

$$r_\mathrm{s} = a\sqrt{\lambda}(t - t_0) . \tag{4.44}$$

Using this fact, we arrive at the approximation

$$Q \approx \begin{cases} \omega_0(t - t_0) & (r > r_\mathrm{s}(t)) , \\ (\omega_0 + g^{-1}a^2\lambda)(t - t_0) - g^{-1} a\sqrt{\lambda} r & (r < r_\mathrm{s}(t)) . \end{cases} \tag{4.45}$$

Let us now turn to a discussion of the corresponding phenomenon in a two-dimensional medium. In this case, in the region outside a circle of radius r_s centered at the origin, the system exhibits a spatially uniform oscillation frequency ω_0, while inside this circle, it is entrained to the high frequency $\tilde{\omega} \equiv \omega_0 + g^{-1}a^2\lambda$. The radius of this entrainment region grows at the constant speed $a\sqrt{\lambda}$. Within the entrainment region, phase waves in the form of a target pattern with wave number $g^{-1}a\sqrt{\lambda}$ also centered at the origin propagate outward with the fixed speed $(\omega_0 g a^{-1} + a\lambda)/\sqrt{\lambda}$ $(> a\sqrt{\lambda})$, and, upon reaching the boundary of the entrainment region, they disappear. The inhomogeneity corresponding to the nonzero region of $\omega(\boldsymbol{r})$ in the center of such an expanding target is regarded as a type of pacemaker.

When two or more pacemakers exist, a target pattern forms corresponding to each, and each of these patterns possesses its own characteristic frequency. When two such patterns, centered at $\boldsymbol{r} = \boldsymbol{r}_1$ and $\boldsymbol{r}_2$ and possessing frequencies ω_1 and ω_2, collide, the behavior differs from the type of interference behavior exhibited in the case of linear wave motion. Let us again think of this as a quantum-mechanical problem. Then we can consider the case in which two potentials are separated by a distance large enough that the mutual overlapping of the bound states of the two potentials (in particular, those of the ground state wave functions) are sufficiently small, and the superposition approximation can be used. As a result, the equation

$$\begin{aligned} \phi &= \omega_0(t - t_0) \\ &\quad + g^{-1} a \max[0, a\lambda_1(t - t_1) \\ &\quad - \sqrt{\lambda_1}|\boldsymbol{r} - \boldsymbol{r}_1| \,,\, a\lambda_2(t - t_2) - \sqrt{\lambda_2}|\boldsymbol{r} - \boldsymbol{r}_2|] \end{aligned} \tag{4.46}$$

is valid as a generalization of (4.45). The structure of the concentration pattern corresponding to this solution can be represented in skeletal form by a set of equal phase lines, each defined by the relation $\phi = 2n\pi$, for integer n. Figure 4.2 illustrates more clearly than (4.45) the form assumed by the equal phase curves. As can be understood from this figure, upon the collision of two target patterns, that corresponding to the higher-frequency pacemaker ultimately destroys that corresponding to the lower-frequency pacemaker.

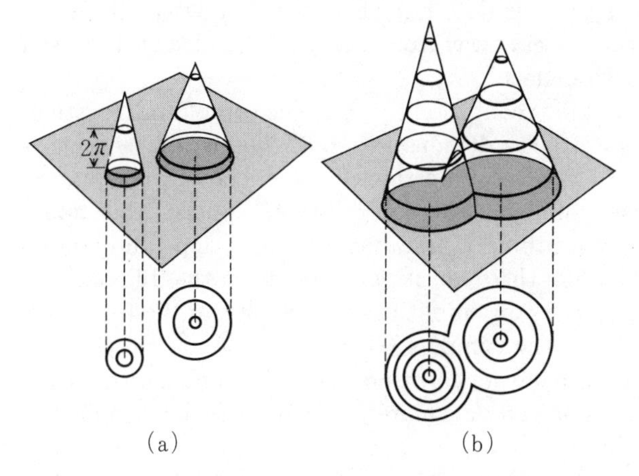

Fig. 4.2. A pair of target patterns described by the approximate solution (4.46) to the phase equation (4.38). In the case depicted, there are two source terms σ corresponding to two different points in space. (**a**) At the initial time, the pattern defined by the phase $\phi(x, y)$ is described by two independently growing circular cones. (**b**) Eventually, these two cones collide, and that with the faster growth rate overwhelms and destroys the other. As shown, if a cone is cut horizontally at 2π intervals (corresponding to 2π intervals in the value of ϕ), the projection on the x–y plane of the resulting circles forms a growing target pattern

4.2.3 Phase Turbulence

In the neighborhood of the phase instability point for a uniformly oscillating field, the KS equation (4.34) displays spatio-temporally chaotic solutions. This kind of spontaneously occurring chaotic behavior in phase fields is termed *phase turbulence*. The phase destabilizing mechanism acting in oscillating reaction–diffusion systems can be qualitatively explained in the following way. The existence of an instability implies that if there exists a region in which the phase is locally advanced, the phase in that region continues to advance to an ever greater extent, while if there is a retarded region, the phase there becomes increasingly retarded. Diffusive coupling serves to push an inhomogeneous system toward uniformity, and thus, for such a system to be unstable, there must exist some mechanism able to negate this tendency toward stabilization. When the diffusion constants for the two components of X differ, such a mechanism in fact does exist. This difference introduces an effect which indirectly leads to phase destabilization, arising from a non-uniformity in the local oscillation amplitude proportional to the curvature of the spatial change of the phase pattern, $\nabla^2\phi$. For example, if the amplitude of a protruding region of advanced phase ($\nabla^2\phi < 0$) decreases slightly, the amplitude in the surrounding region ($\nabla^2\phi > 0$) will tend to increase. Now, in general, the rate of change of the phase, i.e., the frequency, depends on the amplitude. If the nature of the system is such that the frequency increases with a decrease in the amplitude, the local frequency characterizing the protruding region will grow, and the phase in that region will advance further. Then, the stability of the system is determined by the competition between this mechanism pushing the system away from uniformity, and that discussed above possessing the opposite effect.

The Benjamin–Feir instability condition $\alpha < 0$ in the complex GL system, considered in Sect. 2.5, is a concise demonstration of the instability mechanism discussed above. The 1 appearing on the left-hand side of the equation for α, (2.34), expresses the stabilizing effect of diffusion. On the other hand, the second term, c_1c_2, can contribute a destabilizing effect, depending on its sign. As we will see in Sect. 5.4, the quantity c_1 arises from the difference in diffusion coefficients, and c_2 gives an amplitude dependence to the frequency through the term $c_2|W|^2W$ in the complex GL equation. As a consequence of the cooperation between the effects corresponding to c_1 and c_2, the usual stabilizing tendency of diffusion can be counteracted. This is the significance of the BF instability condition.

The behavior of the one-dimensional KS equation has been the subject of a great deal of numerical analysis. If we define the variable v by the relation $v = 2\partial\psi/\partial x$, (4.34) in one dimension becomes

$$\frac{\partial v}{\partial t} = \frac{\partial^2 v}{\partial x^2} - \frac{\partial^4 v}{\partial x^4} + v\frac{\partial v}{\partial x} \,. \tag{4.47}$$

Written this way, the nonlinear term takes on a form similar to that of the inertial term in the Navier–Stokes equation, and we are thus tempted to compare the behavior of the KS equation with that of the NS equation. While it is not possible to make a simple comparison here, we can at least make the following observations. The unstable growth of fluctuations due to negative diffusion occurs in the large-wavelength region, while in the short-wavelength region, fluctuations are attenuated by the $-\partial^4 v/\partial x^4$ term. Then, the nonlinear term facilitates the interaction between the two wavelength regions and insures the global stability of the system. However, for the KS equation, there does not exist, in a definite form, anything that corresponds to the inertial subrange of developed Navier–Stokes turbulence.

One example of the turbulent spatio-temporal patterns exhibited by solutions of the KS equation is displayed in Fig. 4.3. The spatial pattern exhibits an approximately periodic form. The period in question corresponds quite closely to the wave number $k_m = 1/\sqrt{2}$, representing the mode of the largest linear growth rate. It can be seen clearly from the fluctuation spectrum shown in Fig. 4.4 that the amplitude of the fluctuations peaks near this wave number. Many of the various statistical properties of the KS equation in an extended system have not yet been theoretically explained.

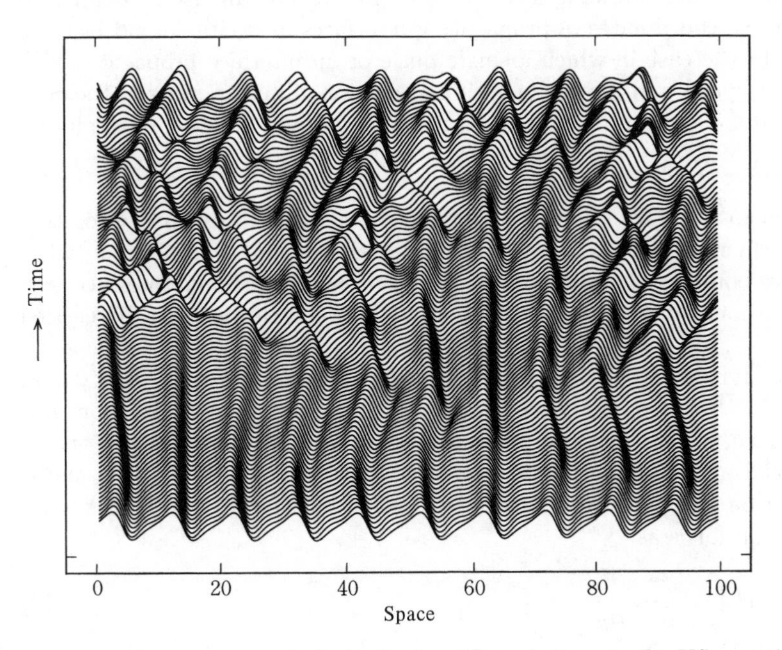

Fig. 4.3. Spatio-temporal chaos displayed by solutions to the KS equation (4.47). The solution represented here begins from a nearly periodic spatial pattern and grows increasingly turbulent with time (figure used by permission of Toshio Aoyagi)

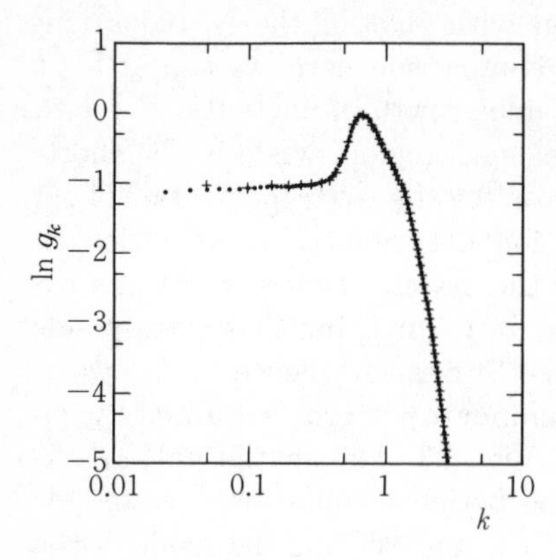

Fig. 4.4. The spectrum of the fluctuations $g_k \equiv \langle |v_k|^2 \rangle$ exhibited by the KS equation (4.47). Here v_k represents the Fourier component of $v(x)$, and $\langle \cdots \rangle$ is a long-time average. There is a distinct increase in the amplitude of fluctuations in the neighborhood of the wave number corresponding to the largest growth rate for the linearized equation (figure used by permission of S. Toh)

4.3 The Phase Dynamics of Interfaces

The context in which the concept of phase can be applied is not limited to systems displaying spatial and temporal oscillation. It can also be defined to represent the coordinates of propagating interfaces in excitable and bistable systems. In the case in which a single pulse or an interface propagates with a constant speed c in a one-dimensional reaction–diffusion system, the concentration of each component can be represented by a function of the form

$$X(x,t) = X_0(x - c_0 t - \psi) , \tag{4.48}$$

where ϕ is an arbitrary phase constant. In two or three dimensions, a solution of this form represents a flat interface. For the sake of simplicity, in what follows we consider the phase dynamics of a one-dimensional interface in a two-dimensional system. Generalization to three dimensions is easily carried out.

Phase dynamics can be applied to situations of the type depicted in Fig. 4.5. In this case, the wave propagates roughly in the x direction, but the interface is not completely straight; it exhibits a gradual bend over a long spatial extent. If we reinterpret ψ as a variable depending on y and t, we should be able to express such a solution in terms of (4.48). Let us write the equation for ψ as

$$\frac{\partial \psi}{\partial t} = H\left(\psi, \frac{\partial \psi}{\partial y}, \frac{\partial^2 \psi}{\partial y^2}, \cdots\right) . \tag{4.49}$$

Equation (4.48) does not contain y, and it thus possesses symmetry with respect to spatial inversion along this direction. In other words, this same symmetry, possessed by the original reaction–diffusion system, is not spontaneously broken by this solution. Thus the phase equation (4.49) must be invariant with respect to the transformation $y \to -y$. These considerations imply the expression

$$\frac{\partial \psi}{\partial t} = a_2 \frac{\partial^2 \psi}{\partial y^2} - a_4 \frac{\partial^4 \psi}{\partial y^4} + g \left(\frac{\partial \psi}{\partial y} \right)^2 + \cdots \tag{4.50}$$

for H, containing only even powers of the operator $\partial/\partial y$. Then, precisely as in the previous section, we obtain the following form for the lowest-order expansion in derivatives:

$$\frac{\partial \psi}{\partial t} = a \frac{\partial^2 \psi}{\partial y^2} + g \left(\frac{\partial \psi}{\partial y} \right)^2 . \tag{4.51}$$

As discussed in Sect. 3.5, when a phase instability of an interface appears, in the neighborhood of the instability point, we have

$$\frac{\partial \psi}{\partial t} = a_2 \frac{\partial^2 \psi}{\partial y^2} - a_4 \frac{\partial^4 \psi}{\partial y^4} + g \left(\frac{\partial \psi}{\partial y} \right)^2 . \tag{4.52}$$

This is also true for the case of an oscillating medium. In this way, the motion of an interface possessing a weak phase instability can be described by the KS equation, and such a system is therefore expected to display phase turbulence.

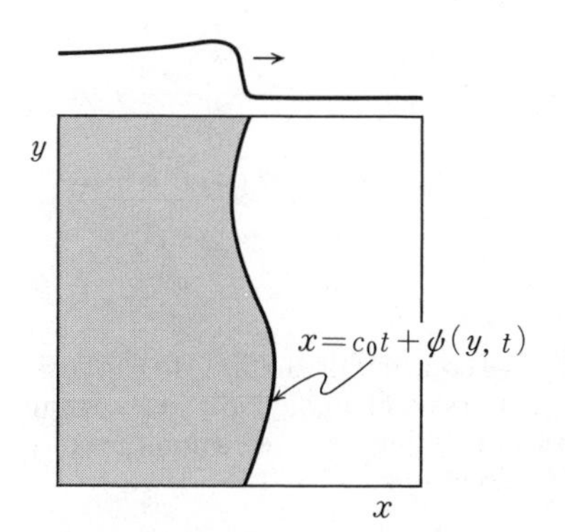

Fig. 4.5. The interface represented by the one-dimensional traveling wave shown above exhibiting slow modulation in a two-dimensional medium. Phase dynamics can be applied to describe an interface undergoing such a time evolution

Considering the particular solution $\psi = sy + gs^2 t$ of (4.51) and (4.52), the meaning of the nonlinear term in these equations becomes clear. This solution represents a steadily propagating straight interface lying along a line making an angle $\theta = \tan^{-1} s$ with the y-axis. Fixing a position on the y-axis and looking from this position in the x direction, the propagation velocity observed is $c_0 + gs^2$. Then, since the propagation velocity in the direction normal to the interface should be c_0, from simple geometric considerations, the approximation $c_0 + gs^2 = c_0 / \cos\theta$ must hold to second order in s. From this, we have

$$g = c_0 / 2 . \tag{4.53}$$

The interface equation of motion (4.51) has solutions that assume the form of (4.36). Such a solution results from the joining of two steadily propagating straight interface solutions making angles $p+|q|$ and $p-|q|$ with respect to the y-axis, as depicted in Fig. 4.6. As a consequence of the propagation of these steady-state waves, the connection between the respective solutions tends to become increasingly sharp, but the curvature effect, represented by the term $\partial^2\psi/\partial y^2$, opposes this tendency, smoothing the form of ψ. The balance of these two effects allows for a fixed local structure to be maintained. In this way, these obliquely colliding interfaces form a type of shock wave. The phase equation (4.51) is restricted to the situation in which the wave front makes an infinitesimal angle with the y-axis everywhere. However, as is also clear from the preceding discussion with regard to the significance of the nonlinear term, the physical import of (4.51) itself is something very general and does not depend on the selection of the coordinate system. That is to say, the local propagation velocity c at some given point, as viewed along the direction normal to the interface, must be given approximately by

$$c = c_0 - a_2\kappa \,, \tag{4.54}$$

from (4.52), where κ is the interface curvature at the point in question. If the interface dynamics are expressed as (4.54), they do not depend on the choice of the coordinate system.

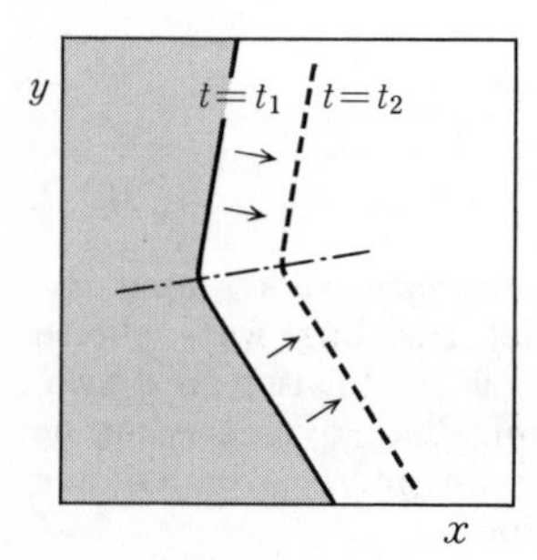

Fig. 4.6. Shock wave solution of the interface phase equation (4.51). This type of behavior results from the collision of two traveling wave solutions propagating in different directions

4.4 Multiple Field Dynamics

To this point in our treatment, we have been considering only the case of a single phase field. In this section we consider the situation in which several coupled fields exist. We encounter problems of this kind when we attempt to apply our phase description to patterns appearing after the occurrence of a number of bifurcations that break the continuous symmetry of the system. For example, when a stationary periodic pattern becomes unstable with respect to oscillation, we can imagine a situation in which the pattern now comes to possess a component corresponding to the time oscillation, in addition to that representing the original pattern. In this case, phases representing both temporal and spatial translation modes coexist, and from their coupling, complicated phenomena can arise. In general, depending on the nature of the continuous symmetry breaking introduced by the original stationary pattern acting as the 'background' and the type of secondary continuous symmetry breaking undergone, there are a variety of cases that we can consider. In this section we take up the relatively simple but important case of a one-dimensional stationary periodic pattern in which a time-oscillation component arises as a secondary mode. As an extension of the previous discussion, we derive a multifield equation based on scaling arguments and the system's inherent symmetry. We then study the the appearance of new properties which cannot be seen in single-field equations.

Let us consider the case in which, in a one-dimensional stationary spatial pattern of wave number k, a nonpropagating limit cycle oscillatory component characterized by the same wave number arises. In general, the field variable for such a situation $X(x,t)$ has the following form:

$$X(x,t) = X_0(\phi_1) + y(\phi_1,\phi_2) \tag{4.55}$$

$$\phi_1 = kx + \psi_1, \qquad \phi_2 = \omega t + \psi_2 \qquad (\psi_1, \psi_2 \text{ arbitrary}) \ .$$

Here, $X_0(\phi+2\pi) = X_0(\phi)$ and $y(\phi_1+2\pi,\phi_2) = y(\phi_1,\phi_2+2\pi) = y(\phi_1,\phi_2)$. As above, we assume that the stationary component possesses symmetry with respect to spatial inversion, i.e., $X_0(-\phi) = X_0(\phi)$.

Now, we can consider the case in which y is spatially symmetric $[y(-\phi_1,\phi_2) = y(\phi_1,\phi_2)]$ and that in which it is spatially asymmetric. In the latter case, we can assume that the pattern is of the 'flip–flop' type, in which the pattern obtained upon spatial inversion coincides with that obtained upon a time translation of a 1/2 period: ϕ_2: $y(-\phi_1,\phi_2) = y(\phi_1,\phi_2+\pi)$. If we do not make this assumption, the plus and minus spatial directions are no longer equivalent, contradicting the assumption that y is a nonpropagating oscillatory component. In the first case, $y(\phi_1,\phi_2) = y(-\phi_1,\phi_2)$, (4.55) is invariant with respect to the simultaneous transformations

(i) $\psi_1 \to \psi_1 + \psi'$, $x \to x - k^{-1}\psi'$ (ψ' arbitrary)
(ii) $\psi_2 \to \psi_2 + \psi'$, $t \to t - \omega^{-1}\psi'$ (ψ' arbitrary)
(iii) $\psi_1 \to -\psi_1$, $x \to -x$,

while in the case of asymmetry, (i) and (ii) above do not change, but (iii)

(iii)′ $\psi_1 \to -\psi_1$, $x \to -x$, $\psi_2 \to \psi_2 + \pi$.

However, as we find below, in either case, the phase equations possess the same form. Let us begin our investigation by writing the coupled phase equations that we are to derive as

$$\frac{\partial \psi_1}{\partial t} = H_1\left(\psi_1, \frac{\partial \psi_1}{\partial x}, \frac{\partial^2 \psi_1}{\partial x^2}, \cdots, \psi_2 \frac{\partial \psi_2}{\partial x}, \frac{\partial^2 \psi_2}{\partial x^2}, \cdots\right) , \tag{4.56a}$$

$$\frac{\partial \psi_2}{\partial t} = H_2\left(\psi_1, \frac{\partial \psi_1}{\partial x}, \frac{\partial^2 \psi_1}{\partial x^2}, \cdots, \psi_2, \frac{\partial \psi_2}{\partial x}, \frac{\partial^2 \psi_2}{\partial x^2}, \cdots\right) . \tag{4.56b}$$

Truncating the expansions in $\partial/\partial x$ at second order, the form satisfying the condition of invariance with respect to (i), (ii), and (iii) (or (iii)′) is

$$\frac{\partial \psi_1}{\partial t} = \nu_1 \frac{\partial^2 \psi_1}{\partial x^2} + \alpha_1 \frac{\partial \psi_2}{\partial x} + h_1 \frac{\partial \psi_2}{\partial x}\frac{\partial \psi_1}{\partial x} \tag{4.57a}$$

$$\frac{\partial \psi_2}{\partial t} = \nu_2 \frac{\partial^2 \psi_2}{\partial x^2} + \alpha_2 \frac{\partial \psi_1}{\partial x} + h_2 \left(\frac{\partial \psi_2}{\partial x}\right)^2 . \tag{4.57b}$$

One feature that makes such coupled phase systems more interesting than single-mode systems is the fact that the interaction between modes can lead to new types of phase instabilities. In the discussion below, we assume that $\nu_1, \nu_2 > 0$; in other words, that each field is stable in the absence of coupling. Then to investigate the stability of the uniform stationary state $\psi_1, \psi_2 =$ const., we assume that the phase disturbances take the form $\psi_1, \psi_2 =\sim \exp(iqx + \lambda t)$, insert this form into the linearized equation, and solve for λ. The two roots of the resulting quadratic equation are given by

$$\lambda_\pm = -\tfrac{1}{2}(\nu_1 + \nu_2)q^2 \pm |q|\sqrt{-\alpha_1\alpha_2 + \tfrac{1}{4}(\nu_1 - \nu_2)^2 q^4} . \tag{4.58}$$

From this we see that if $\alpha_1\alpha_2 < 0$, the system is unstable with respect to long-wavelength phase disturbances. This type of instability in coupled phase systems was discovered by H. Sakaguchi in 1992. It was shown in that study that this instability leads to the formation of local patterns.

As we have seen previously, phase instability often results in the development of the system into a state for which the phase description itself fails. This is certainly also conceivable for the instability mentioned above for this coupled phase system.

The reason the phase description fails in such a situation is that the assumption that the behavior of the system can be described while ignoring the possible coupling between the phase and amplitude of the patterns becomes invalid. Thus it would be very convenient if we could find a simple model equation which includes an amplitude and contains the coupled phase equation (4.57) as a limiting case. Such an evolution equation, incorporating an oscillating field amplitude effect, can in fact be derived in the neighborhood of a Hopf bifurcation point. As we see below, this equation takes the form of a coupled system consisting of the complex Ginzburg–Landau equation and an equation for ψ_1.

In the neighborhood of a Hopf bifurcation, the field variable $X(x,t)$ generally has the following form:

$$X(x,t) = X_0(\phi_1) + v(\phi_1)We^{i\omega t} + \bar{v}(\phi_1)\bar{W}e^{-i\omega t} , \tag{4.59}$$

$$W = Ae^{i\psi_2} .$$

Here $v(\phi_1)$ and $\overline{v}(\phi_2)$ are the Hopf bifurcation critical modes. Thus we can assume that these are symmetric and anti-symmetric functions in ϕ_1, respectively. Equation (4.59) is invariant with respect to (i)–(iii) (or (iii)′) and, in addition,

(iv) $W \to W\exp(\mathrm{i}\psi')$, $t \to t - \omega^{-1}\psi'$ (ψ' arbitrary) .

We write the phase–amplitude equations in the neighborhood of the bifurcation point as

$$\frac{\partial \psi_1}{\partial t} = H_1\left(\psi_1, \frac{\partial \psi_1}{\partial x}, \frac{\partial^2 \psi_1}{\partial x^2}, \cdots, W, \bar{W}, \frac{\partial W}{\partial x}, \frac{\partial \bar{W}}{\partial x}, \cdots\right) , \tag{4.60a}$$

$$\frac{\partial W}{\partial t} = H_2\left(\psi_1, \frac{\partial \psi_1}{\partial x}, \frac{\partial^2 \psi_1}{\partial x^2}, \cdots, W, \bar{W}, \frac{\partial W}{\partial x}, \frac{\partial \bar{W}}{\partial x}, \cdots\right) . \tag{4.60b}$$

Then, enforcing invariance with respect to (i)–(iv), we obtain the following equations, in which only the most important terms (the meaning of this designation to be made clear below) have been retained:

$$\frac{\partial \psi_1}{\partial t} = \frac{\partial}{\partial x}|W|^2 + \mathrm{i}\beta\left(\frac{\partial W}{\partial x}\bar{W} - W\frac{\partial \bar{W}}{\partial x}\right) + \nu\frac{\partial^2 \psi_1}{\partial x^2} , \tag{4.61a}$$

$$\frac{\partial W}{\partial t} = \mu(1 + \mathrm{i}c_0)W - (1 + \mathrm{i}c_2)|W|^2 W + (1 + \mathrm{i}c_1)\frac{\partial^2 W}{\partial x^2} + (\chi_1 + \mathrm{i}\chi_2)\frac{\partial \psi_1}{\partial x}W . \tag{4.61b}$$

Here, by appropriately scaling W, x, and t, several coefficients have been normalized to 1. A solution to the above equations clearly must have the scaling form

$$\psi_1 = |\mu|^{1/2}\tilde{\psi}_1(|\mu|^{1/2}x, |\mu|t) ,$$
$$W = |\mu|^{1/2}\tilde{W}(|\mu|^{1/2}x, |\mu|t) . \tag{4.62}$$

From this, we see that all of the terms in (4.61) are $O(|\mu|^{3/2})$, while the terms ignored in constructing these equations are of higher order in μ, and thus are assumed to be insignificant.

The coupled phase–amplitude equation (4.61) possesses the following plane wave solution as a particular solution:

$$\psi_1 = k_1 x + \omega_1 t , \tag{4.63a}$$

$$W = A\exp[\mathrm{i}\{\mu(c_0 - c_2) + \omega_2\}t + \mathrm{i}k_2 x] . \tag{4.63b}$$

This implies that the wave number of the periodic structure of the full pattern differs from that of the original by an amount k_1 and that the oscillating field has a traveling wave of wave number k_2. Substituting (4.63) into (4.61) yields

$$A = \sqrt{\mu - k_2^2 + \chi_1 k_1} , \tag{4.64a}$$

$$\omega_1 = -2\beta k_2(\mu - k_2^2 + \chi_1 k_1) , \tag{4.64b}$$

$$\omega_2 = (\chi_2 - c_2\chi_1)k_1 + (c_2 - c_1)k_2^2 . \tag{4.64c}$$

It should be clear that a plane wave solution of this type also exists for the coupled phase equation (4.57).

As discussed above in the context of coupled phase equations, the interaction between two fields can cause the uniform state ($k_1 = k_2 = 0$) to become unstable. A similar conclusion can be reached from the linear stability analysis of the phase–amplitude equation (4.61). Because the system possesses three components, the eigenvalues λ corresponding to fluctuations of wave number q in the phase and amplitude are determined by a cubic equation. Expanding the long-wavelength spectrum of the phase branch in powers of q, we obtain the form

$$\lambda = \pm|q|\sqrt{2\mu\beta(\chi_2 - c_2\chi_1)} - \tfrac{1}{2}(1 + c_1c_2 + \nu)q^2 + O(|q|^3) \,. \tag{4.65}$$

Hence, if the field coupling constants are related in such a way as to satisfy the relation $\beta(\chi_2 - \chi_1) > 0$, a phase instability arises.

The coupled phase equation (4.57) can also be derived from the phase–amplitude equation (4.61). Then, with simple considerations similar to those used in Sects. 4.1 and 4.2, explicit formulas for the coefficients $\nu_1, \nu_2, \alpha_1, \alpha_2, h_1$, and h_2 can be determined. First, we consider the special situation in which there is no coupling between the two fields. In this case, the equation describing ψ_2 is expected to be identical to the reduced form of a pure complex GL equation. Thus, from the result in Sect. 4.2, we have $\nu_2 = 1+c_1c_2, h_2 = c_2-c_1$. Then, noting the existence of the plane wave particular solution of the coupled phase equation defined by $\psi_1 = k_1x+\omega_1t$, $\psi_2 = k_2x+\omega_2t$, $\omega_1 = \alpha_1k_2+h_1k_1k_2$, and $\omega_2 = \alpha_2k_1 + h_2k_2^2$, and requiring this to be in agreement, in the long-wavelength regime, with the solution (4.63a,b) of the phase–amplitude equation, we obtain $h_1 = -2\beta\chi_1$, $\alpha_1 = -2\beta\mu$, and $\alpha_2 = \chi_2 - c_2\chi_1$. We can determine the remaining unknown, ν_1, by imposing the condition that the expression for the eigenvalue (4.58) obtained from the linear stability analysis of the coupled phase equation must be, to $O(q^2)$, identical to (4.65). This gives us the identity $\nu_1 = \nu$.

Because of the instability of the coupled phase system, the appearance of a deformation in the periodic structure also causes the phase of the oscillatory field to become inhomogeneous. Let us now attempt to determine the type of state into which the system eventually evolves as a result of such an instability. According to the results of numerical simulations of the phase–amplitude equation (4.61), under appropriate conditions, a singular pattern such as that represented by Fig. 4.7 appears. This figure corresponds to the situation in which the oscillatory region is localized; that is, only inside the localized region has the system undergone the Hopf bifurcation. Outside this region, the system is below the bifurcation point. Note from (4.61b) that if the wave number suffers a change of magnitude k_1, the effective Hopf bifurcation parameter becomes $\mu+\chi_1k_1$, and, in this way, the appearance and disappearance of oscillation comes to be controlled by the change in the wave number. Thus the localization of oscillatory behavior results from the localization of structural distortion. Figure 4.7 clearly displays this point.

The localization of structural distortion of this type causes a simultaneous change in the local frequency of oscillation. The existence of a region such as

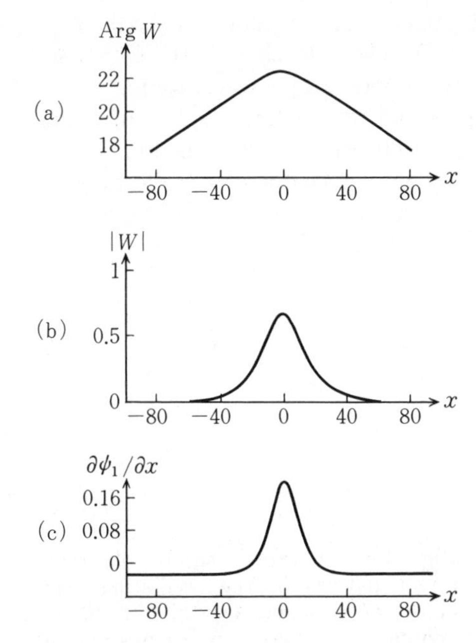

Fig. 4.7. A pattern obtained from numerical simulation of the phase–amplitude equation (4.61) with $\mu = 0.05$, $c_0 = c_2 = 0.5$, $c_1 = 0$, $\chi_1 = 2$, $\chi_2 = 1.05$, $\beta = 2$, and $\nu = 1$. (**a**) The phase of the oscillating field W; (**b**) its amplitude $|W|$; (**c**) the local wave number $\partial\psi_1/\partial x$. [reproduced from Sakaguchi (1992)]

that corresponding to the central portion of Fig. 4.7, in which the frequency is larger than in the region surrounding it, introduces an effect similar to that which would be introduced by the presence of a pacemaker. In fact, the gradient of the phase corresponding to the oscillatory field assumes the form of a single peak superimposed on a constant background extending to $\pm\infty$, and it is known that in this situation a phase wave target pattern like those discussed in Sect. 4.2 (actually, the one-dimensional version of such a pattern) appears. An exact solution to (4.61a,b) corresponding to this type of pattern was found by H. Sakaguchi in 1992.

In a liquid crystal convection experiment, phenomena suggestive of the above theoretical result were discovered by Sano et al. (1993). We treated patterns in liquid crystal convection, and in particular the defect structure and motion that occurs in Williams domains, in Sect. 2.3, but in connection with the theory of multiple fields, the 'grid' phase seen when the applied electric field strength is increased is of particular interest. The grid pattern consists of a two-dimensional square lattice periodic structure, and depending on the conditions present, this pattern can be either stationary or nonstationary. A model equation able to describe the dynamics of such a grid phase destabilized with respect to oscillation has yet to be discovered. However, (4.61) is meaningful as a model of the one-dimensional version of such behavior. With this in mind, the strong similarity between patterns (such as that shown in Fig. 4.8) observed experimentally and the pattern suggested by Fig. 4.7 has attracted a great deal of attention. However, even ignoring questions concerning the two-dimensional nature of the physical pattern, there are several characteristics of observed patterns that cannot be described by (4.61), nor by any other existing theory. For example, there is the fact that, according to experiment, this target pattern does not possess stationary, persistent structure, but, rather, it collapses and reforms repeatedly on a long time scale. When the pattern collapses, defects are created in the underlying periodic structure. A theory able to explain this characteristic life cycle is currently being sought.

Fig. 4.8. An experimentally observed target pattern in the oscillating grid phase of a liquid crystal convection system (figure used by permission of M. Sano)

5. Foundations of Reduction Theory

In the preceding chapters, we developed arguments concerning pattern dynamics based on relatively simple model equations. Many of these equations were derived phenomenologically. In the present chapter, we consider the theoretical foundation of perhaps the most important types of model equations considered in Part I, amplitude equations and phase equations. The degrees of freedom contained in the corresponding reduced equations are generally characterized by slow time development. For this reason, the remaining large number of degrees of freedom are eliminated, so to speak, adiabatically. For dissipative systems, at the foundation of this kind of reduction mechanism is a definite universal structure, and, as will become evident in this chapter, it is possible from the point of view presented here to gain a clear new understanding of the separately developed realizations of reduction theory that we have presented in previous chapters for the study of a variety of outwardly different types of situations. We begin here by considering a simple example through which the fundamental structure of reduction is clarified. We then see how this structure is actually realized in the derivation of amplitude and phase equations. Throughout this entire chapter, we wish to remove emphasis from the presentation of the reduction algorithm and place it, rather, on making clear the physical meaning contained in the reduction approach.

5.1 Two Simple Examples

In this section we apply the reduction procedure to a simple dynamical system which includes no spatial degrees of freedom. As we will discuss later, the question of the presence or absence of continuous spatial degrees of freedom is, from the point of view of the reduction theory presented below, not as important as one may think.

Example 1

$$\mathrm{d}x/\mathrm{d}t = y - x\,, \tag{5.1a}$$
$$\mathrm{d}y/\mathrm{d}t = \epsilon f(x)\,. \tag{5.1b}$$

Here, ϵ is a small parameter, and $f(x)$ is a sufficiently smooth function of x. First, let us describe the normal approach. Because y is the slow variable, we regard this as a constant parameter in the first equation. As a result, x approaches the stable stationary value

$$x = y \ . \tag{5.2a}$$

Then, substituting this into the second equation, we obtain

$$\mathrm{d}y/\mathrm{d}t = \epsilon f(y) \ . \tag{5.2b}$$

Equations (5.2a) and (5.2b) constitute the reduced form of the original system.

Considering the evolution of this system in the x–y plane, the point representing the system first approaches the line defined by (5.2a) with the value of y nearly constant, as depicted in Fig. 5.1. Then, the slow motion along this line is determined by (5.2b). In the application of reduction theory in general, the following procedure is followed:

(1) Find the form of the hypersurface M to which the system is quickly attracted. (In the above example, the relation $x = y$ defines the approximate such form.)
(2) Determine the relation defining the slow motion on M.

In this prescription, as with the above example, the process of the asymptotic approach to M is ignored. If the initial state of the system corresponds to a point on M, the system remains forever on M. The hypersurface M is thus the *invariant manifold* of the system. In the above example, the state approached asymptotically as $t \to \infty$ – the *attractor* – is a fixed point on M (if such a point exists). The power of reduction theory is particularly evident in the situation in which the attractor is complicated and not easily determined from a consideration of the unreduced, original system.

In order to treat the above example more systematically and to higher order, we expand the form of M together with the rate of change of y in powers of ϵ. We have

$$x = x_0(y) + \epsilon x_1(y) + \epsilon^2 x_2(y) + \cdots , \tag{5.3a}$$

$$\mathrm{d}y/\mathrm{d}t = \epsilon G_1(y) + \epsilon^2 G_2(y) + \cdots . \tag{5.3b}$$

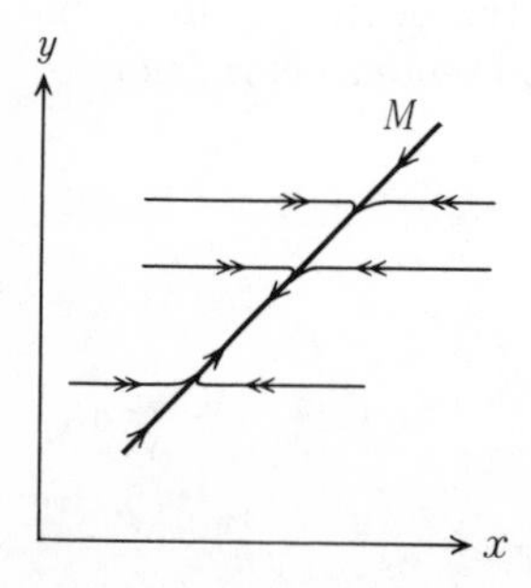

Fig. 5.1. Reduction of the dynamical system (5.1). This reduction is realized by determining the manner in which the evolution can be described as a rapid attraction of the representative point to the invariant curve M followed by slow motion on M

Then, substituting these into the original equation yields

$$\left(\frac{\mathrm{d}x_0}{\mathrm{d}y} + \epsilon\frac{\mathrm{d}x_1}{\mathrm{d}y} + \epsilon^2\frac{\mathrm{d}x_2}{\mathrm{d}y} + \cdots\right)(\epsilon G_1 + \epsilon^2 G_2 + \cdots)$$
$$= y - x_0 - \epsilon x_1 - \epsilon x_2 - \cdots , \qquad (5.4a)$$
$$\epsilon G_1 + \epsilon^2 G_2 + \cdots = \epsilon f(x_0 + \epsilon x_1 + \cdots) \qquad (5.4b)$$

In this way, we have obtained an equality containing only y as an independent variable. The condition that the above equation is satisfied for each power of ϵ gives, at lowest order, the result stated above. At this point it is clear that an approximation of arbitrary order in ϵ can be obtained by successive substitution.

Example 2

$$\frac{\mathrm{d}^2u}{\mathrm{d}t^2} + \omega_0^2 u = \epsilon f\left(u, \frac{\mathrm{d}u}{\mathrm{d}t}\right) . \qquad (5.5)$$

Weakly nonlinear oscillators assume this form. For example, defining $f = (1 - u^2)\dot{u}$ here yields the van der Pol equation. The perturbation theory of weakly nonlinear oscillators has been treated previously by Bogoliubov, Krylov, and Mitropolsky, but we take a slightly different approach in the present treatment.

If we set $v = \mathrm{d}u/\mathrm{d}t$, (5.5) becomes a first-order differential equation in u and v. In this way, this equation can be converted into a two-variable problem, just as in Example 1. In the present case, it is not possible to reduce the number of degrees of freedom any further, and that which corresponds to M in Example 1 is here the entire u–v space itself. In studying this example, emphasis is placed on simplifying the expression for the equation of motion rather than on determining the invariant manifold. We point out here that in Example 1, because M is one-dimensional, there is no need to treat the simplification of the expression for the motion on M. In fact, this amounts to nothing more than making the most natural choice of using y as the variable to designate points on M. However, when the dimension of M is greater than 1, problems related to this expression become nontrivial and cannot be ignored.

In the unperturbed ($\epsilon = 0$) case, the solution to (5.5) can be expressed as follows:

$$u = A\cos\phi, \qquad \phi = \omega_0 t + \psi . \qquad (5.6)$$

The quantities A and ψ here are arbitrary constants. However, we wish to think of these as functions of time for which the relations $\dot{A} = \dot{\psi} = 0$ happen to hold in the $\epsilon = 0$ case. Then, if we add the perturbation and assume the same form for the solution, substitution into (5.5) yields equations of motion for $A(t)$ and $\psi(t)$. Note that this procedure constitutes nothing more than a shift in coordinates from (u, v) to the polar coordinates (A, ϕ). With this shift, we change our point of view and think of u as a function of A and ϕ (rather than t). Then we assume this functional dependence to take a form which differs slightly from that in the unperturbed case, expressed by the first equation in (5.6). With this assumption, the evolution equations for A and ϕ themselves may come to assume simpler forms.

As this discussion suggests, it is our choice whether to subsume the complexity arising from the inclusion of the perturbation into the definition of the state variables or into the form of the evolution equation they obey. In reduction theory, the conciseness of the evolution equation takes priority in all cases. The reason for this is that we expect the most important aspects of the solution in question – for example, its period of oscillation, the topological nature of its attractor and the stability exponents of corresponding trajectories, etc. – to be very insensitive to the choice of the state variables. Rather, we should say that *because* of this insensitivity, these aspects are important.

Following the above thinking, when $\epsilon \neq 0$, we have

$$u = A\cos\phi + \epsilon\rho(A,\phi) \,, \tag{5.7a}$$

$$\mathrm{d}A/\mathrm{d}t = \epsilon G(A) \,, \tag{5.7b}$$

$$\mathrm{d}\phi/\mathrm{d}t = \omega_0 + \epsilon H(A) \,. \tag{5.7c}$$

By substitution, ρ, G, and H can be determined by successive approximation. It is also possible for ρ, G, and H themselves to depend on ϵ. In the above, we have not included ϕ in G and H. For this reason, it is evident that $\rho = 0$ is not allowed. Then, as seen below, with the enforcement of a single condition, ρ, G, and H can be determined uniquely.

Inserting (5.7) into (5.5), we obtain

$$\left[\left\{(\omega_0 + \epsilon H)\frac{\partial}{\partial\phi} + \epsilon G\frac{\partial}{\partial A}\right\}^2 + \omega_0^2\right](A\cos\phi + \epsilon\rho)$$
$$= \epsilon f\left(A\cos\phi + \epsilon\rho, \left\{(\omega_0 + \epsilon H)\frac{\partial}{\partial\phi} + \epsilon G\frac{\partial}{\partial A}\right\}(A\cos\phi + \epsilon\rho)\right) . \tag{5.8}$$

With the exception of the terms on the left-hand side that are linear in the unknown quantities ρ, G, and H, we collect all terms in this equation and lump them into the function $b(A,\phi)$. With this we obtain

$$\omega_0^2 L\rho = 2\omega_0(HA\cos\phi + G\sin\phi) + b(A,\phi) \,, \tag{5.9}$$

$$L = \frac{\partial^2}{\partial\phi^2} + 1 \,.$$

Formally, this is considered as an inhomogeneous linear differential equation in $\rho(\phi)$. The right-hand side contains the inhomogeneous terms. However, since, in general, b contains ρ, the right-hand side is not truly inhomogeneous, but in the lowest-order approximation, we can ignore the small quantities $\epsilon\rho$, ϵG, and ϵH. We thus obtain

$$b \approx f(A\cos\phi, -A\omega_0\sin\phi)\,. \tag{5.10}$$

In this way, b comes to be expressed only in terms of known quantities, containing no dependence on ρ. For example, in the case of the van der Pol equation, $b \approx -\omega_0 A(1 - A^2/4)\sin\phi + (\omega_0 A^3 \sin 3\phi)/4$. Thus, in general, if we can find a general method to solve (5.9) to determine ρ, G, and H uniquely, while considering b as a known quantity, by successive substitution we can determine all of these unknowns to arbitrary order in ϵ.

Implementing this line of reasoning, we assume for the time being that $b(A, \phi)$ is known, and proceed. L is understood as an operator acting on 2π-periodic functions of ϕ. Its eigenfunctions are $\sin m\phi$ and $\cos m\phi$ $(= 1, 2, \ldots)$, where those functions with $m = 1$ are null eigenfunctions (eigenfunctions whose eigenvalues are 0). To solve for ρ, we can expand either side of (5.9) in eigenfunctions of L (in other words, perform Fourier expansions) and compare coefficients of like terms. However, let us note two points here. First, if ρ contains a null eigenfunction component, owing to the action of L, the amplitude of this mode is undetermined. Thus we can assume from the outset that this amplitude is 0 and thus that ρ does not contain such a component. The second point is that the inhomogeneous term cannot contain a contribution from the null eigenfunction. This is known as the solvability condition and is expressed by the equations

$$G = -\frac{1}{2\pi\omega_0}\int_0^{2\pi} b\sin\phi\,\mathrm{d}\phi \tag{5.11a}$$

$$H = -\frac{1}{2\pi\omega_0 A}\int_0^{2\pi} b\cos\phi\,\mathrm{d}\phi\,. \tag{5.11b}$$

If we substitute G and H obtained from the solvability condition into the right-hand side of (5.9) and operate on both sides with $(\omega_0^2 L)^{-1}$, we obtain ρ. For the van der Pol equation, to lowest order,

$$\mathrm{d}A/\mathrm{d}t = \frac{\epsilon}{2}\left(A - \frac{A^3}{4}\right), \tag{5.12a}$$

$$\mathrm{d}\phi/\mathrm{d}t = \omega_0 \tag{5.12b}$$

$$u = A\cos\phi - \frac{\epsilon A^3}{32\omega_0}\sin 3\phi\,. \tag{5.12c}$$

From this approximate solution, b becomes corrected, and we can thus proceed to higher-order approximations.

In the theoretical formalism we have seen here (and in Example 1), the expression of the invariant space M (in the present case, the definition of the (A, ϕ) coordinate system introduced to describe motion on M) and the form of the evolution equation describing motion in this space are progressively and simultaneously determined by successive substitution. This is a general feature of reduction theory.

5.2 The Destabilization of Stationary Solutions

Reduction is generally possible for systems in which there are degrees of freedom that change very slowly with time. In Example 1 of the previous section, there is one such degree of freedom, y, while in Example 2 there are two, A and ψ. This type of situation appears in its most typical form in the neighborhood of a bifurcation point. Amplitude equations accompanying the destabilization of stationary solutions in this context were derived phenomenologically in Chaps. 1 and 2. In this and the following two sections, we apply the thinking detailed in the previous section to systems in the neighborhood of bifurcation points. In the process, we make clear the basis of the phenomenological derivations. First, in the present section, we organize the main points regarding the linear stability analysis of stationary solutions to ordinary differential equations. In the next section we carry out the reduction for systems of ordinary differential equations, and in Sect. 5.4 we treat spatially extended systems and discuss the reduction method as applied to partial differential equations.

We consider a system of n state variables $(X_1, X_2, \ldots, X_n) \equiv \boldsymbol{X}$ whose time evolution obeys the equation

$$\mathrm{d}\boldsymbol{X}/\mathrm{d}t = \boldsymbol{F}(\boldsymbol{X};\mu)\ . \tag{5.13}$$

Unless specifically stated otherwise, we will take all quantities appearing in the formulas below to be real. The quantity $\boldsymbol{F}$ in (5.13) is a sufficiently smooth nonlinear function of $\boldsymbol{X}$ containing the parameter μ. We assume that, for some range of values of μ, a stationary solution $\boldsymbol{X_0}(\mu)$ exists. Let us then define the difference between the actual solution and this stationary solution by $\boldsymbol{u} = \boldsymbol{X} - \boldsymbol{X_0}$. Now, we split $\boldsymbol{F}$ into pieces that are linear and nonlinear in $\boldsymbol{u}$ and write these as $\boldsymbol{Lu}$ and $\boldsymbol{N}(\boldsymbol{u})$, respectively:

$$\mathrm{d}\boldsymbol{u}/\mathrm{d}t = \boldsymbol{L}\boldsymbol{u} + \boldsymbol{N}(\boldsymbol{u})\ . \tag{5.14}$$

The i,jth element of the Jacobi matrix L is given by $L_{ij} = \partial F_i(\boldsymbol{X_0})/\partial X_{0j}$, and we assume that $\boldsymbol{N}(\boldsymbol{u})$ can be expanded in powers of $\boldsymbol{u}$. Thus we write

$$\boldsymbol{N}(\boldsymbol{u}) = N_2\boldsymbol{uu} + N_3\boldsymbol{uuu} + \cdots\ , \tag{5.15}$$

where the ith components of the terms on the right-hand side are given by

$$\begin{aligned}
(N_2\boldsymbol{uv})_i &= \sum_j\sum_k \frac{1}{2!}\frac{\partial^2 F_i(\boldsymbol{u})}{\partial u_j \partial u_k}\bigg|_{\boldsymbol{u}=0} u_j v_k\ ,\\
(N_3\boldsymbol{uvw})_i &= \sum_j\sum_k\sum_l \frac{1}{3!}\frac{\partial^3 N_i(\boldsymbol{u})}{\partial u_j \partial u_k \partial u_l}\bigg|_{\boldsymbol{u}=0} u_j v_k w_l\ ,
\end{aligned}$$

and so on.

The linear stability of the trivial solution $\boldsymbol{u} = 0$ is determined by the spectrum of eigenvalues λ of the operator L. The values of λ are given as the roots of an nth degree polynomial with real coefficients. In the case in which complex roots exist, they appear in conjugate pairs. The necessary and sufficient condition for the stability of the trivial solution is that the relation $\mathrm{Re}\,\lambda < 0$ holds for all values of λ. That is to say, in the complex λ plane, all values of λ lie in the left half-plane. Therefore, in the case in which by changing μ, the stability of this solution becomes destroyed at some value, in general the following two situations are possible:

(A) A single real eigenvalue passes through the origin.
(B) A single pair of complex conjugate eigenvalues cross the imaginary axis.

Case (B) is referred to as a Hopf bifurcation. While these may appear to be the only important cases, we should point out, however, that for physical systems it is often the case that $\boldsymbol{F}(\boldsymbol{X})$ possesses some type of special symmetry, and thus it cannot be said that situations in which, for example, multiple real eigenvalues pass through the origin simultaneously or cases (A) and (B) occur simultaneously are necessarily exceptional. Despite this fact, in the present treatment, for the sake of simplicity we will assume that there is no degeneracy among the eigenvalues. In other words, we consider only so-called 'simple bifurcations' in our presentation.

Let us suppose that, for the model in question, when μ changes from negative to positive, either a type (A) or type (B) instability is realized. Then, in the neighborhood of the bifurcation point $\mu = 0$, we split L and λ as

$$L = L_0 + \mu L_1 \,, \tag{5.16a}$$

$$\lambda = \lambda_0 + \mu \lambda_1 \,, \tag{5.16b}$$

expressing them as the sum of their realizations at the critical point and some appropriate remainder term. In general, L_1 and λ_1 too can depend on μ, but for the discussion that follows, this dependence is unimportant, and we therefore ignore it. For the same reason, we also ignore the μ dependence of the nonlinear operator $\boldsymbol{N}(\boldsymbol{u})$.

Writing the real parts of λ_0 and λ_1 as σ_0 and σ_1 and their imaginary parts as ω_0 and ω_1, by assumption we have the following relations:

$$\sigma_0 = 0 \,, \qquad \sigma_1 > 0 \,, \tag{5.17a}$$

$$\omega_0 = \omega_1 = 0 \qquad \text{[in case (A)]} \,,$$

$$\omega_0, \omega_1 \neq 0 \qquad \text{[in case (B)]} \,. \tag{5.17b}$$

Now, let us write the right and left eigenvectors of the operator L_0 corresponding to the critical eigenvalue λ_0 as $\boldsymbol{U}$ and $\boldsymbol{U}^*$, respectively. Explicitly, we have

$$L_0 \boldsymbol{U} = \lambda_0 \boldsymbol{U} \,, \qquad \boldsymbol{U}^* L_0 = \lambda_0 \boldsymbol{U}^* \,. \tag{5.18}$$

In case (A), $\boldsymbol{U}$ and $\boldsymbol{U}^*$ are real vectors, and in case (B) they are complex. We normalize these vectors so that they satisfy $(\boldsymbol{U}, \boldsymbol{U}^*) = 1$. In case (B), we additionally have $(\boldsymbol{U}^*, \overline{\boldsymbol{U}}) = (\overline{\boldsymbol{U}}, \boldsymbol{U}^*) = 0$. Also, we note that the following equalities hold:

$$\lambda_0 = (\boldsymbol{U}^*, L_0\boldsymbol{U}), \qquad \lambda_1 = (\boldsymbol{U}^*, L_1\boldsymbol{U}) . \tag{5.19}$$

For the situation under consideration, we can now write the evolution equation (5.14) as

$$\left(\frac{\mathrm{d}}{\mathrm{d}t} - L_0\right)\boldsymbol{u} = \mu L_1\boldsymbol{u} + \boldsymbol{N}(\boldsymbol{u}) . \tag{5.20}$$

In the next section this will be considered as the main equation in our treatment of reduction. Before we proceed, however, there is an additional point which must be made. The situation in that the above equation can be used to describe the system after the stability of the stationary solution $\boldsymbol{X}_0$ has been destroyed is limited to those cases in which, although unstable, this solution continues to exist. For cases in which the stationary solution disappears together with this loss of stability, this equation does not provide a valid description of the system. We will discuss this point further in the latter half of the following section.

In the treatment below, we consider both terms on the right-hand side of (5.20) as perturbations. In other words, at the bifurcation point, the linear system represents the unperturbed system, and the effect introduced by μ being shifted from its value at the bifurcation point, together with the effect of nonlinear terms, is treated as a perturbation. The validity of this treatment follows from the assumption that both μ and $\boldsymbol{u}$ are small quantities. This assumption is justified because, physically, the smallness of $\boldsymbol{u}$ results from the smallness of μ, and in fact it is expected that a fixed formula exists relating their sizes. However, this formula comes to be known from information concerning the physically meaningful solution obtained from the reduced equation, and thus, at the present stage, we must think of $\boldsymbol{u}$ and μ as independent objects.

5.3 Foundations of the Amplitude Equation

In this section, we carry out the reduction of (5.13) for both types of destabilization of the stationary solution mentioned above.

Case (A)

First, as a special situation, we consider the case in which the equation of motion is invariant under a change in sign of $\boldsymbol{X}$; that is,

$$\boldsymbol{F}(-\boldsymbol{X}) = -\boldsymbol{F}(\boldsymbol{X}) . \tag{5.21}$$

We will later investigate the case in which this symmetry is lost. In the case with this symmetry, the trivial solution $\boldsymbol{X}_0 = 0$ exists for all values of the bifurcation parameter, and we can use (5.20) as the fundamental equation. In the case under consideration, of course the relation $\boldsymbol{N}(-\boldsymbol{u}) = -\boldsymbol{N}(\boldsymbol{u})$ is also satisfied. Now, let us use $\boldsymbol{u}_0$ to represent the solution to this equation in the $t \to \infty$ limit for the case in which the right-hand side of (5.20) vanishes (the unperturbed case). Then, since all the components of this solution other than that corresponding to the null eigenvector decay, it is evident that

$$\boldsymbol{u}_0 = A\boldsymbol{U} , \tag{5.22}$$

where A is a free real parameter.

As we saw in the two examples of Sect. 5.1, when a free parameter appears in the asymptotic (i.e., $t \to \infty$) solution of the unperturbed system, reinterpretation of this parameter as a state variable leads us to determine an equation governing its slow motion in the perturbed case. This is a universally exercised concept in reduction theory. We thus reinterpret A here in this way and note that our starting point is the unperturbed system, where we have

$$\mathrm{d}A/\mathrm{d}t = 0 . \tag{5.23}$$

Equations (5.22) and (5.23) for the unperturbed system represent, respectively, the trivial invariant space (critical eigenspace) and the trivial motion in this space.

In response to perturbation, this invariant space will suffer a small deformation, and at the same time, gentle motion should develop in this space. We thus expect that the perturbed system will possess the following reduced form:

$$\boldsymbol{u} = \boldsymbol{u}_0 + \boldsymbol{\rho}(A) , \tag{5.24a}$$

$$\mathrm{d}A/\mathrm{d}t = G(A) . \tag{5.24b}$$

Here, ρ and G are both small quantities. Equation (5.24a) represents the deformed invariant space M, while (5.24b) represents the slow motion in this deformed space.

Unfortunately, ρ and G cannot be uniquely determined as a result of simply making the assumption leading to (5.24). The reason for this is that the coordinate A, introduced to describe motion on M, has not yet been defined. Each point in the critical eigenspace corresponds to a particular value of A and, in this sense, A represents this space, but we have not yet thought about how the correspondence between these points and those in M should be made. The arbitrariness existing at this stage in the correspondence between A and M is equivalent to an arbitrariness in the manner in which $\boldsymbol{u}$ is split into $\boldsymbol{u}_0$ and $\boldsymbol{\rho}$. Figure 5.2 expresses this equivalence. In particular, when $\boldsymbol{u}$ is decomposed into component eigenvectors of L_0, depending on whether or not the $\boldsymbol{U}$ component so obtained is completely accounted for by the $\boldsymbol{u}_0$ term, the meaning inherent in the quantity A can be changed. Hence, in order to remove this indeterminacy, an additional condition is necessary.

As the most natural additional condition, let us require the nature of $\boldsymbol{\rho}$ to be such that it does not contain a $\boldsymbol{U}$ component:

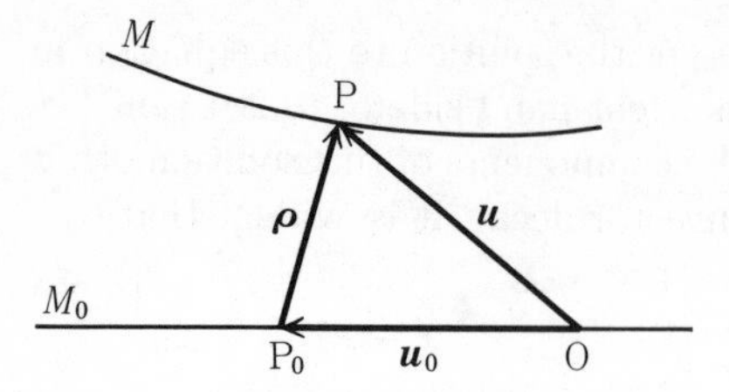

Fig. 5.2. The introduction of the coordinate A on the invariant surface M. The parameter A is initially defined on the eigenvalue space M_0. Then, since we can define a one-to-one correspondence between each point P_0 on M_0 and each point P on M, if we stipulate that corresponding points on M_0 and M possess the same value of A, A comes to be defined on M as well. Note that there is an arbitrariness involved in defining the correspondence between points on M_0 and M. This follows from the arbitrariness inherent in the definition of $\boldsymbol{u}$ in terms of $\boldsymbol{u}_0$ and $\boldsymbol{\rho}$

$$(\boldsymbol{U}^*, \boldsymbol{\rho}) = 0 . \tag{5.25}$$

A similar requirement was enforced in Example 2 of Sect. 5.1. Now, inserting (5.24) into the equation of motion (5.20), we obtain

$$L_0\boldsymbol{\rho} = G\boldsymbol{U} + \boldsymbol{b} , \tag{5.26a}$$

$$\boldsymbol{b} = G\frac{\mathrm{d}\boldsymbol{\rho}}{\mathrm{d}A} - \mu L_1 \boldsymbol{u} - \boldsymbol{N}(\boldsymbol{u}) . \tag{5.26b}$$

Note the close resemblance between (5.26a) and (5.9).

Using the solvability condition, we obtain a relation similar to that found for Example 2:

$$G = -(\boldsymbol{U}^*, \boldsymbol{b}) . \tag{5.27}$$

Under conditions (5.25) and (5.27), formally solving (5.26b) for $\boldsymbol{\rho}$ yields

$$\boldsymbol{\rho} = L_0^{-1}(\boldsymbol{b} - (\boldsymbol{U}^*, \boldsymbol{b})\boldsymbol{U}) . \tag{5.28}$$

The lowest approximation for $\boldsymbol{b}$ is obtained by substituting the unperturbed solution $G = 0$, $\boldsymbol{\rho} = 0$ into (5.26b). This gives

$$\boldsymbol{b} \approx -\mu A L_1 \boldsymbol{U} - \boldsymbol{N}(A\boldsymbol{U}) . \tag{5.29}$$

Using this expression, we can determine G and $\boldsymbol{\rho}$ from (5.27) and (5.28). Then, by successive substitution, we can raise the order of the approximation, following reasoning identical to that given in Sect. 5.1. In fact, in many cases it is sufficient to determine only G, which contains the most important terms in μ and A. We will now carry out this determination.

By the symmetry assumption (5.21),

$$\boldsymbol{N} = A^3 N_3 \boldsymbol{UUU} + O(A^5) . \tag{5.30}$$

We ignore terms of fifth and higher order in $\boldsymbol{U}$. Then inserting (5.30) into (5.29), we obtain, as the starting point of our succession of substitutions, the expression

$$\boldsymbol{b} \approx \mu A L_1 \boldsymbol{U} - A^3 N_3 \boldsymbol{U}\boldsymbol{U}\boldsymbol{U} . \tag{5.31}$$

From this, we obtain the lowest approximation of G, and therefore of $\dot{A}$,

$$\begin{aligned} \mathrm{d}A/\mathrm{d}t &= \mu\lambda_1 A - gA^3 , \\ g &= -(\boldsymbol{U}^*, N_3\boldsymbol{U}\boldsymbol{U}\boldsymbol{U}) . \end{aligned} \tag{5.32}$$

The function $\boldsymbol{\rho}$ can be obtained similarly. The first equation in (5.32) has the same form as (1.12), where the real amplitude W plays the role of A. Proceeding with the successive orders of substitution in general yields $\dot{A}$ as an expansion in odd orders of A.

Physically, we are interested in the situation in which the two terms on the right-hand side of (5.32) are 'balanced', and thus, the assumption that $A \sim O(|\mu|^{1/2})$ is *physically* justified. Anticipating this situation, we can efficiently carry out the perturbative calculation of ρ and G. For this purpose, we introduce the $O(|\mu|^{1/2})$ indicator ϵ, and redefine A and μ according to $A \to \epsilon A$ and $\mu \to \epsilon^2\mu$. (In the end, we will set ϵ to 1.) Now let us expand $\boldsymbol{\rho}$, G and $\boldsymbol{b}$ in the single parameter ϵ as

$$\boldsymbol{\rho} = \epsilon\boldsymbol{\rho}_1 + \epsilon^2\boldsymbol{\rho}_2 + \cdots , \tag{5.33a}$$

$$G = \epsilon G_1 + \epsilon^2 G_2 + \cdots , \tag{5.33b}$$

$$\boldsymbol{b} = \epsilon\boldsymbol{b}_1 + \epsilon^2\boldsymbol{b}_2 + \cdots , \tag{5.33c}$$

and substitute these forms into (5.26a). We then require the individual equations consisting of all terms of equal order in ϵ to hold. We then immediately obtain the equalities $\boldsymbol{\rho}_1 = \boldsymbol{\rho}_2 = 0$, $G_1 = G_2 = 0$, and $G_3 = \mu\lambda_1 A - gA^3$. Thus we see that, at the level of the $G \approx G_3$ approximation, (5.32) holds. For the more complicated problems discussed below, the kind of expansion given in (5.33) is very convenient in that it allows one to gain a good perspective on the calculation – although, of course, one should be careful not to get caught in this kind of mechanical formalism and lose sight of the essential meaning of reduction that we have been emphasizing.

In the above analysis, we assumed the system to possess the special symmetry (or, more precisely, anti-symmetry) expressed by (5.21). We now comment on the case in which such a symmetry does not exist. To see the connection between these two types of systems, we consider the situation in which the symmetry is only weakly broken.

Let us write $\boldsymbol{F}(\boldsymbol{X})$ as the sum of an antisymmetric part $\boldsymbol{F}_0(\boldsymbol{X})$ and the remaining part $\boldsymbol{f}(\boldsymbol{X})$:

$$\boldsymbol{F}(\boldsymbol{X};\mu) = \boldsymbol{F}_0(\boldsymbol{X};\mu) + \boldsymbol{f}(\boldsymbol{X}) . \tag{5.34}$$

In the case considered presently, $\boldsymbol{f}$ is a small term. For this reason, and because its μ dependence is of secondary importance (since μ itself is small), this dependence is ignored. As a result of the existence of $\boldsymbol{f}$, the true stationary state (assuming that it continues to exist) is in general shifted slightly from $\boldsymbol{X} = 0$. Now, considering the term $\boldsymbol{f}$ as a perturbation, the present case differs from the symmetric case only in the addition of this term to the original perturbative terms $\mu L_1 \boldsymbol{u}$ and $\boldsymbol{N}(\boldsymbol{u})$. Thus, with the exception that the expression for $\boldsymbol{b}$, (5.26b), is appended with the term $-\boldsymbol{f}$, the above analysis can be applied as it stands to the present case. Taking into account the lowest order for each of these three small terms, we immediately obtain

$$\begin{aligned} \mathrm{d}A/\mathrm{d}t &= \mu\lambda_1 A - A^3 + h \\ h &= (\boldsymbol{U}^*, \boldsymbol{f}(0)) . \end{aligned} \tag{5.35}$$

As displayed in Fig. 5.3, disappearance of the symmetry causes the bifurcation diagram to become topologically altered. Focusing on the branch B_1 containing the reflection point, we see that in response to changing μ, the stable stationary solution and the unstable stationary solution represented by this branch can be made to appear and disappear together. Generally, for an asymmetric system in which the stability of the originally unique stationary solution is lost as a result of one eigenvalue attaining the value 0, the bifurcation diagram will be topologically equivalent to the branch B_1. In this case, we see that the bifurcation is characterized by the mutual disappearance and appearance of the stable and unstable stationary solutions in response to changing μ. This type of bifurcation is known as a *saddle-node bifurcation.*

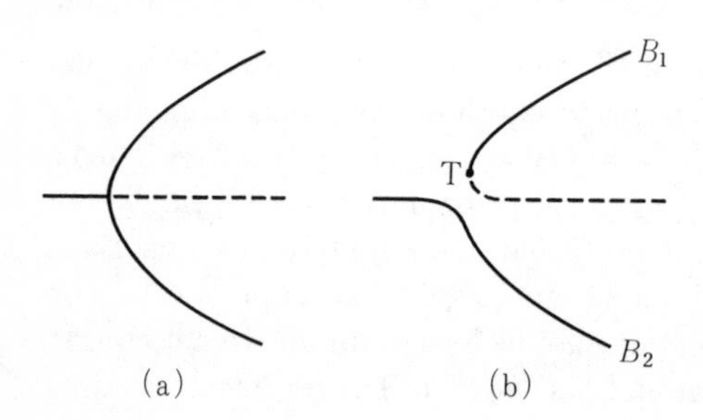

Fig. 5.3. Topological change in the bifurcation diagram resulting from loss of symmetry. Bifurcation diagrams for **(a)** symmetric and **(b)** asymmetric systems. The branch B_1 possesses the reflection point T

Case (B)

Here we assume that the stationary state $\boldsymbol{X}_0$ continues to exist after it loses stability and base our study on the evolution equation (5.20). If we once again view the right-hand side as a perturbation, the solution to the unperturbed system in the limit as $t \to \infty$ (the asymptotic unperturbed solution) is

$$\boldsymbol{u}_0 = W\mathrm{e}^{\mathrm{i}\omega_0 t}\boldsymbol{U} + \mathrm{c.c.}\,, \tag{5.36}$$

where W is an arbitrary complex number. Writing the amplitude and phase of W as A and ϕ, the above equation can be written as

$$\boldsymbol{u}_0 = A\mathrm{e}^{\mathrm{i}\phi}\boldsymbol{U} + \mathrm{c.c.}, \qquad \phi = \omega_0 t + \psi\,. \tag{5.37}$$

For the unperturbed system, a single harmonic oscillator constitutes the only degree of freedom surviving in the limit as $t \to \infty$. Thus we treat this problem as that of a perturbed harmonic oscillator. This treatment closely resembles that of Example 2 in Sect. 5.1.

Let us write the reduced form of (5.20) as

$$\boldsymbol{u} = \boldsymbol{u}_0 + \boldsymbol{\rho}(A, \phi) , \tag{5.38a}$$

$$\mathrm{d}A/\mathrm{d}t = G(A) , \tag{5.38b}$$

$$\mathrm{d}\phi/\mathrm{d}t = \omega_0 + H(A) , \tag{5.38c}$$

where $\boldsymbol{\rho}$ is a 2π-periodic function of ϕ. We can combine (5.38b) and (5.38c) into the single equation

$$\mathrm{d}W/\mathrm{d}t = (G + \mathrm{i}HA)\mathrm{e}^{\mathrm{i}\psi} . \tag{5.39}$$

Now, inserting (5.38) into the fundamental equation (5.20),

$$\tilde{L}_0\boldsymbol{\rho} = (G + \mathrm{i}HA)\mathrm{e}^{\mathrm{i}\phi}\boldsymbol{U} + \mathrm{c.c.} + \boldsymbol{b} \tag{5.40}$$

is obtained, where the following identities hold:

$$\tilde{L}_0 = L_0 - \omega_0\frac{\partial}{\partial\phi} , \tag{5.41a}$$

$$\boldsymbol{b} = G\frac{\partial\boldsymbol{\rho}}{\partial A} + H\frac{\partial\boldsymbol{\rho}}{\partial\phi} - \mu L_1\boldsymbol{u} - \boldsymbol{N}(\boldsymbol{u}) . \tag{5.41b}$$

In order to determine $\tilde{L}_0$, G, and H perturbatively from (5.40), it is natural to expand both sides in the eigenvectors of $\tilde{L}_0$. This operator is defined to act only on n-dimensional vectors that are 2π-periodic in ϕ. Let us define the inner product of two such vectors $\boldsymbol{f}(\phi)$ and $\boldsymbol{g}(\phi)$ by

$$[\boldsymbol{f}, \boldsymbol{g}] = \frac{1}{2\pi}\int_0^{2\pi}(\boldsymbol{f}(\phi), \boldsymbol{g}(\phi))\mathrm{d}\phi . \tag{5.42}$$

The adjoint of $\tilde{L}_0$, $\tilde{L}_0^*$, is defined by

$$[\tilde{L}_0^*\boldsymbol{f}, \boldsymbol{g}] = [\boldsymbol{f}, \tilde{L}_0\boldsymbol{g}] . \tag{5.43}$$

We note that $\exp(\mathrm{i}l\phi)\boldsymbol{U}$ and $\exp(-\mathrm{i}l\phi)\boldsymbol{U}$ (l integer) are both eigenvectors of $\tilde{L}_0$. In particular, for $l = 1$, these represent null eigenvectors. Then, note that these eigenvectors appear on the right-hand side of (5.40). If we take the inner product of both sides of this equation with $\boldsymbol{U}^*\exp(-\mathrm{i}\phi)$ (the null eigenvector of $\tilde{L}_0^*$), from the solvability condition, we can obtain the formula for $G + \mathrm{i}HA$; in other words, the reduced equation. This treatment exactly parallels that for Example 2 considered above and case (A) in this section. We will now carry out this treatment.

As can be seen immediately from the definition (5.24), when taking the inner product $[\boldsymbol{U}^*\exp(-\mathrm{i}\phi), \boldsymbol{f}(\phi)]$, all 2π-periodic vector Fourier components of $\boldsymbol{f}(\phi)$ vanish, with the exception of the fundamental mode $\exp(\mathrm{i}\phi)$. More explicitly, Fourier expanding $\boldsymbol{f}(\phi) = \sum_l \boldsymbol{f}^{(l)}\exp(\mathrm{i}l\phi)$, we have $[\boldsymbol{U}^*\exp(-\mathrm{i}\phi), \boldsymbol{f}]$ $= (\boldsymbol{U}^*, \boldsymbol{f}^{(1)})$. As in case (A), we require here that the null eigenvector of

$\tilde{L}_0$ should not appear as a component of $\boldsymbol{\rho}$. In other words, we require the equality

$$[\boldsymbol{U}^* \exp(-\mathrm{i}\phi), \boldsymbol{\rho}] = [\bar{\boldsymbol{U}}^* \exp(\mathrm{i}\phi), \boldsymbol{\rho}] = 0 \tag{5.44}$$

to hold. This equation can also be expressed as

$$(\boldsymbol{U}^*, \boldsymbol{\rho}^{(1)}) = (\bar{\boldsymbol{U}}^*, \boldsymbol{\rho}^{(-1)}) = 0\ . \tag{5.45}$$

Then, from the condition that the left-hand side of (5.40) does not contain the null eigenvector of the operator $\tilde{L}_0$ (the solvability condition), we obtain

$$G + \mathrm{i}HA = -(\boldsymbol{U}^*, \boldsymbol{b}^{(1)})\ . \tag{5.46}$$

Under the two conditions (5.45) and (5.46), (5.40) can be solved formally. In particular, the expression

$$\boldsymbol{\rho}^{(l)} = (L_0 - \mathrm{i}l\omega_0)^{-1}\boldsymbol{b}^{(1)} \qquad (l \neq \pm 1) \tag{5.47}$$

is obtained.

Let us now explicitly determine the formula for $G+\mathrm{i}HA$ (i.e., the evolution equation for W) from the pair of equations (5.46) and (5.47). $G + \mathrm{i}HA$ and $\boldsymbol{\rho}$ should be obtainable together as power expansions in the small quantities μ and $|W|$. In the following, we again employ the mechanical expansion in ϵ applied in Case A. Thus we rewrite W and μ according to the rules $W \to \epsilon W$ and $\mu \to \epsilon^2\mu$, and expand $\boldsymbol{\rho}$, G, H and $\boldsymbol{b}$ as in (5.33). In this way, both sides of (5.46) and (5.47) also become expansions in ϵ. From these, it is then possible to write down equations of successive orders in ϵ (balanced equations). For the first two terms in the expansion of $\boldsymbol{b}$ we have

$$\boldsymbol{b}_1 = 0\ , \tag{5.48a}$$
$$\boldsymbol{b}_2 = -N_2\boldsymbol{u}_0\boldsymbol{u}_0\ . \tag{5.48b}$$

From (5.46), these imply that $G_1 = G_2 = H_1 = 0$ and, in addition, from (5.47), that $\boldsymbol{\rho}_1 = 0$. Hence,

$$\boldsymbol{b}_3 = -\mu L_1\boldsymbol{u}_0 - N_3\boldsymbol{u}_0\boldsymbol{u}_0\boldsymbol{u}_0 - 2N_2\boldsymbol{u}_0\boldsymbol{\rho}_2\ , \tag{5.49}$$

and the lowest order nontrivial equations become

$$G_3 + \mathrm{i}H_2A = -(\boldsymbol{U}^*, \boldsymbol{b}_3^{(1)})\ , \tag{5.50a}$$
$$\boldsymbol{\rho}_2^{(l)} = (L_0 - \mathrm{i}l\omega_0)^{-1}\boldsymbol{b}_2^{(1)} \qquad (l \neq \pm 1)\ . \tag{5.50b}$$

Then, multiplying (5.50a) by $\exp(\mathrm{i}\phi)$ produces the lowest-order expression for $\dot{W}$:

$$\mathrm{d}W/\mathrm{d}t = -(\boldsymbol{U}^*, \boldsymbol{b}_3^{(1)})\mathrm{e}^{\mathrm{i}\phi}\ . \tag{5.51}$$

Therefore, if $\boldsymbol{\rho}_2$ contained in $\boldsymbol{b}_3$ can be expressed in terms of W, we can obtain a closed equation of motion for W.

Now, from (5.50b) and (5.48b), the following relations hold for $\boldsymbol{\rho}_2^{(l)} (l \neq \pm 1)$:

$$\rho_2^{(0)} = -2|W|^2 \boldsymbol{V}_0, \qquad \boldsymbol{V}_0 = L_0^{-1} N_2 \boldsymbol{U}\overline{\boldsymbol{U}} \,, \tag{5.52a}$$

$$\rho_2^{(\pm 2)} = -|W|^2 \boldsymbol{V}_\pm \,, \tag{5.52b}$$

$$\boldsymbol{V}_+ = (L_0 - 2\mathrm{i}\omega_0)^{-1} N_2 \boldsymbol{U}\boldsymbol{U}, \quad \boldsymbol{V}_- = (L_0 + 2\mathrm{i}\omega_0)^{-1} N_2 \bar{\boldsymbol{U}}\bar{\boldsymbol{U}},$$

$$\rho_2^{(l)} = 0 \qquad (l \neq 0, \pm 1, \pm 2) \,. \tag{5.52c}$$

Substituting these into (5.49) yields

$$\begin{aligned} b_3^{(1)} &= -\mu |W| L_1 \boldsymbol{U} - 3|W|^3 N_3 \bar{\boldsymbol{U}}\boldsymbol{U}\boldsymbol{U} \\ &\qquad - 2|W| N_2 \boldsymbol{U} \rho_2^{(0)} - 2|W| N_2 \bar{\boldsymbol{U}} \rho_2^{(2)} \\ &= -\mu |W| L_1 \boldsymbol{U} + |W|^3 (-3 N_3 \bar{\boldsymbol{U}}\boldsymbol{U}\boldsymbol{U} \\ &\qquad + 4 N_2 \boldsymbol{U} \boldsymbol{V}_0 + 2 N_2 \bar{\boldsymbol{U}} \boldsymbol{V}_+) \,, \end{aligned} \tag{5.53}$$

and, substituting this into (5.51), we obtain the amplitude equation

$$\mathrm{d}W/\mathrm{d}t = \mu\lambda_1 W - g|W|^2 W \,. \tag{5.54}$$

The coefficient g here is, as λ_1, a complex constant:

$$g = -3(\boldsymbol{U}^*, N_3 \bar{\boldsymbol{U}}\boldsymbol{U}\boldsymbol{U}) + 4(\boldsymbol{U}^*, N_2 \boldsymbol{U}\boldsymbol{V}_0) + 2(\boldsymbol{U}^*, N_2 \bar{\boldsymbol{U}}\boldsymbol{V}_+) \,. \tag{5.55}$$

5.4 The Introduction of Continuous Spatial Degrees of Freedom

We have been describing the reduction of ordinary differential equations. However, as represented by the Navier–Stokes equation and reaction-diffusion equations, many important evolution equations in the study of dissipative structures are partial differential equations containing continuous spatial degrees of freedom, and often, in an idealized limit, these spatial dimensions extend to infinity. Now we wish to ask whether the type of reduction analysis we have been presenting in the context of ordinary differential equations can be applied to such partial differential equations as well. This may seem to be a difficult problem, since in this case the idea of separating and removing a small number of long time-scale degrees of freedom from all others does not apply. For example, the eigenvalue of the critical mode at the bifurcation point is not isolated; here, the eigenvalue whose real part is zero exists as the endpoint of a continuous branch of eigenvalues. There is no definite way to separate such a spectrum into slow modes and fast modes. When the reduction of partial differential equations is viewed in such a manner, these points represent very serious technical difficulties. However, we are able to avoid these difficulties by treating the spatially extended nature of the systems we study from a slightly different point of view.

5.4.1 The Hopf Bifurcation

In this subsection, we demonstrate the effectiveness of the type of thinking used in the reduction of partial differential equations by carrying out the derivation of the complex Ginzburg–Landau equation from a reaction–diffusion equation. The same type of idea can be applied to the derivation of the Newell–Whitehead equation, which is considered in the following subsection.

Adding a diffusion term to (5.13), we obtain the reaction–diffusion equation

$$\partial \boldsymbol{X}/\partial t = \boldsymbol{F}(\boldsymbol{X};\mu) + D\nabla^2 \boldsymbol{X} \,. \tag{5.56}$$

The diffusion coefficient D is assumed to be a nonzero diagonal matrix, and the system is considered as extending to infinity in the spatial dimension. Let us suppose that the stationary solution of the system obtained by removing the diffusion term, $\dot{\boldsymbol{X}} = \boldsymbol{F}$, displays type (B) instability (with respect to oscillation) for $\mu > 0$. In precisely the same manner then, quite obviously, for (5.56) the spatially uniform stationary solution is unstable with respect to spatially uniform oscillation. Such oscillation can be induced through the application of a spatially uniform perturbation to this solution.

We carry out the reduction of (5.56) in the neighborhood of $\mu = 0$, with the goal of obtaining a generalization of (5.54). The most important point in this process is that we treat the diffusion term here as a type of perturbation. Physically, we are always thinking in terms of a spatially extended field, but in the mathematical treatment, the diffusion term is regarded as acting at an individual spatial point as an externally applied perturbation on the dynamical system $\dot{\boldsymbol{X}} = \boldsymbol{F}$. In the treatment of the previous section, we regarded the linear system at $\mu = 0$ as the unperturbed system and the effect of the shift from $\mu = 0$, as well as that of the nonlinear terms, as a perturbation. In the present context, the diffusion term is considered as an additional perturbation. Implicit in such a treatment is the assumption that the diffusion term is small; in other words, that the spatial modulation of $\boldsymbol{X}$ is gentle. The validity or invalidity of this assumption can only be demonstrated by the nature of the solutions to the reduced equations themselves. (More precisely, this validity depends on whether or not solutions of this nature are physically meaningful in the neighborhood of the bifurcation point.) As discussed in previous sections in a different context, the assumptions concerning the nonlinear effect, as well as the smallness of the amplitude, too can be justified only after the fact.

With the inclusion of the diffusion term, the equation of motion for the quantity $\boldsymbol{u}$ representing the difference between the stationary solution $\boldsymbol{X}_0$ and X corresponding to (5.20) becomes

$$\left(\frac{\partial}{\partial t} - L_0\right)\boldsymbol{u} = \mu L_1 \boldsymbol{u} + \boldsymbol{N}(\boldsymbol{u}) + D\nabla^2 \boldsymbol{u} \,. \tag{5.57}$$

We seek the reduced form of this equation analogous to (5.38), but in the present case we must assume that $\boldsymbol{\rho}$ contains not only A and ϕ, but also various spatial derivatives of these functions. Similar statements hold for G

and H. Here, as in case (B) studied previously, the invariant space M is two-dimensional. However, we should note that the various spatial derivatives of A and ϕ act as parameters in the expression defining the shape taken by M.

The equation for $\boldsymbol{\rho}$ once again takes the form of (5.40) [with the definition (5.41a)], but in the present case, $\boldsymbol{b}$ is somewhat more complicated than (5.41b). Here we must use the form

$$\begin{aligned} \boldsymbol{b} &= \sum_{j=0}^{\infty}\left\{\nabla^j G \frac{\partial \boldsymbol{\rho}}{\partial(\nabla^j A)} + \nabla^j H \frac{\partial \boldsymbol{\rho}}{\partial(\nabla^j \phi)}\right\} \\ &\qquad - \mu L_1 \boldsymbol{u} - \boldsymbol{N}(\boldsymbol{u}) - D\nabla^2 \boldsymbol{u} \,. \end{aligned} \tag{5.58}$$

Of course, for the amplitude equation subsuming only the lowest-order effects from the three types of perturbations discussed above, among the new terms in this expression for $\boldsymbol{b}$, only the last term is represented. Furthermore, for this term, we are allowed to replace $\boldsymbol{u}$ by $\boldsymbol{u}_0$ and thus we arrive at the generalization of (5.54),

$$\partial W/\partial t = \mu\lambda_1 W - g|W|^2 W + d\nabla^2 W \,, \tag{5.59}$$

the complex Ginzburg–Landau equation. Here, d is a complex constant defined by $d = (\boldsymbol{U}^*, D\boldsymbol{U})$. In the case in which the diffusion coefficient corresponding to each component is equal to some value D_0 (i.e., D is a scalar), we have $d = D_0$.

5.4.2 The Turing Instability

In this subsection we consider the simple situation in which a one-dimensional stationary structure of wave number k_c forms as a result of the Turing instability discussed in Sect. 3.4, and we make clear the foundation of the amplitude equation (equivalent to the NW equation) describing the slow modulation of this periodic structure. In this case, the diffusion process participates in the formation of the periodic structure as well as in the time development of the gentle spatial modulation. Thus, in this case, it is not possible to think of diffusion as a purely perturbative effect, as we did in the previous subsection.

If we separate the two roles played by diffusion, it is appropriate to express the spatial dependence of the variable $\boldsymbol{X}(x,t)$ in two parts; namely, that corresponding to the periodic change and that corresponding to the slow modulation. We express this explicitly by writing $\boldsymbol{X}(x,t)$ as $\boldsymbol{X}(x,t,\xi)$. Physically, x and ξ refer to the same spatial coordinate, but mathematically we consider these as mutually independent variables. The function $\boldsymbol{X}$ is an exact $2\pi/k_c$-periodic function of x, while it is in general assumed to be a slowly varying function of ξ.

Following the above line of reasoning, we must rewrite the spatial derivative in the reaction–diffusion equation as $\partial/\partial x \to \partial/\partial x + \partial/\partial\xi$. We therefore have

$$\frac{\partial \boldsymbol{X}}{\partial t} = \boldsymbol{F}(\boldsymbol{X};\mu) + D\left(\frac{\partial}{\partial x} + \frac{\partial}{\partial \xi}\right)^2 \boldsymbol{X}\,, \tag{5.60}$$

or, for the function representing the 'distance' from the stationary uniform solution, $\boldsymbol{u}$, this becomes

$$\begin{aligned}\left(\frac{\partial}{\partial t} - \tilde{L}_0\right)\boldsymbol{u} &= \mu L_1 \boldsymbol{u} + \boldsymbol{N}(\boldsymbol{u}) + 2D\frac{\partial^2 \boldsymbol{u}}{\partial x \partial \xi} + D\frac{\partial^2 \boldsymbol{u}}{\partial \xi^2}\,, \\ \tilde{L}_0 &= L_0 + D\frac{\partial^2}{\partial x^2}\,.\end{aligned} \tag{5.61}$$

The operator $\tilde{L}_0$ here is understood to act only on 2π-periodic functions of x. We write the null eigenvector of this operator as $\exp(\pm \mathrm{i}k_\mathrm{c}x)\boldsymbol{U}$, where $\boldsymbol{U}$ is a real vector. If we make the definition $(\boldsymbol{U}^*, (L_0 - Dk^2)\boldsymbol{U}) \equiv \lambda(k)$, since in the neighborhood of k_c we have the relation

$$\lambda(k) \propto (k - k_\mathrm{c})^2\,, \tag{5.62}$$

we obtain $d\lambda(k_\mathrm{c})/dk_\mathrm{c} = 0$, or

$$(\boldsymbol{U}^*, D\boldsymbol{U}) = 0\,. \tag{5.63}$$

We assume that all other eigenvalues of $\tilde{L}_0$ have negative real parts.

Let us define the inner product of the two $2\pi/k_\mathrm{c}$-periodic n-dimensional vectors $\boldsymbol{f}(x)$ and $\boldsymbol{g}(x)$ in analogy to (5.42):

$$[\boldsymbol{f}, \boldsymbol{g}] = \frac{k_\mathrm{c}}{2\pi}\int_0^{2\pi/k_\mathrm{c}} \mathrm{d}x(\boldsymbol{f}(x), \boldsymbol{g}(x))\,. \tag{5.64}$$

Note that the $t \to \infty$ solution of the unperturbed system is

$$\boldsymbol{u}_0 = W\mathrm{e}^{\mathrm{i}k_\mathrm{c}x}\boldsymbol{U} + \text{c.c.}\,. \tag{5.65}$$

Then, writing $W = A\exp(\mathrm{i}\psi)$, (5.65) can be expressed as $\boldsymbol{u}_0 = A\exp(\mathrm{i}\phi)\boldsymbol{U}+$ c.c., with $\phi = k_\mathrm{c}x + \psi$. In the present case, the arbitrary nature of the phase ψ is due to the symmetry of the system with respect to spatial translation.

We next rewrite W as a function of ξ and t and assume the following system for the reduced form of the perturbed system (5.61):

$$\boldsymbol{u} = \boldsymbol{u}_0 + \boldsymbol{\rho}\left(A, k_\mathrm{c}x + \psi, \frac{\partial A}{\partial \xi}, \frac{\partial \psi}{\partial \xi}, \cdots\right), \tag{5.66a}$$

$$\frac{\partial A}{\partial t} = G\left(A, \frac{\partial A}{\partial \xi}, \frac{\partial \psi}{\partial \xi}, \cdots\right), \tag{5.66b}$$

$$\frac{\partial \psi}{\partial t} = H\left(A, \frac{\partial A}{\partial \xi}, \frac{\partial \psi}{\partial \xi}, \cdots\right). \tag{5.66c}$$

Here, $\boldsymbol{\rho}$ is a $2\pi/k_\mathrm{c}$-periodic function of x. Then, substituting (5.66) into (5.61), we obtain a formal expression for the linear ordinary differential equation obeyed by $\boldsymbol{\rho}$. The form of this equation is similar to that of (5.40):

$$\tilde{L}_0\boldsymbol{\rho} = \{(G + \mathrm{i}HA)\mathrm{e}^{\mathrm{i}(k_\mathrm{c}x+\psi)} + \text{c.c.}\}\boldsymbol{U} + \boldsymbol{b}\,. \tag{5.67}$$

In this equation, $\boldsymbol{b}$ is given by

$$\boldsymbol{b} = \sum_{j=0}^{\infty} \left\{ \frac{\partial^j G}{\partial \xi^j} \frac{\partial \boldsymbol{\rho}}{\partial(\partial_\xi^j A)} + \frac{\partial^j H}{\partial \xi^j} \frac{\partial \boldsymbol{\rho}}{\partial(\partial_\xi^j \psi)} \right\} - \mu L_1 \boldsymbol{u} - \boldsymbol{N}(\boldsymbol{u}) - 2D \frac{\partial^2 \boldsymbol{u}}{\partial x \partial \xi} - D \frac{\partial^2 \boldsymbol{u}}{\partial \xi^2} . \qquad (5.68)$$

The function $\boldsymbol{b}$, like $\boldsymbol{\rho}$, is 2π-periodic in x. Let us write these functions in Fourier expanded forms:

$$\boldsymbol{\rho} = \sum_{l=-\infty}^{\infty} \boldsymbol{\rho}^{(l)} \exp(\mathrm{i} l k_c x), \qquad \boldsymbol{b} = \sum_{l=-\infty}^{\infty} \boldsymbol{b}^{(l)} \exp(\mathrm{i} l k_c x) .$$

Then we require that $\boldsymbol{\rho}$ does not contain as a component the null eigenvector of $\tilde{L}_0$; that is,

$$(\boldsymbol{U}^*, \boldsymbol{\rho}^{(1)}) = (\bar{\boldsymbol{U}}^*, \boldsymbol{\rho}^{(-1)}) = 0 . \qquad (5.69)$$

From the same condition applied to the right-hand side of (5.67) (the solvability condition), the relation

$$(G + \mathrm{i} H A)\mathrm{e}^{\mathrm{i}\psi} = -(\boldsymbol{U}^*, \boldsymbol{b}^{(1)}) \qquad (5.70)$$

is required. Regarding the right-hand side of (5.67) formally as an inhomogeneous term, this equation can be thought of as a linear ordinary differential equation for $\boldsymbol{\rho}$. Hence, under conditions (5.69) and (5.70), (5.67) can be solved uniquely for $\boldsymbol{\rho}$. Explicitly, we have

$$\boldsymbol{\rho}^{(\pm 1)} = (L_0 - k_c^2 D)^{-1} \{ -\boldsymbol{U}^*, \boldsymbol{b}^{(\pm 1)} \boldsymbol{U} + \boldsymbol{b}^{(\pm 1)} \} \qquad (5.71a)$$

$$\boldsymbol{\rho}^{(l)} = (L_0 - l^2 k_c^2 D)^{-1} \boldsymbol{b}^{(l)} \qquad (l \neq \pm 1) . \qquad (5.71b)$$

From (5.70) and (5.71), $\boldsymbol{\rho}$, G, and H can be obtained through successive approximation. Introducing expansions in ϵ allows us to carry out this procedure systematically, as seen in previous sections. We thus rewrite the various quantities in the following manner: $\mu \to \epsilon^2 \mu$, $A \to \epsilon A$, and $\partial/\partial\xi \to \epsilon \partial/\partial\xi$. Then we expand $\boldsymbol{\rho}$, G, H, and $\boldsymbol{b}$ as in (5.33). Trivially, we have $\boldsymbol{b}_1 = 0$, and from this we obtain $G_1 = 0$ and $\boldsymbol{\rho}_1 = 0$. These relations yield

$$\boldsymbol{b}_2 = -\boldsymbol{N}_2 \boldsymbol{u}_0 \boldsymbol{u}_0 - 2D \frac{\partial^2 \boldsymbol{u}_0}{\partial x \partial \xi} , \qquad (5.72)$$

and this gives

$$\begin{aligned} (G_2 + \mathrm{i} H_1 A)\mathrm{e}^{\mathrm{i}\psi} &= -(\boldsymbol{U}^*, \boldsymbol{b}_2^{(1)}) \\ &= 2\mathrm{i} k_c \frac{\partial W}{\partial \xi} (\boldsymbol{U}^*, D\boldsymbol{U}) = 0 , \end{aligned} \qquad (5.73)$$

where we have used (5.63). Continuing, we obtain the lowest-order nontrivial balanced equation,

$$(G_3 + \mathrm{i}H_2A)\mathrm{e}^{\mathrm{i}\psi} = -(\boldsymbol{U}^*, \boldsymbol{b}_3^{(1)}) , \tag{5.74a}$$

$$\boldsymbol{\rho}_2^{(l)} = (L_0 - l^2k_c^2D)^{-1}\boldsymbol{b}_2^{(l)} , \tag{5.74b}$$

where $\boldsymbol{b}_3$ is given by

$$\boldsymbol{b}_3 = -\mu L_1\boldsymbol{u}_0 - N_3\boldsymbol{u}_0\boldsymbol{u}_0\boldsymbol{u}_0 - 2N_2\boldsymbol{u}_0\boldsymbol{\rho}_2 - 2D\frac{\partial^2\boldsymbol{\rho}_2}{\partial x\partial\xi} . \tag{5.75}$$

Each Fourier component of $\boldsymbol{\rho}_2$ can be obtained from (5.72) and (5.74). The results are as follows:

$$\boldsymbol{\rho}_2^{(0)} = -2|W|^2\boldsymbol{V}_0, \qquad \boldsymbol{V}_0 = L_0^{-1}N_2\boldsymbol{U}\boldsymbol{U} , \tag{5.76a}$$

$$\boldsymbol{\rho}_2^{(\pm1)} = \mp 2\mathrm{i}k_c\frac{\partial W}{\partial\xi}\boldsymbol{V}_{\pm1}, \qquad \boldsymbol{V}_{\pm1} = (L_0 - k_c^2D)^{-1}D\boldsymbol{U} , \tag{5.76b}$$

$$\boldsymbol{\rho}_2^{(2)} = -W^2\boldsymbol{V}_2, \qquad \boldsymbol{V}_2 = (L_0 - 4k_c^2D)^{-1}N_2\boldsymbol{U}\boldsymbol{U} , \tag{5.76c}$$

$$\boldsymbol{\rho}_2^{(-2)} = -\bar{W}^2\boldsymbol{V}_{-2}, \qquad \boldsymbol{V}_{-2} = \boldsymbol{V}_2 . \tag{5.76d}$$

All components not appearing here vanish. By inserting these results for $\boldsymbol{\rho}_2$ into (5.75), and substituting the expression for $\boldsymbol{b}_3$ thus obtained into (5.74a), we can find $(G_3 + \mathrm{i}H_2A)\exp(\mathrm{i}\psi)$; that is, the lowest order expression for $\dot{W}$. Then, re-identifying ξ with x, the coordinate from which it originated, this equation becomes

$$\frac{\partial W}{\partial t} = \mu\lambda_1 W - g|W|^2W + d\frac{\partial^2 W}{\partial x^2} . \tag{5.77}$$

In this expression, the quantities λ_1, d, and g are all real numbers. They are defined by the following expressions: $\lambda_1 = (\boldsymbol{U}^*, L_1\boldsymbol{U})$, $d = 4k_c^2(\boldsymbol{U}^*, D\boldsymbol{V}_1)$, and $g = -3(\boldsymbol{U}^*, N_3\boldsymbol{U}\boldsymbol{U}\boldsymbol{U}) + 4(\boldsymbol{U}^*, N_2\boldsymbol{U}\boldsymbol{V}_0) + 2(\boldsymbol{U}^*, N_2\boldsymbol{U}\boldsymbol{V}_2)$.

5.5 Fundamentals of Phase Dynamics

In this section we detail the theoretical foundation for the phase equation that we derived phenomenologically in Chap. 4. This will be done by faithfully following both the derivation of the phase equation given there and the basic reduction formalism presented in Sect. 5.1. Just as in the derivation of the amplitude equation performed in the previous section, we here regard the effect of the slow spatial modulation as a perturbation. However, in the several cases considered below, this is the unique perturbation to which the system is subject.

In this section, we again consider reaction–diffusion equations. Paralleling the treatments given in Sects. 5.4.1 and 5.4.2, we first consider the phase dynamics of a uniform oscillating field, and then we consider a system possessing a one-dimensional stationary periodic structure. Finally, we discuss the phase equation for an interface. These correspond to the three representative physical situations considered in Chap. 4.

5.5.1 Phase Dynamics in a Uniform Oscillating Field

We write the uniform oscillatory solution of the reaction–diffusion equation (5.56) as $\boldsymbol{X}_0(\phi)$ $[= \boldsymbol{X}_0(\phi + 2\pi)]$, where $\phi = \omega_0 t + \psi$, and ψ is an arbitrary constant. Writing

$$\mathrm{d}\boldsymbol{X}_0/\mathrm{d}\phi = \boldsymbol{u}_0(\phi)\,, \tag{5.78}$$

$\boldsymbol{u}_0$ clearly satisfies the equality $\omega_0 \boldsymbol{u}_0(\phi) = \boldsymbol{F}(\boldsymbol{X}_0(\phi))$. Taking the derivative of this expression with respect to ϕ, we obtain

$$\begin{aligned} \tilde{L}_0 \boldsymbol{u}_0 &= 0\,, \\ \tilde{L}_0 &= L_0(\phi) - \omega_0 \frac{\mathrm{d}}{\mathrm{d}\phi}\,, \end{aligned} \tag{5.79}$$

where $L_0(\phi)$ is the Jacobi matrix corresponding to $\boldsymbol{F}$ (i.e., its components satisfy $(L_0)_{ij} = \partial F_i(\boldsymbol{X}_0(\phi))/\partial X_{0j}(\phi)$.) The operator $\tilde{L}_0$ is understood to act only on 2π-periodic vector valued functions, and, according to (5.79), $\boldsymbol{u}_0(\phi)$ is its null eigenvector. As before, the vector inner product is defined by (5.42), and $\tilde{L}_0^*$ is used to denote the adjoint of $\tilde{L}_0$. The null vector of this adjoint operator is represented by $\boldsymbol{u}_0^*$. We assume the normalization condition $[\boldsymbol{u}_0^*, \boldsymbol{u}_0] = 1$.

The phase ϕ is also considered to be a slowly changing function of the spatial variable, and we now set out to determine its evolution equation. On the basis of the experience we have gained in previous sections, we assume the following equations defining the reduced form for the reaction–diffusion equation:

$$\boldsymbol{X}(\boldsymbol{r},t) = \boldsymbol{X}_0(\phi(\boldsymbol{r},t)) + \boldsymbol{\rho}(\phi, \nabla\phi, \nabla^2\phi, \cdots)\,, \tag{5.80a}$$

$$\partial\phi/\partial t = \omega_0 + H(\nabla\phi, \nabla^2\phi, \cdots)\,. \tag{5.80b}$$

Substituting (5.80a) into (5.56) yields

$$\tilde{L}_0 \boldsymbol{\rho} = H\boldsymbol{u}_0 + \boldsymbol{b}\,, \tag{5.81a}$$

$$\boldsymbol{b} = -D\nabla^2 \boldsymbol{X}_0 - \boldsymbol{N}(\boldsymbol{\rho}) + \sum_{j=0}^{\infty} \nabla^j H \frac{\partial \boldsymbol{\rho}}{\partial(\nabla^j \phi)}\,. \tag{5.81b}$$

Let us note here that the procedure involved in this example closely parallels that for the derivation of the amplitude equation, but for the sake of conciseness we will not make note of each of the corresponding points.

The conditions that $\boldsymbol{\rho}$ and the right-hand side of (5.81a) do not contain the null eigenvector of $\tilde{L}_0$ imply

$$[\boldsymbol{u}_0^*, \boldsymbol{\rho}] = 0 \tag{5.82}$$

and

$$H = -[\boldsymbol{u}_0^*, \boldsymbol{b}]\,, \tag{5.83}$$

respectively. Under these conditions, it is possible to uniquely solve (5.81a) for $\boldsymbol{\rho}$:

$$\rho = \tilde{L}_0^{-1}(H\boldsymbol{u}_0 + \boldsymbol{b}) . \tag{5.84}$$

Then, starting from the lowest-order formula for $\boldsymbol{b}$,

$$\begin{aligned} \boldsymbol{b} &\approx -D\nabla^2 \boldsymbol{X}_0 = -D\frac{\mathrm{d}\boldsymbol{X}_0}{\mathrm{d}\phi}\nabla^2\phi - D\frac{\mathrm{d}^2\boldsymbol{X}_0}{\mathrm{d}\phi^2}(\nabla\phi)^2 \\ &= -D\boldsymbol{u}_0\nabla^2\phi - D\frac{\mathrm{d}\boldsymbol{u}_0}{\mathrm{d}\phi}(\nabla\phi)^2 , \end{aligned} \tag{5.85}$$

ρ and H can be obtained from the pair of equations (5.82) and (5.84) by successive substitution. From the lowest-order approximation of H, the phase equation (4.26) is produced:

$$\partial\phi/\partial t = \omega_0 + a\nabla^2\phi + g(\nabla\phi)^2 . \tag{5.86}$$

Here, $a = [\boldsymbol{u}_0^*, D\boldsymbol{u}_0]$ and $g = [\boldsymbol{u}_0^*, D\mathrm{d}\boldsymbol{u}_0/\mathrm{d}\phi]$. Finally, we can facilitate a systematic small-parameter expansion by taking $\nabla \to \epsilon\nabla$ and developing arguments similar to those given from the derivation of the amplitude equation.

5.5.2 Phase Dynamics for a System with Periodic Structure

If the reaction–diffusion equation (5.56) possesses a stationary periodic solution $\boldsymbol{X}_0(kx+\psi)$ $[= \boldsymbol{X}_0(kx+\psi+2\pi)]$, this solution satisfies

$$\boldsymbol{F}(\boldsymbol{X}_0(\phi)) + k^2 D\frac{\mathrm{d}^2\boldsymbol{X}_0}{\mathrm{d}\phi^2} = 0, \qquad \phi = kx + \psi . \tag{5.87}$$

From the equations obtained by taking the derivative of both sides of this expression with respect to ϕ,

$$\tilde{L}_0\boldsymbol{u}_0 = 0 , \tag{5.88a}$$

$$\tilde{L}_0 = L_0 + k^2 D\frac{\mathrm{d}^2}{\mathrm{d}\phi^2} \tag{5.88b}$$

$$\boldsymbol{u}_0 = \mathrm{d}\boldsymbol{X}_0/\mathrm{d}\phi , \tag{5.88c}$$

it is clear that $\boldsymbol{u}_0$ is the null eigenvector of $\tilde{L}_0$. Again, we use (5.64) to define the inner product $[\boldsymbol{f}, \boldsymbol{g}]$.

Then, substituting the pair of equations

$$\boldsymbol{X}(x,y,t) = \boldsymbol{X}_0(\phi) + \rho\left(\phi, \frac{\partial\psi}{\partial x}, \frac{\partial\psi}{\partial y}, \frac{\partial^2\psi}{\partial x^2}, \cdots\right) , \tag{5.89a}$$

$$\frac{\partial\psi}{\partial t} = H\left(\frac{\partial\psi}{\partial x}, \frac{\partial\psi}{\partial y}, \frac{\partial^2\psi}{\partial x^2}, \cdots\right) , \tag{5.89b}$$

into the reaction–diffusion equation (5.56) yields

$$\tilde{L}_0\rho = H\boldsymbol{u}_0 + \boldsymbol{b} , \tag{5.90a}$$

$$\boldsymbol{b} = -D\frac{\mathrm{d}^2\boldsymbol{X}_0}{\mathrm{d}\phi^2}\left\{2k\frac{\partial\psi}{\partial x} + \left(\frac{\partial\psi}{\partial x}\right)^2 + \left(\frac{\partial\psi}{\partial y}\right)^2\right\}$$

$$
\begin{aligned}
&- D\frac{\mathrm{d}\boldsymbol{X}_0}{\mathrm{d}\phi}\left(\frac{\partial^2\psi}{\partial x^2}+\frac{\partial^2\psi}{\partial y^2}\right) - \boldsymbol{N}(\boldsymbol{\rho}) \\
&-D\frac{\partial^2\boldsymbol{\rho}}{\partial\phi^2}\left\{2k\frac{\partial\psi}{\partial x}+\left(\frac{\partial\psi}{\partial x}\right)^2\right\} \\
&-D\frac{\partial\boldsymbol{\rho}}{\partial\psi}\partial^2\psi\partial x^2 + \boldsymbol{b}'\left(\phi,\frac{\partial\psi}{\partial x},\cdots\right),
\end{aligned} \tag{5.90b}
$$

where $\boldsymbol{b}'$ is related to the time and space derivatives through the derivatives $\partial\psi/\partial x$, $\partial\psi/\partial y$, $\partial^2\psi/\partial x^2$, etc., contained in $\boldsymbol{\rho}$. Its general expression is quite complicated, so we do not include it here.

The solvability condition corresponding to (5.90a) is

$$
H = -[\boldsymbol{u}_0^*, \boldsymbol{b}] . \tag{5.91}
$$

This condition and that on $\boldsymbol{\rho}$,

$$
[\boldsymbol{u}_0^*, \boldsymbol{\rho}] = 0 , \tag{5.92}
$$

give

$$
\boldsymbol{\rho} = \tilde{L}_0^{-1}(H\boldsymbol{u}_0 + \boldsymbol{b}) . \tag{5.93}
$$

Then, beginning from the lowest-order approximation for $\boldsymbol{b}$,

$$
\begin{aligned}
\boldsymbol{b} \;\approx\; & -D\frac{\mathrm{d}^2\boldsymbol{X}_0}{\mathrm{d}\phi^2}\left\{2k\frac{\partial\psi}{\partial x}+\left(\frac{\partial\psi}{\partial x}\right)^2+\left(\frac{\partial\psi}{\partial y}\right)^2\right\} \\
& -D\frac{\mathrm{d}\boldsymbol{X}_0}{\mathrm{d}\phi}\left(\frac{\partial^2\psi}{\partial x^2}+\frac{\partial^2\psi}{\partial y^2}\right),
\end{aligned} \tag{5.94}
$$

we can use (5.91) and (5.93) to determine H and $\boldsymbol{\rho}$ through successive substitution.

The lowest-order approximation for H gives

$$
\frac{\partial\psi}{\partial t} = a_2\frac{\partial^2\psi}{\partial x^2} + b_2\frac{\partial^2\psi}{\partial y^2} . \tag{5.95}
$$

From the symmetry considerations discussed in Chap. 4, it is known that $\partial\psi/\partial x, (\partial\psi/\partial x)^2, \partial\psi/\partial y \ldots$ cannot appear in $\dot{\psi}$. Actually, the coefficients of these terms are proportional to $[\boldsymbol{u}_0^*, D, \mathrm{d}^2\boldsymbol{X}_0/\mathrm{d}\phi^2]$, but from the assumption of the spatial inversion symmetry of $\boldsymbol{X}_0$ $[\boldsymbol{X}_0(\phi) = \boldsymbol{X}_0(-\phi)]$ and the fact that $\boldsymbol{u}_0 = \mathrm{d}\boldsymbol{X}/\mathrm{d}\phi$ and $\boldsymbol{u}_0^*$ have the same parity, these coefficients become identically zero.

5.5.3 Interface Dynamics in a Two-Dimensional Medium

We consider the reaction–diffusion system (5.56) in two spatial dimensions and study the situation in which a kink solution $\boldsymbol{X}_0(\phi)$ propagates in the x direction. Let us write $\phi = x - ct - \psi$. The analysis which follows can be applied in essentially unaltered form to the case of a pulse solution.

The solution $\boldsymbol{X}_0$ obviously satisfies the equation

$$-c\frac{\mathrm{d}\boldsymbol{X}_0}{\mathrm{d}\phi} = \boldsymbol{F}(\boldsymbol{X}_0) + D\frac{\mathrm{d}^2\boldsymbol{X}_0}{\mathrm{d}\phi^2} . \tag{5.96}$$

Taking the derivative of this equation with respect to ϕ gives

$$\tilde{L}_0\boldsymbol{u}_0 = 0 , \tag{5.97a}$$

$$\tilde{L}_0 = L_0 + c\frac{\mathrm{d}}{\mathrm{d}\phi} + D\frac{\mathrm{d}^2}{\mathrm{d}\phi^2} , \tag{5.97b}$$

$$\boldsymbol{u}_0 = \mathrm{d}\boldsymbol{X}_0/\mathrm{d}\phi . \tag{5.97c}$$

Let us assume that the operator $\tilde{L}_0$ acts on functions ϕ that decay to zero sufficiently quickly in the $\phi \to \pm\infty$ limits. Equation (5.97a) expresses the fact that $\boldsymbol{u}_0$ is the null eigenvector of this operator. We define the inner product of the functions $\boldsymbol{f}(\phi)$ and $\boldsymbol{g}(\phi)$ by

$$[\boldsymbol{f}, \boldsymbol{g}] = \int_{-\infty}^{\infty} \mathrm{d}\phi(\boldsymbol{f}(\phi), \boldsymbol{g}(\phi)) . \tag{5.98}$$

The entire eigenvalue spectrum of $\tilde{L}_0$ is discrete, and we assume that, with the exception of the null eigenvalue, the real part of each of these is negative.

To describe the dynamics of a deformed interface like that shown in Fig. 4.5, we reinterpret ψ as the function $\psi(y,t)$, and substitute the form

$$\boldsymbol{X}(x,y,t) = \boldsymbol{X}_0\left(x - ct - \psi(y,t)\right) + \boldsymbol{\rho}\left(\phi, \frac{\partial\psi}{\partial y}, \frac{\partial^2\psi}{\partial y^2}, \cdots\right) \tag{5.99a}$$

into the reaction–diffusion equation (5.56). Then, in the resulting expansion, using

$$\frac{\partial\psi}{\partial t} = H\left(\frac{\partial\psi}{\partial y}, \frac{\partial^2\psi}{\partial y^2}, \cdots\right) , \tag{5.99b}$$

we obtain

$$\tilde{L}_0\boldsymbol{\rho} = H\boldsymbol{u}_0 + \boldsymbol{b} , \tag{5.100a}$$

$$\boldsymbol{b} = -D\left\{\frac{\mathrm{d}\boldsymbol{u}_0}{\mathrm{d}\phi}\left(\frac{\partial\psi}{\partial y}\right)^2 + \boldsymbol{u}_0\frac{\partial^2\psi}{\partial y^2} + \frac{\partial^2\boldsymbol{\rho}}{\partial\phi^2}\left(\frac{\partial\psi}{\partial y}\right)^2 + \frac{\partial\boldsymbol{\rho}}{\partial\phi}\frac{\partial^2\psi}{\partial y^2}\right\}$$
$$+H\frac{\partial\boldsymbol{\rho}}{\partial\phi} - \boldsymbol{N}(\boldsymbol{\rho}) + \boldsymbol{b}' . \tag{5.100b}$$

Here, as above, $\boldsymbol{b}'$ is related to the time and space derivatives through the derivatives $\partial\psi/\partial x$, $\partial\psi/\partial y$, $\partial^2\psi/\partial x^2$, etc., contained in $\boldsymbol{\rho}$, but, again, since its form is very complicated, we refrain from including it.

The solvability condition corresponding to (5.100a) is

$$H = -[\boldsymbol{u}_0^*, \boldsymbol{b}] . \tag{5.101}$$

Then we enforce the condition that $\boldsymbol{\rho}$ should contain no component of the null eigenvector of $\tilde{L}_0$:

$$[\boldsymbol{u}_0^*, \boldsymbol{\rho}] = 0 \,. \tag{5.102}$$

Under the conditions (5.101) and (5.102), it is possible to rewrite (5.100a) as

$$\boldsymbol{\rho} = \tilde{L}_0^{-1}(H\boldsymbol{u}_0 + \boldsymbol{b}) \,. \tag{5.103}$$

In the lowest-order approximation, we have

$$\boldsymbol{b} \approx -D\left\{ \frac{\mathrm{d}\boldsymbol{u}_0}{\mathrm{d}\phi}\left(\frac{\partial\psi}{\partial y}\right)^2 + \boldsymbol{u}_0\frac{\partial^2\psi}{\partial y^2} \right\} \,. \tag{5.104}$$

Substituting this into (5.101), we obtain the lowest-order form for the phase equation:

$$\begin{aligned} \frac{\partial\psi}{\partial t} &= a\frac{\partial^2\psi}{\partial y^2} + g\left(\frac{\partial\psi}{\partial y}\right)^2 \,, \\ a &= [\boldsymbol{u}_0^*, D\boldsymbol{u}_0], \qquad g = \left[\boldsymbol{u}_0^*, D\frac{\mathrm{d}\boldsymbol{u}_0}{\mathrm{d}\phi}\right] \,. \end{aligned} \tag{5.105}$$

Supplement I: Dynamics of Coupled Oscillator Systems

The discussion of dissipative structures appearing in the first five chapters was given in the context of spatio-temporal patterns existing in continuous media. In the present chapter we consider another important class of dissipative systems consisting of a large number of degrees of freedom, those composed of aggregates of isolated elements. A neural network consisting of intricately coupled excitable oscillators (neurons) is one example of such a system. In addition, there are many systems of this type composed of groups of cells exhibiting physiological activity. As the subject of study in nonlinear dynamics has broadened in recent years to include phenomena found in living systems, interest in coupled oscillator systems has grown. In what follows, we consider a relatively simple example forming the subject of a great deal of present study, that of a collection of limit cycles oscillators, and basing our investigation on the method of phase dynamics, we discuss the fundamental points regarding synchronization phenomena.

SI.1 The Phase Dynamics of a Collection of Oscillators

In Chaps. 4 and 5 we touched upon the subject of the phase dynamics of an oscillator field. However, in the present context of a collection of isolated oscillators, it is convenient to use a phase description that differs from that appearing in previous chapters. We first consider a single limit cycle oscillator (with angular frequency ω) represented by an n-dimensional dynamical system, $\dot{\boldsymbol{X}} = \boldsymbol{F}(\boldsymbol{X})$. We then define the phase ϕ corresponding to an arbitrary X in the state space representing the system. The most natural treatment is to define ϕ so that, with the time development of $\boldsymbol{X}$ given by the above equation, ϕ satisfies $\dot{\phi} = \omega$. This implies that $\phi(\boldsymbol{X})$ satisfies

$$\mathrm{grad}_{\boldsymbol{X}}\phi \cdot \boldsymbol{F}(\boldsymbol{X}) = \omega \, . \tag{SI.1}$$

The phase dynamics method is extremely effective in describing systems of weakly coupled oscillators. In this case, we can treat the interaction effect on a single oscillator as being created by a weak external force $\boldsymbol{p}(t)$. Let us proceed in this manner, writing

$$\mathrm{d}\boldsymbol{Z}/\mathrm{d}t = \boldsymbol{F}(\boldsymbol{X}) + \boldsymbol{p}(t) \, . \tag{SI.2}$$

Then, we consider the correspondingly perturbed motion of ϕ. Formally, we have

$$\mathrm{d}\phi/\mathrm{d}t = \omega + \mathrm{grad}_{\boldsymbol{X}}\phi \cdot (\boldsymbol{F} + \boldsymbol{p}) . \tag{SI.3}$$

As shown in Fig. SI.1a, the vector $\mathrm{grad}_{\boldsymbol{X}}\phi$ is perpendicular to the surface of constant phase I_ϕ. Then, if the perturbation in question is sufficiently weak, we expect that the point $\boldsymbol{X}$ representing the system will be close to the limit-cycle orbit C of the unperturbed system. Thus, as described by Fig. SI.1b, we can replace $\mathrm{grad}_{\boldsymbol{X}}\phi$ by its realization at the intersection of I_ϕ and C, $\mathrm{grad}_{\boldsymbol{X}_0}\phi \equiv \boldsymbol{Z}(\phi)$. In other words, we write

$$\mathrm{d}\phi/\mathrm{d}t = \omega + \boldsymbol{Z}(\phi) \cdot \mathbf{p} . \tag{SI.4}$$

Clearly, we have $\boldsymbol{Z}(\phi + 2\pi) = \boldsymbol{Z}(\phi)$. Let us now consider the situation in which our system consists of only two identical oscillators and in which the interaction between oscillators can be represented by a perturbation of the form $\boldsymbol{p} = \boldsymbol{V}(\boldsymbol{X}, \boldsymbol{X}')$, where $\boldsymbol{X}'$ is the state vector of the second oscillator. Now, with $\boldsymbol{X}$ and $\boldsymbol{X}'$ traversing paths along the equal-phase surfaces $I(\phi)$ and $I(\phi')$, with the same reasoning as used above, we can replace $\boldsymbol{X}$ and $\boldsymbol{X}'$ in $\boldsymbol{V}$ by their values at the intersections of C with $I(\phi)$ and C with $I(\phi')$, $\boldsymbol{X}_0$ and $\boldsymbol{X}_0'$. We thus obtain the following evolution equation involving only phases:

$$\begin{aligned} \mathrm{d}\phi/\mathrm{d}t &= \omega + G(\phi, \phi') , \\ G(\phi, \phi') &= \mathbf{Z}(\phi)\mathbf{V}(\mathbf{X}_0(\phi), \mathbf{X}_0(\phi')) . \end{aligned}$$

The function $G(\phi, \phi')$ is 2π-periodic in both ϕ and ϕ'. We further simplify this equation as follows. First, writing $\phi = \omega t + \psi$, and $\phi' = \omega t + \psi'$, and replacing ϕ and ϕ' in (SI.5) with these expressions, we have

$$\mathrm{d}\psi/\mathrm{d}t = G(\omega t + \psi, \omega t \psi') , \tag{SI.5}$$

but since the coupling between oscillators is weak, G is small, and ψ and ψ' change slowly in time. Thus, in one period $2\pi/\omega$, ψ and ψ' change only very slightly, and we can approximate them as constants. We thus average the right-hand side of (SI.6) over one period and obtain

$$\begin{aligned} \mathrm{d}\psi/\mathrm{d}t &= \Gamma(\psi, \psi') , \\ \Gamma(\psi - \psi') &= \frac{1}{2\pi}\int_0^{2\pi} \mathrm{d}(\omega t) G(\omega t + \psi, \omega t + \psi') , \end{aligned}$$

or, returning to a description in terms of ϕ,

$$\mathrm{d}\phi/\mathrm{d}t = \omega + \Gamma(\phi - \phi') , \tag{SI.6}$$

The function Γ is 2π-periodic in $\phi - \phi'$.

In the above discussion, we assumed the two oscillators to be identical, but similar analysis can be applied to the case represented by

$$\boldsymbol{F}(\boldsymbol{X}) = \boldsymbol{F}_0(\boldsymbol{X}) + \delta\boldsymbol{F}(\boldsymbol{X}), \quad \boldsymbol{F}(\boldsymbol{X}' = \boldsymbol{F}_0(\boldsymbol{X}') + \delta\boldsymbol{F}(\boldsymbol{X}') , \tag{SI.7}$$

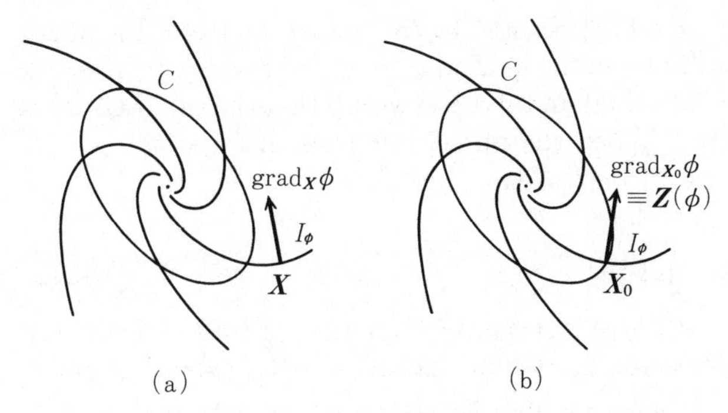

Fig. SI.1. (**a**) The geometrical meaning of the gradient vector $\mathrm{grad}_x\phi$. (**b**) When the state point X is in the neighborhood of the limit cycle, the gradient can be evaluated approximately by assuming that X is on the limit cycle trajectory C

in which the vector field corresponding to each deviates slightly from some common $\boldsymbol{F}_0(\boldsymbol{X})$. Then, writing the full equation of motion for each oscillator in the form $\mathrm{d}\boldsymbol{X}/\mathrm{d}t = \boldsymbol{F}(\boldsymbol{X})$, the vector field $\boldsymbol{p}$ in (SI.4) can be included in $\delta\boldsymbol{F}(\boldsymbol{X})$, and under the approximation made above, in which $\boldsymbol{X}$ is replaced by $\boldsymbol{X}_0(\phi)$, G in (SI.5) comes to possess the additional term $\boldsymbol{Z}(\phi)\delta\boldsymbol{F}(\boldsymbol{X}_0(\phi))$. Again, taking the average over one period, this additional term yields the constant value $\delta\omega$. This represents a constant shift from the characteristic frequency. With this, (SI.8) comes to assume the form

$$\mathrm{d}\phi/\mathrm{d}t = \omega + \delta\omega + \Gamma(\phi - \phi') \,. \tag{SI.8}$$

Then, if the interaction is symmetric in the two oscillators, we have

$$\mathrm{d}\phi'/\mathrm{d}t = \omega + \delta\omega' + \Gamma(\phi' - \phi) \tag{SI.9}$$

for the second oscillator. Generalizing this discussion to the case of an N-oscillator system, we have

$$\mathrm{d}\phi_j/\mathrm{d}t = \omega_j + \sum_{j'=1}^{N} \Gamma(\phi_j - \phi_{j'}) \quad (j = 1, 2, \cdots, N) \,, \tag{SI.10}$$

where $\omega_j = \omega + \delta\omega_j$.

SI.2 Synchronization Phenomena

The phase model (SI.11) has been widely used in the investigation of, among other things, synchronous/asynchronous phenomena among coupled oscillators and oscillator network dynamics. In what follows, after giving an explanation based on this model of the concept of synchronization between two oscillators, we briefly consider the corresponding behavior of a large collection of oscillators.

In the two-oscillator model represented by (SI.9) and (SI.10), if the phase difference $|\phi - \phi'|$ remains finite in the limit as $t \to \infty$, the average frequencies for the two oscillators are identical. In this case, we say that the two oscillators are mutually synchronized. When the phase difference diverges as $t \to \infty$, this synchronization is said to be broken.

Writing $\phi - \phi' = \Delta\phi$, we have

$$\mathrm{d}\Delta\phi/\mathrm{d}t = \Delta\omega + 2\Gamma_{\mathrm{odd}}(\Delta\phi) . \tag{SI.11}$$

Here, $\Delta\omega = \delta\omega + \delta\omega'$, and $\Gamma_{\mathrm{odd}}(y)$ is the anti-symmetric part of $\Gamma(y)$. Because Γ_{odd} is both anti-symmetric and 2π-periodic, it satisfies $\Gamma_{\mathrm{odd}}(0) = \Gamma_{\mathrm{odd}}(\pi) = 0$. Thus, in the case of identical oscillators ($\Delta\omega = 0$), we have at least two synchronized solutions, one with $\Delta\phi = 0$ and one with $\Delta\phi = \pi$. In order for these solutions to be stable, their corresponding differential coefficients, $\Gamma'(0)$ and $\Gamma'(\pi)$, must be negative. For example, when $\Gamma(y) = -K\sin(y+\alpha)$ ($K > 0$), if $|\alpha| < \pi/2$, the motion of the oscillators becomes synchronized with phase difference 0, and if $|\alpha| > \pi/2$ it becomes synchronized with phase difference π. For more complicated forms of $\Gamma(y)$, states in which the phase difference is neither 0 nor π can result. It is also clear that stable synchronized solutions exist over some finite range of $\Delta\omega$ around 0.

We now investigate the collective behavior displayed by a system of a large number of oscillators, as determined by the fundamental two-oscillator synchronous/asynchronous system described above. This problem has been widely studied using the so-called 'mean-field coupled model' in which each pair of oscillators is coupled with equal strength. In such a system there exists a *group entrainment transition*. This is a phase transition that accompanies the reversal of the predominance characterizing the relation between the tendency toward synchronization resulting from the oscillator-oscillator interactions and the tendency toward asynchronization resulting from the variance in the distribution of characteristic oscillator frequencies. This transition has been studied using the following model equation:

$$\frac{\mathrm{d}\phi_j}{\mathrm{d}t} = \omega_j + \frac{K}{N}\sum_{j'=1}^{N}\sin(\phi_{j'} - \phi_j)\ (j = 1, 2, \cdots, N) . \tag{SI.12}$$

Here, ω_j is taken to be a random parameter whose distribution consists of a single symmetric peak centered at some ω_0.

Let us now introduce the complex order parameter $z = \omega \mathrm{e}^{\mathrm{i}\Psi}$ through the relation

$$z = \frac{1}{N}\sum_{j=1}^{N}\mathrm{e}^{\mathrm{i}\phi_j} = \frac{1}{2\pi}\int_0^{2\pi} n(\psi, t)\mathrm{e}^{\mathrm{i}\psi}\mathrm{d}\psi . \tag{SI.13}$$

Here we have taken the $N \to \infty$ limit, and using the phase distribution function $n(\psi, t)$, replaced the sum by an integral. Using the parameter z, (SI.13) becomes

$$d\phi_j/dt = \omega_j - K\sigma \sin(\phi_j - \Psi) , \qquad \text{(SI.14)}$$

and the problem has taken on the appearance of that for a single oscillator. In particular, if σ is independent of time, and assuming that $\Psi = \omega t$, (SI.15) can be explicitly solved for each ϕ_j in terms of the unknown quantity σ. This implies that the distribution $n(\psi)$ becomes expressed in a form containing σ. Substituting this form into the equation defining z, (SI.14), we obtain an equation $\omega = s(\omega)$ which is to be solved self-consistently to obtain ω. According to analysis of this system, there exists a critical coupling strength $K_c = 2/(\pi g(\omega_0)$. For $K < K_c$, $\omega = 0$ represents the unique solution, and macroscopic oscillation is not displayed. However, when $K > K_c$, a finite value of σ (proportional to $\sqrt{K - K_c}$) arises. This marks the appearance of collective oscillatory motion.

The type of group entrainment transition that has been discussed here has been generalized through the introduction and study of a number of solvable models.

In the models considered in this short chapter, the interactions are such that they tend to cause a uniformity among phases. In addition to these, there also exist models in which oscillator–oscillator couplings tend to stabilize opposite-phase relationships. In this case, even when the ω_j are all identical, the behavior of a system with a large number of coupled oscillators is not necessarily simple. Here, the segregation of the system into a number of smaller groups of synchronized oscillators, and other complex collective behavior, can appear.

Part II

The Structure and Physics of Chaos

Introduction

Study of the dynamics of chaos began in 1899 with the three-body problem (nonintegrable system) of Poincaré. However, not until recently have random and unpredictable solutions – chaotic orbits – come to be widely understood as universal phenomena in nonlinear dynamical systems. As we understand it now, chaos can be thought of as the main cause of the diversity that we see displayed in Nature's perpetually changing panorama.

Let us consider the post-transient dynamics exhibited by dynamical systems. In most general terms, this behavior can be grouped into three categories: periodic, multiperiodic, and aperiodic. In the chaotic region of phase space (that region in which chaos appears), all orbits are unstable, and there is a countably infinite set of coexisting periodic orbits. However, there also exists an uncountable set of aperiodic orbits, and, therefore, in nearly all cases, the system's motion is aperiodic, and it exhibits chaos. The form of chaos displayed by a given system is determined by its coexisting invariant sets (periodic orbits, tori, Cantor repellors, etc.), while the types of coexisting invariant sets differ with the nature of the system in question, as well as with the values of its bifurcation parameters. The great multitude of possible types of invariant sets can be thought of as the source of the variety of chaotic behavior that we observe.

In Part II we consider the manner in which the various types of chaos come into being, the form and structure that they possess, the methods used in their description, and the types of statistical laws to which they are subject. One important physical quantity that we will employ to characterize chaotic orbits is the coarse-grained expansion rate of the distance between neighboring orbits. This quantity describes the dynamics of chaotic orbits moving along unstable manifolds. In addition to this quantity, we will use the local dimension describing the self-similar nested structure of strange attractors. The long-time averages and fluctuations exhibited by these quantities along the chaotic orbits afford a statistical characterization of the geometrical form and structure of these orbits. Such quantities also provide a means of investigating the qualitative change in the structure and form of chaotic behavior resulting from bifurcation as well as the anomalous nature of transport phenomena found in chaos. The universality exhibited by chaotic systems is expressed by, among other things, similarity laws (scaling laws) and q-phase transitions.

6. A Physical Approach to Chaos

The term 'chaos' is used in reference to unstable, aperiodic motion in dynamical systems and also to the state of a system which exhibits such motion. Almost any nonequilibrium open system will, when some bifurcation parameter characterizing the system is made sufficiently large, display chaotic behavior. It can be said that chaos is Nature's universal dynamical form. Chaos is characterized by the coexistence of an infinite number of unstable periodic orbits that determine the form and structure of the chaotic behavior exhibited by any given system. In this chapter we consider the problem of identifying the descriptive signature of such a set of orbits, and we establish the point of view from which we will elucidate the nature of chaos.

A chaotic orbit is unstable with respect to small disturbances and is thus irreproducible. However, the long-time averages of physical quantities taken with respect to such orbits are stable and reproducible, and, in fact, there exist clear statistical laws regarding the behavior of such quantities. We thus wish to consider descriptive physical quantities whose behavior characterizes the variety of coexisting periodic orbits. We then attempt to clarify the fundamental nature of chaotic behavior by studying the long-time averages and fluctuations of these quantities.

6.1 The Phase Space Structure of Dissipative Dynamical Systems

Let us consider an often studied system in the context of dissipative dynamics, the periodically forced pendulum. The equation of motion for the angle ϕ representing this system can be written

$$\ddot{\phi} + \gamma\dot{\phi} + \sin\phi = a\cos(\omega t) , \tag{6.1}$$

where the parameter γ is the coefficient of friction. The term on the right-hand side represents a driving force with period $T \equiv 2\pi/\omega$ and amplitude a. Fig. 6.1 displays the quantity $X_i \equiv (\phi(t_i), \dot{\phi}(t_i))$ of a solution to this equation with $\gamma = 0.22$, $\omega = 1$, and $a = 2.7$ at times $t_i \equiv iT$ $(i = 0, 1, 2, \ldots)$. These data were obtained through numerical integration of (6.1), using the initial values $\phi = -1.5$ and $\dot{\phi} = 1.3$. The first 100 data points (corresponding to the time interval $[-100T, 0]$) were regarded as representing transient behavior and discarded. Figure 6.1a gives the time series of $\dot{\phi}(t_i)$ for $i = 0$–300, while Fig. 6.1b displays the orbit X_i for the time interval $i = 0$–5×10^4. Here, values for the angle ϕ were calculated mod 2π, with $-\pi \le \phi < \pi$. This system provides an example of nonlinear rotational motion characterized by a random time series as depicted in Fig. 6.1a. A chaotic orbit corresponding to this random rotational motion follows a path like that shown in Fig. 6.1b. If we set $x = \phi$ and $z = \dot{\phi}$, (6.1) becomes

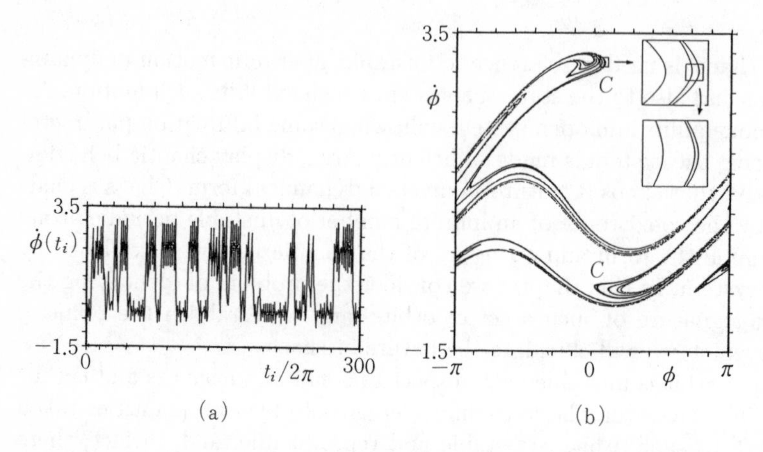

Fig. 6.1. The time series (**a**) and the chaotic attractor (**b**) of the forced pendulum (6.1). The inserted figures in (**b**) are schematic representations of the self-similar nested structure (fractal structure) exhibited here. Expanding the small boxes as shown reveals, at each successive level, that the apparently single exterior line is seen to appear, upon closer examination, as two lines

$$\dot{x} = z, \qquad \dot{z} = -\gamma z - \sin x + a\cos(\omega t), \qquad \dot{t} = 1\,. \tag{6.2}$$

The phase space of this system is the three-dimensional space (x, z, t) shown in Fig. 6.2. Noting the periodicity of the driving force, we take the x-z plane corresponding to each $t = t_i \equiv iT$ to coincide with that for $t = 0$. This x–z plane defines the Poincaré section for this system. The plot of X_i in Fig. 6.1b determines the *Poincaré map*

$$X_{i+1} = F(X_i) \qquad (i = 0, \pm 1, \pm 2, \cdots) \tag{6.3}$$

composed of the intersection points $X_i = (x_i, z_i)$ of the orbit $(x(t), y(t), t)$ and the Poincaré section. With $J(X_i) \equiv \det\{DF(X_i)\} = \partial(x_{i+1}, z_{i+1})/\partial(x_i, z_i)$, where $DF(X_i)$ is the Jacobian matrix, we have the relation $S_{i+1} = |J(X_i)|S_i$, where $S_i = \mathrm{d}x_i \mathrm{d}z_i$ is a surface element of the x–z plane and S_{i+1} is its image under F. Then $\dot{S}_i/S_i = \operatorname{div} \boldsymbol{V} \equiv (\partial\dot{x}/\partial x) + (\partial\dot{z}/\partial z) = -\gamma$, and

$$\ln|J(X_i)| = \int_{t_i}^{t_{i+1}} \mathrm{d}t \operatorname{div}\boldsymbol{V} = -\gamma T < 0\,. \tag{6.4}$$

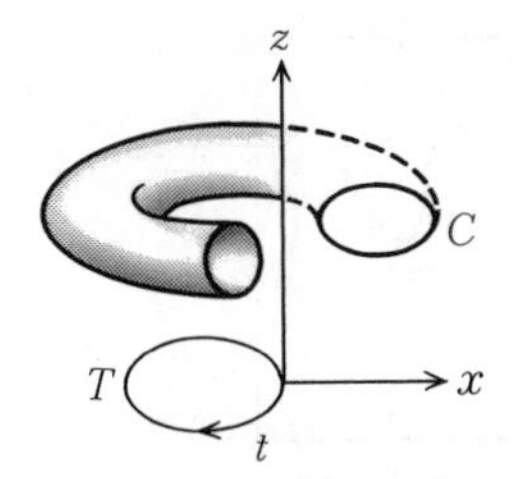

Fig. 6.2. The cross-section C of a torus lying in the x–z plane of the (x, z, t) phase space. The time axis is drawn as an ellipse of circumference T, the fundamental period of the system in question, and thus the plane corresponding to each of the times $\pm T, \pm 2T, \dots$ coincides with that for $t = 0$. The closed curve C represents the intersection of the torus shown with the Poincaré section

Thus we obtain $|J(X_i)| = \mathrm{e}^{-\gamma T}$. Then, writing the ith image of a point X_0 as $X_i = F^i(X_0)$, the surface area μ of the ith image of an arbitrary region R in the x–z plane, $F^i R$, decreases to 0 in accordance with $\mu(F^i R) \propto \mathrm{e}^{-\gamma T i}$. This behavior is characteristic of a *dissipative dynamical system.*

Fluctuations in the time series $\dot{\phi}(t_i)$ are random. The curve in Fig. 6.1b, however, is not random, displaying a rather clean and elegant form. In addition, for almost all initial conditions, an orbit X_i is attracted to this curve; after the passing of some transient time, such an orbit moves along a path composed of points contained in the set defined by this curve. This kind of bounded set is called an *attractor.* The attractor in the present system possesses a multiplex string-like structure, and if we magnify a part of this attractor, as exhibited in the insert of Fig. 6.1b, a nested self-similar structure is revealed. This kind of structure is referred to as a *fractal.* (Although a treatment of the concept of the *fractal dimension* is not given until Sect. 8.2, we note here that the fractal dimension for the present example is $D \approx 1.4$.) An attractor composed of a fractal structure is known as a *strange attractor.* A chaotic orbit moves along such a limitlessly complex attractor in a random manner. This randomness and complexity are reflected in the behavior displayed by the system represented by such an orbit. However, in addition to this randomness, the motion displayed by such a system is also characterized by an intricate self-organization.

In order to understand the process that produces this form within chaos, consider Fig. 6.3. Here we display the manner in which a set of chaotic orbits, moving in the attractor, behave during one period of time $T = 2\pi$ starting at $t = 0$. The region (called a 'cell') shaded black at $t = 0$ becomes stretched and elongated during the first $T/3$ units of time, and in the next $T/3$ it becomes folded. At $t = T$, taking the mod 2π value of ϕ and returning to the interval $-\pi \le \phi < \pi$, we see that the shaded region has been folded several times and is now spread throughout the entire attractor. These actions of *stretching* and *folding* are fundamental processes in chaos, and their repeated application causes the development of fine structure within the system. They also induce a *mixing* in the system in which (almost) any set of orbits singled out for observation becomes mixed with other orbits, and a region defined as in the above situation spreads in phase space. To precisely capture such a mixing process and to understand how a simple equation such as (6.1) produces such complex geometrical forms is one of the goals of Part II of this book.

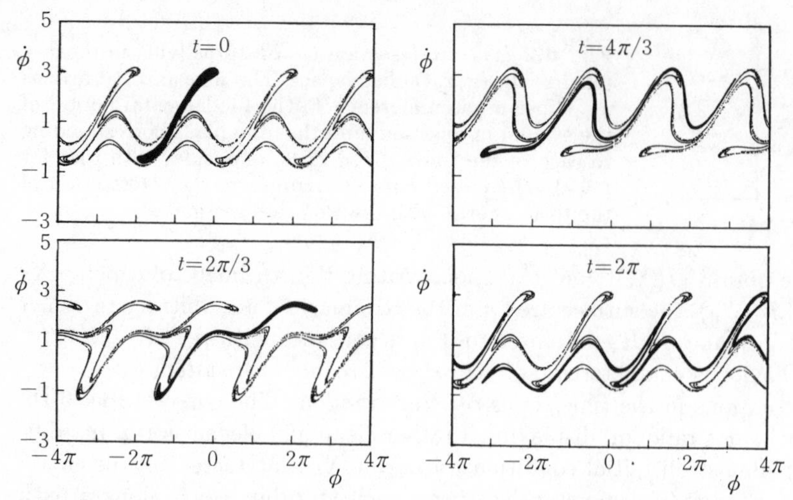

Fig. 6.3. The form of the attractor and the stretching and folding of a particular cell (darkly shaded region) over a time interval of length $T = 2 < \pi$. Note that the area of this cell decreases with time

In general, an attractor is the limit set traced out as $t \to \infty$ by an orbit representing a solution of the system in question. The types of such sets include fixed points, limit cycles (periodic orbits), and tori (multiperiodic orbits), in addition to strange attractors. When $a = 0$, an orbit representing a solution of (6.1) will converge to the fixed point $\phi = \dot{\phi} = 0$ in the limit as $t \to \infty$. When a is small (but nonzero), in this limit, such an orbit represents a forced oscillator solution. This is a period $T = 2\pi/\omega$ limit cycle. If γ is not too small, increasing a further results in the repeated occurrence of a phenomenon referred to as *period doubling*, and for values $\hat{a}_n$ of a ($n = 1, 2, \ldots, \infty$), period $2^n T$ limit cycles appear in succession. After this period-doubling cascade reaches completion (i.e., for $a > \hat{a}_\infty$, where $\hat{a}_\infty$ represents the value corresponding to an infinite period limit cycle), chaotic orbits appear. This process is studied in Sect. 7.3.

A characteristic property of dissipative systems is the production of heat. The amount of heat produced per unit time due to friction is measured by the *energy dissipation rate*, $W_{\mathrm{f}}(t) \equiv \gamma\dot{\phi}^2$. The energy supplied by the driving force, $W_{\mathrm{d}}(t)$, is distributed into kinetic and potential energy, and the rate of dissipation of this energy, $W_{\mathrm{f}}(t)$, fluctuates in time. The long-time average of this quantity is given by

$$\overline{W}_{\mathrm{f}} = \langle w(X) \rangle \equiv \lim_{N\to\infty} \frac{1}{N} \sum_{i=0}^{N-1} w(X_i) \tag{6.5a}$$

$$w(X_i) \equiv \frac{1}{T}\int_{t_i}^{t_{i+1}} \mathrm{d}t\, W_{\mathrm{f}}(t) = \frac{\gamma}{T}\int_{t_i}^{t_{i+1}} \dot{\phi}(t)\, \mathrm{d}\phi(t)\,, \tag{6.5b}$$

where $\overline{W}_{\rm f} = \overline{W}_{\rm d}$. In particular, for a period QT limit cycle, $\overline{W}_{\rm f} = \gamma I/QT$, where $I \equiv \oint \dot{\phi}(t) d\phi(t)$ is the area of the region inside the path defined by the limit cycle in the ϕ–$\dot{\phi}$ phase space. In this way, $w_i(X_i)$ becomes the action integral per period T. As we will see in Sect. 9.1, the fluctuations of this quantity provide useful geometrical information concerning the motion of chaotic orbits in phase space.

For many systems, phase space possesses two or more attractors [we denote these $A^\alpha (\alpha = 1, 2, \ldots)$], and it is divided into *basins* B^α corresponding to these attractors. For such a basin, we have $A^\alpha = \cap_{i=0}^{\infty} F^i B^\alpha$. Let us consider some region R^α of nonzero area in some B^α. The ith image of R^α, $F^i R^\alpha$, is attracted to A^α as $t \to \infty$, and in this limit it comes to cover A^α. However, in the same limit, the area of the region in question converges to 0 (since we are considering dissipative systems), and thus the area of A^α is 0. Hence the dimension D of such an attractor satisfies $2 > D \geq 0$. The chaotic attractor ($D \approx 1.4$) shown in Fig. 6.1b is composed of a bundle of string-like structures. A 'string' itself is a one-dimensional curve, while here the 'bundle' constitutes a 0.4-dimensional Cantor set.

The geometrical structure of chaotic attractors is studied further in the next chapter, but we give here a summary of the overall picture obtained there:

(1) A chaotic attractor A contains a multitude of unstable *invariant sets* $S^\beta = F(S^\beta)$ $(\beta = 1, 2, \ldots)$. In particular, it possesses a countably infinite number of unstable periodic orbits which can be of arbitrarily long periods. (These orbits are known as 'saddles'.)
(2) A contains an uncountably infinite number of unstable aperiodic orbits (chaotic orbits). Furthermore, there exists a chaotic orbit that passes through every neighborhood of every point in A, and therefore A is indecomposable.
(3) This type of attractor A is identical to the closure[1] of the unstable manifolds of the saddles that it contains.
(4) The structure of an attractor A as described above is not destroyed (qualitatively altered) by small disturbances. It is thus 'statistically stable'.

In what follows, we will discuss this type of geometrical structure in chaos from various points of view. The goal of these chapters is to obtain a statistical characterization of this structure and an elucidation of transport phenomena in chaos. An easily understood illustration of (1) above is seen in the situation that chaos emerges as the result of a period-doubling $2^n T$ cascade. In this case the chaotic attractor will possess an infinite number of unstable periodic orbits of period $2^n T$. Increasing a, $m \times 2^n T$-periodic orbits are generated for successively larger (odd) values of m. Increasing a beyond the value at which a

[1] The closure of a set W consists of the set of all points in W together with the set of all accumulation points of W which are not in W. The closure of W is denoted $[W]$ and is the smallest closed set containing W.

given such orbit appears, this orbit eventually becomes unstable and comes to be contained within the attractor. For such a system, the form and structure of the attractor can be characterized according to the type of saddles that it possesses. The periodic motion, the self-similar fractal structure, and the various kinds of similarity that we discuss below represent the so-called ‘order in chaos’.

6.2 The Phase Space Structure of Conservative Dynamical Systems

Let us consider the motion of fluid particles in a Bénard convection system confined to a two-dimensional surface defined by $(-\infty < x < \infty, -0.5 < z < 0.5)$. Here, z is the vertical direction, and x is the horizontal direction along which are aligned infinitely many rolls. Let us consider the situation in which drops of dye are placed in the system to facilitate the observation of the fluid motion. The particles of dye can be followed to observe the mixing and diffusion of the dyed portion of the fluid. The orbit of a fluid particle parameterized in terms of time t, $(x(t), z(t))$, can be obtained by numerical integration of

$$\dot{x} = u(x, z, t), \qquad \dot{z} = w(x, z, t) , \tag{6.6}$$

where (u, w) represents the velocity of the fluid at (x, y, t). Let us suppose that the fluid is incompressible and that it obeys the continuity equation $(\partial u/\partial x) + (\partial w/\partial z) = 0$. In this case, a stream function $\Psi(x, z, t)$ exists, and we can write (6.6) as

$$\dot{x} = -\partial\Psi/\partial z, \qquad \dot{z} = \partial\Psi/\partial x . \tag{6.7}$$

This is precisely Hamilton's equation for the ‘Hamiltonian’ Ψ, with x playing the role of the particle coordinate and z the role of the corresponding momentum. Thus, with dynamics governed by (6.7), the phase space (x, z) will possess various properties seen in phase spaces of Hamiltonian dynamical systems. Among other properties, the phase space for this system is characterized by: (1) conservation of area (under time evolution); (2) hierarchical structure of islands of invariant tori and chaotic seas; and (3) a variety of structures produced by the stable and unstable manifolds of unstable periodic points. Let us begin by briefly describing these properties.

As a concrete example, we consider the often used stream function

$$\Psi(x, z, t) = (A/\pi) \sin[\pi\{x + B \sin(2\pi t)\}]W(z) . \tag{6.8}$$

This contains an oscillating term of amplitude B, periodic in time (period $T = 1$). In this equation, A represents the maximal value of the dimensionless velocity w, and $W(z)$ is an even function of z. Let us stipulate that at points in the horizontal planes defined by $z = \pm 0.5$ the velocity is 0. The case

$B = 0$ is that of steady Bénard convection. For $B > 0$, the rolls become unstable with respect to lateral oscillations, and all rolls exhibit non-steady convective motion in which they oscillate with period 1 along the x direction. This system corresponds to a three-dimensional phase space (x, z, t) like that shown in Fig. 6.2. Here, because of the period $T = 1$ periodicity, the x–z planes corresponding to $t = \pm 1, \pm 2, \ldots$ are considered as coinciding with that for $t = 0$. Thus, an orbit in this three-dimensional phase space crosses the $t = 0$ plane (x–z plane) at times $t = 0, \pm 1, \pm 2, \ldots$ This plane constitutes the Poincaré section for this system. The intersection points $X_t \equiv (x_t, z_t)$ define the Poincaré map

$$X_{t+1} = F(X_t) \qquad (t = 0, \pm 1, \pm 2, \cdots)\,. \tag{6.9}$$

If we use (6.6), we can write the continuity equation as $(\partial \dot{x}/\partial x) + (\partial \dot{z}/\partial z) = 0$, and therefore the Jacobian becomes $J(X_t) \equiv \det\{DF(X_t)\} = 1$. A system of this type is called a *conservative dynamical system.* The stability and structure of the neighborhoods of periodic points in such systems are briefly analyzed in Sect. A.1 of the Appendix.

Figure 6.4 displays plots of the Poincaré map (6.9) consisting of the points $X_t = (x_t, z_t)$ representing values of the orbit $(x(t), z(t))$ at times $t = 0, 1, 2, \ldots$. This figure exhibits data collected using many different sets of initial values with $B = 0.01$ for (a) and $B = 0.05$ for (b). The many orbits from which these graphs were generated reveal the phase space structure of the system and how this structure depends on the roll oscillation amplitude, B. Let us, for the time being, define the cell of the jth roll to be the region $(j < x < j + 1, -0.5 < z < 0.5)$. (We will precisely define the cell boundary W_j later.) Equation (6.7) is invariant with respect to the shift $(x, z) \to (x + j, (-1)^j z)$ and time reversal $t \to -t$, $x \to -x$, $z \to -z$, and each cell possesses a similar structure, reflecting this symmetry. In the center of each cell lies a set of concentric elliptically shaped curves, and either above or below this is situated a set of concentric crescent moon-shaped curves. The centers of these sets of curves, $X_j^\alpha (\alpha = 1, 2)$ are *neutral periodic points* (see Sect. A.1 of the Appendix) which satisfy the relation $X_j^\alpha = F(X_j^\alpha)$ for the map (6.9). These are the points at which the neutral, period $T = 1$ periodic orbits in the three-dimensional phase space (x, z, t) intersect the x–z cross-section. Each closed curve C_j^β contained in these sets of curves is an invariant set satisfying $C_j^\beta = F(C_j^\beta)$. Each such curve rotates with period $T'_\beta (> 1)$ about the neutral periodic orbit and in fact constitutes the cross-section C of a torus defined by a bi-periodic orbit, as in the situation described in Fig. 6.2. The outermost such curve corresponds to a so-called *critical torus.* This is a closed curve on the verge of collapse. The entire set of such a collection of concentric curves is referred to as an *island of tori.* In the case of Fig. 6.4a, the elliptical and crescent-shaped sets of curves have merged to form a single large island, while in Fig. 6.2b the outer curves, shared by the two sets in Fig. 6.2a, have collapsed, leaving two distinct sets of smaller islands. An is-

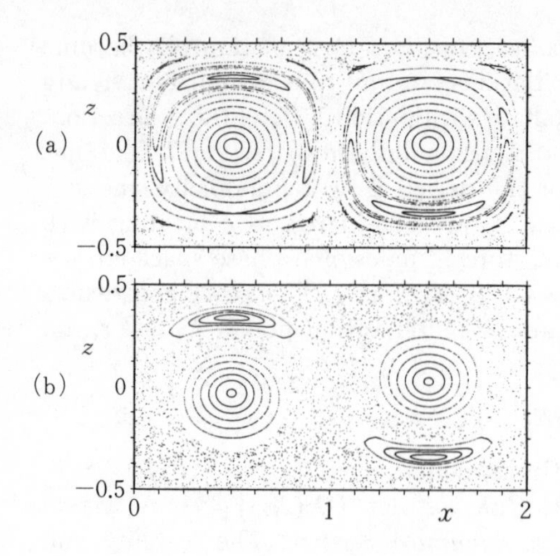

Fig. 6.4. The phase space structure of the conservative system (6.7) for (a) $B = 0.01$ and (b) $B = 0.05$. These plots consist of points on a variety of orbits

land of such closed curves represents a region in which fluid particles become trapped in the roll's vortex streamlines.

The many data points existing outside these islands are situated in the region through which the unstable aperiodic orbits (chaotic orbits) pass in this x–z cross-section. This is the chaotic region, or the the *chaotic sea*. In the case of nonsteady convection, streamline oscillation can cause fluid particles to jump between streamlines and, in particular, can cause particles to jump from one cell to another. The chaotic orbits represent the orbits of fluid particles which, by virtue of this type of jumping between streamlines, are not trapped in the vorticies of a roll. These orbits diffuse from cell to cell, and the displacements x_t of the x coordinate of a set of such orbits behave according to the variance

$$\langle (x_t - x_0)^2 \rangle_{\mathrm{E}} = 2D_\eta t^\eta \qquad (1 \le \eta < 2) \tag{6.10}$$

in the limit as $t \to \infty$. Hence the root mean square of this displacement increases without bound as $t^{\eta/2}$. The angle brackets $\langle \cdots \rangle_{\mathrm{E}}$ here represent the operation of averaging over the set of some large number of dye particles located in the middle of the chaotic sea at $t = 0$. Computation reveals that for $B = 0.01$, we have $\eta = 1$ and $D_1 \approx 0.095$. (D_1 here is the diffusion coefficient.) Thus diffusion of dye particles with $\eta = 1$ is realized through advection.

In order to describe the appearance of this kind of chaotic orbit, the corresponding mixing and diffusion, and the manner in which these phenomena depend on the value of the amplitude B from a dynamical system point of view, it is necessary to understand the structure of the chaotic sea. Figure 6.5 describes this structure in the x–z plane between the fixed surfaces at $z = \pm 0.5$ inside which our Bénard system is confined. Depicted here are the invariant manifolds $W^{\mathrm{s}}(p_j^{\pm})$ and $W^{\mathrm{u}}(p_j^{\mp})$ emerging from the unstable fixed points $p_j^{\pm} \equiv (x = j, z = \pm 0.5)$, the lobes $L_{j,j\pm 1} (j = 0, \pm 1)$, and the boundaries W_j separating the roll cells R_{j-1} and R_j. The curve $W^s(p_j^+)$ represents the *stable manifold* emerging from the point p_j^+ (j even), and $W^s(p_j^-)$ from the point p_j^- (j odd). The curves $W^u(p_j^-)$ and $W^u(p_j^+)$ represent *unstable manifolds* emerging from p_j^- (j even) and p_j^+ (j odd), respectively. In general terms, each of these is an *invariant manifold* [i.e., $W^\alpha = F(W^\alpha)$ $(\alpha = s, u)$] corresponding to an unstable periodic point p. Defining $|X|$ to be the length of the vector X, the invariant manifold $W^s(p)$ is defined as the set of points X for which $|F^n(X) - F^n(p)| \to 0$ as $n \to \infty$ (where F^n represents n iterations of the map F). Then, defining the inverse of the map (6.9) by $X_t = F^{-1}(X_{t+1})$, the invariant manifold $W^u(p)$ is defined as the set of points X for which $|F^{-n}(X) - F^{-n}(p)| \to 0$ as $n \to \infty$ (where $F^{-n} = F^{-1} \circ F^{-(n-1)}$). Manifolds $W^s(p_j^\pm)$ and $W^u(p_j^\mp)$ pass through every open set in the chaotic region, and in almost all of these neighborhoods they intersect transversely. Excluding neighborhoods of points at which some $W^s(p)$ and $W^u(p')$ contact tangentially, an arbitrary segment S of $W^s(p)$ will become shortened, and an arbitrary segment U of W^u will become lengthened under the iteration of the map F acting in the direction of increasing time.

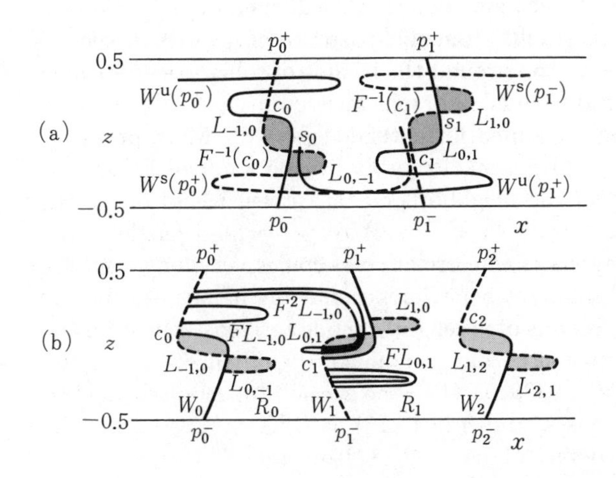

Fig. 6.5. Schematic of the structure of the chaotic sea. **(a)** The lobes $L_{\mp 1,0}$ and $L_{0,\pm 1}$ (*shaded regions*) determined by the invariant manifolds W^u (*solid curves*) and W^s (broken curves). **(b)** The cell boundaries W_j and the stretching and folding of the lobe $L_{-1,0}$

The points c_j and s_j in Fig. 6.5 are the consecutive intersection points (primary intersection points) of $W^s(p_j^\pm)$ and $W^u(p_j^\mp)$. The lobe $L_{j-1,j}$ consists of the region bounded by the segments $S(c_j, s_j)$ and $U(c_j, s_j)$ of the stable and unstable invariant manifolds, respectively, lying between these two points, while $L_{j,j-1}$ is the region bounded by $S(s_j, F^{-1}(c_j))$ and $U(s_j, F^{-1}(c_j))$. The image under $F^{\pm n}$ of $L_{j,j\pm 1}$ for arbitrary n, $F^{\pm n}L_{j,j\pm 1}$, is also referred to as a lobe. The area of each such lobe is identical to that of $L_{j,j\pm 1}$. The cell boundary W_j consists of a curve joining p_j^+ and p_j^-. When j is even, this is the curve composed of $S(p_j^+, c_j)$ and $U(c_j, p_j^-)$, and when j is odd, it is the curve composed of $U(p_j^+, c_j)$ and $S(c_j, p_j^-)$. In the limit as $B \to 0$, the boundary W_j approaches the separatrix ($x = j, -0.5 \le z \le 0.5$), and the corresponding lobes disappear.

The nth images of c_j and s_j, $F^n(c_j)$ and $F^n(s_j)$, are also intersection points of the two manifolds (*heteroclinic points*) situated on the boundary W_j. Thus the set of intersection points of $W^s(p_j^\pm)$ and $W^u(p_j^\mp)$ contains an infinite number of heteroclinic points on W_j. These points define a so-called heteroclinic *tangle*, and they accumulate at $p_j^\pm$ along $W^s(p_j^\pm)$. Thus lobes $F^n L_{j-1,j}$ and $F^n L_{j,j-1}$ become progressively more elongated and folded as n increases (as depicted by the sequences $L_{-1,0} \to FL_{-1,0} \to F^2 L_{-1,0}$ and $L_{0,1} \to FL_{0,1}$ in Fig. 6.5b). However, as dictated by the conservation law, these lobes all have equal area. This type of *stretching* and *folding* introduces mixing and diffusion into the fluid occupying the chaotic sea. The region mapped from the cell R_{j-1} to the cell R_j under the map F is limited to the lobe $L_{j-1,j}$, and, conversely, that mapped from R_j to R_{j-1} is limited to $L_{j,j-1}$. Because these lobes are shifted to opposite sides of the boundary W_j under F, they are referred to as *turnstiles*. Through such turnstiles, fluid particles diffuse from cell to cell[2].

Even fluid particles undergoing diffusive migration will become temporarily trapped in the vortices of rolls, and as a result will spend long periods of time in their vicinities. This results from the existence of a series of successively smaller islands in the region around the islands of roll vortices forming a self-similar, 'islands around islands' hierarchical structure (see Fig. 10.4) in which chaotic orbits become trapped for extended periods. More precisely, the probability $W(t)$ that a given chaotic orbit will be trapped for a time greater than some value t in the neighborhood of a given island obeys the inverse power law $W(t) \propto t^{-(\beta-1)}$ $(1 < \beta < 2)$. When a fluid particle does become trapped in this way, its time correlation becomes very long, and the corresponding mixing and diffusive behavior become anomalous. As we will discuss in Sect. 10.4, when islands of accelerator mode tori exist, the diffusion coefficient diverges and diffusion becomes anomalously enhanced. For example, when B is in the range $(0.0398, 0.0419)$, the accelerator mode defined by $x_{t+2} = x_t \pm 2$, $z_{t+2} = z_t$ exists, and η in (6.10) satisfies $\eta = 3 - \beta > 1$. It follows that the diffusion coefficient $D_1 = D_\eta t^{\eta-1}$ diverges as $t \to \infty$.

As described above, the chaotic sea in the phase space of a conservative dynamical system is composed of a myriad of islands of tori. To understand the chaos exhibited by such systems, it is important to explain and describe the structure of the physical processes of mixing and diffusion of the fluid particles found in the chaotic sea in terms of the stretching and folding of lobes and the hierarchical structure of 'islands around islands', which imparts

[2] Note that, to some extent, the definition of a cell boundary, and thus the definition of its turnstile, is arbitrary. However, we wish to define a boundary to be composed of such intervals of the invariant manifolds in question that it is 'closest' to the $B = 0$ separatrix as $B \to 0$. This makes both the definition of the boundary and of its turnstile unique.

a decisive influence on this stretching and folding behavior. This problem will be considered in Chap. 10 from the point of view presented in Sect. 6.4.

6.3 Orbital Instability and the Mixing Nature of Chaos

6.3.1 The Liapunov Number

In a chaotic attractor of a dissipative system or a chaotic sea of a conservative system, there exist a set of a countably infinite number of periodic orbits and a set of an uncountably infinite number of aperiodic orbits. All of these orbits are unstable. What kind of physical quantities should be considered in order to describe and characterize this diverse collection of unstable orbits? We now address this question.

To determine the stability of the orbit $X_n = F^n(X_0)$ with respect to small disturbances, let us consider a neighboring orbit whose initial point differs slightly from X_0, $X'_n = F^n(X_0 + y_0)$. The orbit X_n is unstable if the difference between X_n and X'_n grows with n. The time development of this distance, $y_n \equiv X'_n - X_n$, is given by the variational equation

$$y_{i+n} = DF^n(X_i) \cdot y_i \qquad (n, i = 0, 1, 2, \cdots) \tag{6.11}$$

around the nth image of X_i, $X_{i+n} = F^n(X_i)$. Here, $|y_{i+n}| \ll 1$, and $DF^n(X_i)$ is the Jacobian matrix of $F^n(X_i)$.

First, let us consider the orbit $X'_n = X^*_n + y_n$ in the neighborhood of the Q-periodic orbit $X^*_{i+Q} = X^*_i$ $(i = 1, 2, \ldots, Q)$. Each point in this periodic orbit constitutes a fixed point of the Qth iterate: $X^*_i = F^Q(X^*_i)$. The Jacobian matrix for this map is

$$\begin{aligned} DF^Q(X^*_i) &= D\{F(F^{Q-1}(X^*_i))\} = DF(F^{Q-1}(X^*_i))DF^{Q-1}(X^*_i) \\ &= DF(F^{Q-1}(X^*_i)) \cdots DF(F(X^*_i))DF(X^*_i) \,. \end{aligned} \tag{6.12}$$

The eigenvalues ν_1^Q and ν_2^Q $(|\nu_1^Q| \geq |\nu_2^Q|)$ of this matrix are common to the Q distinct periodic points X^*_i. If we define $\overline{\lambda}_\alpha \equiv (1/Q) \ln |\nu_\alpha^Q|$, taking the logarithm of the absolute value of the determinant of (6.12) yields

$$\overline{\lambda}_1(X^*_i) + \overline{\lambda}_2(X^*_i) = \frac{1}{Q} \sum_{t=1}^{Q} \ln |J(X^*_t)| \leq 0 \,, \tag{6.13}$$

where $J(X^*_t) = \det\{\mathrm{DF(X^*_t)}\}$. Then, using $\overline{\lambda}_1(X^*_i) \geq \overline{\lambda}_2(X^*_i)$ (from the assumption above), in the $m \to \infty$ limit we have

$$|y_{i+Qm}| = |\{DF^Q(X^*_i)\}^m \cdot y_i| = \exp[\overline{\lambda}_1(X^*_i)Qm]g_i \,, \tag{6.14}$$

where we have assumed that y_i is not parallel to the eigenvector E_2 corresponding to the eigenvalue ν_2^Q. The value of g_i is determined by y_i and satisfies $0 < g_i \ll 1$. When $\overline{\lambda}_1 > 0$, $|y_{i+Qm}|$ increases exponentially with

m. In this case, it thus follows that orbits neighboring X_i^* will move away from X_i^* exponentially. Hence X_i^* is *unstable*. When $\overline{\lambda}_1 < 0$, $|y_{i+Q_m}|$ decreases exponentially, and thus X_i^* is *stable*. When $\overline{\lambda}_1 = 0$, X_i^* is *neutral*. In the chaotic region, all periodic points X_i^* are *saddle points*, satisfying $\overline{\lambda}_1(X_i^*) > 0 > \overline{\lambda}_2(X_i^*)$.

For aperiodic orbits, the concept of the eigenvalue does not apply. However, a generalization of this concept, the *Liapunov number* (or Liapunov exponent), plays a similar role. In the case of a one-dimensional map $x_{t+1} = f(x_t)$, this quantity is defined using the nth iterate of f by

$$\overline{\lambda}_1(x_0) \equiv \lim_{n\to\infty} \frac{1}{n} \ln \left| \frac{\mathrm{d}f^n(x_0)}{\mathrm{d}x_0} \right| , \tag{6.15a}$$

$$= \lim_{n\to\infty} \frac{1}{n} \sum_{t=0}^{n-1} \ln |f'(x_t)| . \tag{6.15b}$$

A two-dimensional map is represented in three-dimensional space as a two-dimensional surface. The vector y_n $(i = 0)$ in (6.11) represents the vector (*tangent vector*) defining the tangent plane at each point of the orbit $X_n = F^n(X_0)$ on this two-dimensional surface. Here, we define

$$\overline{\lambda}(X_0, y_0) \equiv \lim_{n\to\infty} \frac{1}{n} \ln |DF^n(X_0) \cdot y_0| , \tag{6.16}$$

where the quantity $|\cdots|$ is the length (norm) of the tangent vector. According to (6.11), we can interpret $DF^n(X)$ as moving the tangent plane at X to that at $F^n(X)$. For this reason, $DF(X)$ is known as the *tangent map*. With a periodic orbit $X_{i+Q}^* = X_i^*$, for the eigenvector E_α of the Jacobian matrix (6.12) we have the relation $DF^{Qm}(X_i^*) \cdot E_\alpha = (\nu_\alpha^Q)^m E_\alpha$. Hence, if we take $y_0 = E_\alpha$ and $n = Qm$, then (6.16) becomes $\overline{\lambda}(X_i^*, E_\alpha) \equiv \overline{\lambda}_\alpha(X_i^*)$, and the Liapunov number and eigenvalue are identical. According to Oseledec (1968), a basis e_α exists for each tangent plane even for chaotic orbits, and the Liapunov number is given by $\overline{\lambda}_\alpha(X_0) \equiv \overline{\lambda}(X_0, e_\alpha)$. Here we have $\overline{\lambda}_1 + \overline{\lambda}_2 = \langle \ln |J(X)| \rangle \leq 0$. [Note that the angle brackets represent the long-time average (6.5a).] This is a generalization of (6.13). For chaotic orbits, the long time average $\langle G(X) \rangle$ can be thought of as a constant, assuming the same value for almost all choices of the initial point X_0. As we will discuss in Sect. 6.3.3, this follows from the 'loss of memory' associated with chaotic orbits. This can be seen by noting that for almost all chaotic orbits there is a unique and common positive Liapunov exponent, $\Lambda^\infty \equiv \overline{\lambda}_1(X_0)$, and for $n \to \infty$, we can write

$$|y_n| = |DF^n(X_0) \cdot y_0| \propto \exp[\Lambda^\infty n] , \tag{6.17}$$

where we stipulate that the directions of y_0 and e_2 are not coincident. However, in general, the Liapunov numbers of periodic orbits, $\overline{\lambda}_1(X_i^*)$, differ for each such orbit.

6.3.2 The Expansion Rate of Nearby Orbits, $\lambda_1(X_t)$

To facilitate investigation of the local structure of the region surrounding a chaotic orbit X_t $(t = 0, 1, 2, \ldots)$, let us introduce the expansion rate of neighboring orbits, $\lambda_1(X_t)$. When the eigenvalues of the Jacobian matrix (6.12) for a periodic point X_i^* are not on the unit circle, this point is termed *hyperbolic*. In the chaotic region, each X_i^* is hyperbolic and a saddle, satisfying $\overline{\lambda}_1(X_i^*) > 0 > \overline{\lambda}_2(X_i^*)$. The lines determined by the corresponding eigenvectors E_1 and E_2 are shown in Fig. 6.6 (here simply labeled E_1 and E_2). The curves $W^{\mathrm{u}}(X_i^*)$, tangent to E_1 at X_i^*, and the curves $W^{\mathrm{s}}(X_i^*)$, tangent to E_2 at X_i^*, are the unstable and stable manifolds of X_i^*. These are the invariant manifolds described in Fig. 6.5. If we define $\phi(X) \equiv F^Q(X)$, because $|\nu_1^Q(X_i^*)| > 1$, in the neighborhood of X_i^*, all points q on the unstable manifold $W^{\mathrm{u}}(X_i^*)$ approach X_i^* exponentially under the backward map $\phi^{-n}(q)$, and, similarly, all points p on $W^{\mathrm{s}}(X_i^*)$ approach X_i^* exponentially under the forward map $\phi^n(p)$. Thus any set of points in the neighborhood of W^{u} is stretched along W^{u} under the action of ϕ, and as the orbit containing the numbered circles in Fig. 6.6, all orbits passing through the neighborhood of X_i^* (which are not on W^{s}) are attracted to $W^{\mathrm{u}}(X_i^*)$. Hence all chaotic orbits are attracted to $W^{\mathrm{u}}(X_i^*)$, and in the asymptotic time limit they lie on its closure. The following points will be discussed in Sect. 7.1:

(a) A chaotic attractor coincides with the closures of the unstable manifolds $[W^{\mathrm{u}}(X_i^*)]$ corresponding to the saddles X_i^* contained within this attractor[3].
(b) The two invariant manifolds $W^{\mathrm{u}}(X_i^*)$ and $W^{\mathrm{s}}(X_i^*)$ for each such saddle point X_i^* intersect repeatedly within the attractor, creating an infinite number of homoclinic points and an infinitely fine net-like structure which is dense in the attractor.

It follows from the above two observations that through any open neighborhood of any point X_t on a chaotic orbit there pass two invariant manifolds W^{s} and W^{u}. Additionally, given an arbitrary segment of some W^{u}, for sufficiently large n, its length will grow exponentially under the action of the forward map F^n, while the length of a given segment of some W^{s} will shrink exponentially under the same map. Thus a chaotic orbit is unstable along an unstable manifold W^{u} and stable along a stable manifold W^{s}.

---

[3] Here, the closures of the unstable manifolds corresponding to any two saddles contained within a given chaotic attractor are identical. To see this, let us consider two such saddles, X_i^* and Y_i^*. The manifolds $W^{\mathrm{u}}(Y_i^*)$ and $W^{\mathrm{s}}(X_i^*)$ intersect, and the points representing the forward mappings F^n of any such intersection point accumulate at X_i^*. This situation is depicted in Fig. A.1(b) of the Appendix. Thus $W^{\mathrm{u}}(Y_i^*)$ extends into an arbitrary open neighborhood of $W^{\mathrm{u}}(X_i^*)$, and hence $[W^{\mathrm{u}}(Y_i^*)] \supset W^{\mathrm{u}}(X_i^*)$. Similarly, $W^{\mathrm{u}}(X_i^*)$ and $W^{\mathrm{s}}(Y_i^*)$ intersect, and the points representing the forward mappings of any such intersection point accumulate at Y_i^*, implying $[W^{\mathrm{u}}(X_i^*)] \supset W^{\mathrm{u}}(Y_i^*)$. Therefore $[W^{\mathrm{u}}(Y_i^*)] = [W^{\mathrm{u}}(X_i^*)]$.

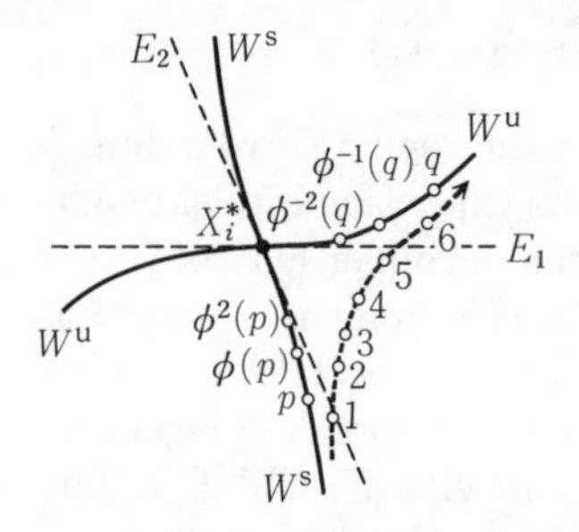

Fig. 6.6. The neighborhood of the saddle $X_i^* = \phi(X_i^*)$: the axes determined by the eigenvectors E_1 and E_2 of the matrix $D\phi(X_i^*)$, the invariant manifolds W^u and W^s, and the time series $\phi^n(X_0)(n = 1, 2 \ldots, 6)$ of an arbitrary orbit. Under the map ϕ, the length of a segment of W^u increases and the length of a segment of W^s decreases exponentially

In a chaotic region, for any point X_t on a chaotic orbit, there exists a sequence of points of intersection of the type in question which converge to X_t. (This follows from the fact that these intersection points form a dense set in the attractor.) Then, corresponding to a sequence of such intersection points converging to X_t there is a sequence of local structures. Thus, assuming the existence of the limits of some properties of this local structure, we can say that such structure is implied at X_t by virtue of this limit. In this way, the neighborhood of any point X_t on a chaotic orbit corresponds to either hyperbolic structure (Fig. 6.7a), or tangency structure (Figs. 6.7b and c). The unstable manifold W^u and stable manifold W^s passing through the point X_t cross in case (a) and intersect without crossing in cases (b) and (c) of Fig. 6.7. Under the map F, the length of a segment of W^u grows exponentially, and the length of a segment of W^s shrinks exponentially. Hence W^u is stretched in cases (a) and (c) and bent and folded in case (b). By defining the local structure at each point in the manner discussed above, it can be stated that a chaotic orbit possesses hyperbolic structure at almost all points. A system consisting entirely of hyperbolic structure is referred to as a *hyperbolic dynamical system.*

Physical systems possess tangency structure in addition to hyperbolic structure. In the attractor displayed in Fig. 6.1b for the pendulum, the curved regions labeled C above and below, as well as their forward mappings $F^n(C)$, possess the type I tangency structure described in Fig. 6.7b, while their backward mappings $F^{-n}(C)$ possess the type II tangency structure of Fig. 6.7c. Designating the width of the cell in the W^u direction at X_t in Fig. 6.7 by $2|y_t^u|$, we represent the expansion rate of this quantity resulting from the stretching and bending of W^u by

$$\lambda_1(X_t) = \ln \left| y_{t+1}^u / y_t^u \right| . \tag{6.18}$$

For the hyperbolic structure shown in Fig. 6.7a, $\lambda_1(X_t) > 0$, for the type I tangency structure of Fig. 6.7b, $\lambda_1(X_t) < 0$, and for the type II tangency structure of Fig. 6.7c, $\lambda_1(X_t) > 0$. A positive value of this quantity implies stretching, and a negative value implies bending and folding. We can also write this quantity in the following way, using the unstable component of the tangent mapping $DF(X_t)$:

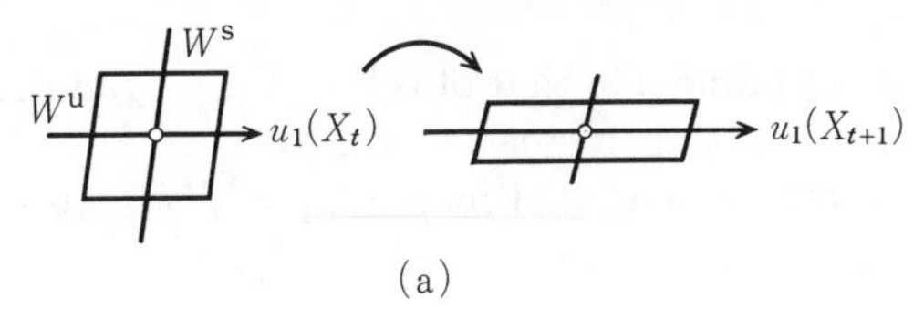

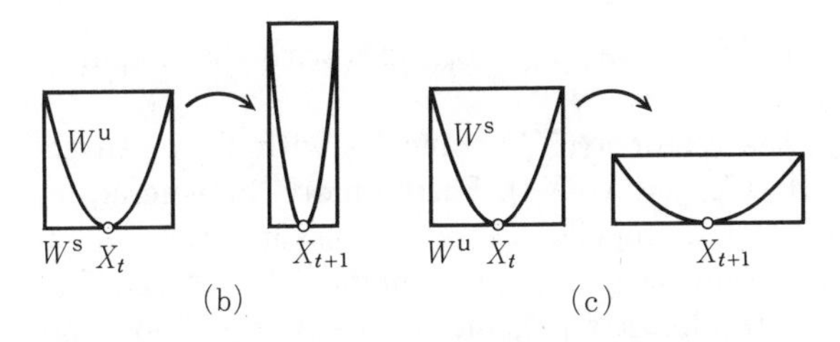

Fig. 6.7. The local structure of chaos and the stretching and folding of W^u. Here, (**a**) represents the case of hyperbolic structure ($\lambda_1 > 0$), (**b**) represents type I tangency structure ($\lambda_1 < 0$), and (**c**) represents type II tangency structure ($\lambda_1 > 0$). The quantity $|\lambda_1|$ expresses the local rate of stretching of W^u when $\lambda_1 > 0$ and its local rate of folding when $\lambda_1 < 0$. The quantity $\lambda_1(X_t)$ describes the basic process of mixing of chaotic orbits

$$\lambda_1(X_t) = \ln |DF(X_t) \cdot u_1(X_t)| \,. \tag{6.19}$$

Here, $u_1(X_t)$ is the unit vector tangent to W^u at X_t, and in a dissipative system this vector is also tangent to the attractor in question at X_t.

Let us consider the matrix equation $y_{t+1} = DF(X_t) \cdot y_t$. The map $DF(X_t)$ transforms the vector $u_1(X_t)$ into a vector in the direction of the tangent vector $u_1(X_{t+1})$, leading to $DF(X_t) \cdot u_1(X_t) = \nu_1(X_t) u_1(X_{t+1})$ and $|\nu_1(X_t)| = \exp[\lambda_1(X_t)]$, and, therefore, for the component of y_t in the direction of $u_1(X_t)$, y_t^u, the equality $y_{t+1}^u = \nu_1(X_t) y_t^u$ holds, implying (6.18). Then, the orbit $X_t' = X_t + y_t$ in the neighborhood of X_t is attracted to W^u, and if $y_0^u \neq 0$ holds, y_t will become parallel to $u_1(X_t)$ as t increases. Therefore, for almost all orbits in the neighborhood of X_t, we have $\lambda_1(X_t) = \ln[|y_{t+1}|/|y_t|]$ $(t \gg 1)$. This formula is useful for calculating or measuring $\lambda_1(X_t)$.

The orbital expansion rate $\lambda_1(X_t)$ fluctuates in time, assuming a variety of values as the chaotic orbit X_t is traced out. As we see below, during its motion, a chaotic orbit acts as a 'probe', investigating the local structure of phase space, and this structure is revealed by the fluctuations of $\lambda_1(X_t)$. Following Fujisaka (1983) and Morita et al. (1988), let us introduce the time-averaged value of $\lambda_1(X_t)$:

$$\Lambda_n(X_0) \equiv \frac{1}{n} \sum_{t=0}^{n-1} \lambda_1(X_t) \,. \tag{6.20}$$

Taking the $n \to \infty$ limit of this expression, we obtain the long-time average $\Lambda_\infty(X_0) = \langle \lambda_1(X) \rangle$. Comparing this with the Liapunov number (6.16), if we assume that the direction of y_0 is not parallel to that of $W^{\rm s}$ at X_0, y_t gradually turns to point in the direction of $W^{\rm u}$ as time increases, and thus we can write $\Lambda_\infty(X_0) = \overline{\lambda}_1(X_0)$. In fact, for a one-dimensional map $x_{t+1} = f(x_t)$, (6.19) becomes

$$\lambda_1(x_t) = \ln |f'(x_t)| , \tag{6.21}$$

and the long-time average, $\Lambda_\infty(x_0) = \langle \lambda_1(x) \rangle$, coincides with the Liapunov number, (6.15).

For a periodic orbit X_i^*, the vector $u_1(X_i^*)$ coincides with the eigenvector $E_1(X_i^*)$ and, therefore, $\Lambda_\infty(X_i^*) = \overline{\lambda}_1(X_i^*)$. Furthermore, in general, invariant sets (periodic orbits, Cantor repellors, etc.) possess a variety of Liapunov numbers which differ from the Liapunov number $\Lambda^\infty \equiv \Lambda_\infty(X_0)$ ($X_0 \in$ chaotic orbit) common to almost all chaotic orbits. For a chaotic orbit, as n increases, the value of $\Lambda_n(X_0)$ fluctuates as it reflects in succession the Liapunov numbers of the invariant sets and the orbital expansion rates of local structure characterizing the neighborhoods of successive X_t. The time average defined in (6.20) is termed the *coarse-grained expansion rate.* It represents a physical quantity which, by virtue of the fluctuations that it experiences along a chaotic orbit, describes the local structure of the many kinds of invariant sets and invariant manifolds coexisting in phase space.

6.3.3 Mixing and Memory Loss

In a chaotic region, each orbit is unstable along the direction of an unstable manifold, and the corresponding Liapunov number $\Lambda_\infty(X_0) = \overline{\lambda}_1(X_0)$ is thus positive. Therefore almost all neighboring orbits move apart exponentially, in accordance with (6.14) or (6.17). This implies that an 'error' in the initial conditions (or, more precisely, an uncertainty in their specification or measurement) also grows exponentially. A system characterized by such an instability is described as possessing a 'sensitive dependence on the initial point'. If we designate the scale of some measuring precision by l and the initial error by y_0, even if $|y_0| \ll l$, after a time $t_* \equiv (1/\Lambda_\infty) \ln[l/|y_0|]$, this error becomes larger than l, and prediction regarding the orbit (based on knowledge concerning the initial point) becomes impossible. Then, since the precision inherent in any type of measurement is finite, when considering large times, such orbits are experimentally irreproducible. In addition, the behavior of a chaotic orbit cannot be predicted by any theoretical means, be they analytical or numerical.

The stretching undergone by a segment of an unstable manifold $W^{\rm u}$ can be regarded as the source of both the instability and the mixing nature of the chaotic orbits attracted to $W^{\rm u}$. As a result of this stretching, for chaotic orbits there is a loss of memory with regard to initial conditions, and the motion of a chaotic orbit on an attractor is in some sense random. This randomness implies the *indeterminism* characterizing such systems and results in the random behavior of the time series corresponding to such an orbit, as in the case of the forced pendulum described in Fig. 6.1a.

Mixing in a fluid system is caused by a stretching and folding of local regions. This can be understood by observing a dyed region in the fluid (for example, a drop of milk in a cup of coffee). This is similar to the mixing displayed by chaotic orbits. Here, focusing on a set of points contained within some small cell, as a consequence of the stretching and folding of this cell (as in the cells of Fig. 6.3 and the lobes of Fig. 6.5), the chaotic orbits to which these points belong eventually become mixed with other orbits. This is a result of the stretching and folding of W^{u}. The quantity $|\Lambda_n(X_0)|$ represents the rate of stretching when $\Lambda_n(X_0) > 0$ and the rate of folding when $\Lambda_n(X_0) < 0$. The expansion rate $\Lambda_n(X_0)$ expresses the coarse-grained rate of mixing in the contexts of both fluids and chaotic orbits.

6.4 The Statistical Description of Chaos

6.4.1 The Statistical Stability of Chaos

It has been stated [Sinai (1981)] that when the orbits of a system are unstable, the system behaves according to statistical laws. It is important to note that the existence of such an instability requires neither a large number of degrees of freedom nor an external source of indeterminacy.

Let us first consider how the presence of stable manifolds W^{s} insures the stability of the probability distribution characterizing a system. Any given chaotic orbit exists (after some transient) on the closure of an unstable manifold W^{u}, and along this manifold it is unstable. However, such an orbit is stable along the directions of the stable manifolds that pass through every neighborhood of every point along it.

Now let us take a typical chaotic orbit $X_i = F^i(X_0)$ $(i = 0, 1, 2 \ldots)$ and consider the set of an infinite number of points $\{X_0, X_1, \ldots, X_{N-1}\}$ $(N \to \infty)$, which we refer to as $\Xi \equiv \{X_i\}$. Then, the average value of some quantity $G(X)$ over this set Ξ is given by

$$\langle G(X) \rangle \equiv \lim_{N\to\infty} \frac{1}{N} \sum_{i=0}^{N-1} G(X_i) \,. \tag{6.22}$$

This is simply the long-time average (6.5a) of $G(X)$. Since memory of the initial conditions is lost in some finite time τ, this time-averaged quantity is independent of the initial point X_0 for almost all choices of X_0.

Note that since the chaotic orbit X_i passes through every neighborhood of any point on the unstable manifold W^u, the set Ξ is densely distributed on the closure of W^u.

Now let us apply a small disturbance, localized in time and space, to the chaotic orbit X_i and denote the disturbed orbit by X_i'. This disturbance could represent, for example, errors inevitably arising in computer simulation or experimental measurement. Here we may assume that this kind of disturbance does not alter the structure of the closure of W^u appreciably. The effect of such a disturbance is to shift the chaotic orbit X_i along W^u. (Even if X_i is caused to deviate from W^u, it converges back to W^u exponentially along the stable manifolds it encounters.) Thus the disturbed orbit X'_i also lies on W^u. Hence, since the sets $\Xi = \{X_i\}$ and $\Xi' = \{X'_i\}$ are dense in W^u, for any sufficiently long time after the disturbance, given any point $X'_j \in \Xi'$ there exists a point $X_k \in \Xi$ arbitrarily close to X'_j, and given any point $X_k \in \Xi$ there exists a point $X'_j \in \Xi'$ arbitrarily close to X_k. Thus we see that the set Ξ and the average quantity (6.22) are not altered appreciably by a disturbance of the type in question, although the disturbed orbit X'_i differs greatly from the original orbit X_i for large time i ($\gg \tau$) due to orbital instability.

We often take a chaotic orbit $X_t = F^t(X_0)$ $(t = 0, 1, \ldots, n-1)$ with a finite n ($\lesssim \tau$) and consider the coarse-grained quantity $G_n(X_0) = (1/n)\sum_{t=0}^{n-1} G(X_t)$. Here we may assume that $G_n(X_0)$ is a unique function of X_0 for a given length n. Then, the average value $\langle G_n(X)\rangle$ over the set Ξ is not altered appreciably by a small disturbance. This is a result of the fact that the chaotic orbit is dense in W^u and its memory of the initial point is lost after some finite time τ.

From the above discussion, we see that the average quantity $\langle G_n(X)\rangle$ is stable with respect to small disturbances, and hence it is experimentally reproducible. Let us refer to this property as *statistical stability*. In light of these considerations, we see that experiments (both real and numerical) have meaning only with regard to averages over a sufficiently large number N of orbits (say, $N \gg 10^5$). Thus reproducible average quantities such as that represented by (6.22) are the observables that we can compute and measure.

6.4.2 Time Coarse-Graining and the Spectrum $\psi(\Lambda)$

As an example of the kind of observable discussed above, let us consider the probability density of the coarse-grained expansion rate of nearby orbits

$$P(\Lambda; n) \equiv \langle \delta(\Lambda_n(X) - \Lambda)\rangle \ , \tag{6.23}$$

where $\delta(g)$ is the delta-function. For a given value of the length n, $\Lambda_n(X_i)$ assumes various values, as determined by the initial point X_i. The probability that this value falls in the range $(\Lambda, \Lambda + \mathrm{d}\Lambda)$ is $P(\Lambda; n)\mathrm{d}\Lambda$. Let us write the n dependence of $P(\Lambda; n)$ in the $n \to \infty$ limit as

$$\ln P(\Lambda; n) = -\psi(\Lambda)n - \phi(\Lambda)\ln n + \gamma(\Lambda) \ . \tag{6.24}$$

This expression represents the statistical law of time coarse-graining. Here, the coefficient $\psi(\Lambda)$ is defined by

$$\psi(\Lambda) \equiv \lim_{n\to\infty} \psi_n(\Lambda) \tag{6.25}$$

$$\psi_n(\Lambda) \equiv -(1/n)\ln[P(\Lambda;n)/P(\Lambda^\infty;n)] , \tag{6.26}$$

where Λ^∞ is the Liapunov number for a chaotic orbit. The function $P(\Lambda;n)$ is maximal at $\Lambda = \Lambda^\infty$ for a given n ($\gg 1$) and decreases when moving away from this point, implying that $\psi(\Lambda)$ is a downward convex function satisfying $\psi(\Lambda) \geq \psi(\Lambda^\infty) = 0$. The function $\psi(\Lambda)$ is called the *spectrum* of Λ. This probability density is a stable and experimentally reproducible observable that can be obtained through calculation or observation of $\Lambda_n(X_i)$ for various X_i (provided that $N \gg n \gg 1$). In numerical experiments considered in later chapters, we have $n \sim 10$–10^2 and $N \sim 10^6$–10^8.

The function $P(\Lambda;n)$ determines the distribution of the mixing rate $\Lambda_n(X_i)$ representing the stretching and folding of segments of W^{u} and thus describes the mixing nature of chaos. For $\Lambda > 0$, $P(\Lambda;n)$ is characterized by the distribution of the Liapunov numbers $\Lambda_\infty(X_i^*)$ of the coexisting periodic orbits, while for $\Lambda < 0$, it is characterized by the type I tangency structure depicted in Fig. 6.7b. We will investigate the form of $P(\Lambda;n)$ in Chap. 8, but in general, when there is no region in which $\psi(\Lambda)$ is identically 0, from (6.24), we can write

$$P(\Lambda;n) = \exp[-n\psi(\Lambda)]P(\Lambda^\infty;n) . \tag{6.27}$$

Even in a conservative dynamical system, the spectrum $\psi(\Lambda)$ reflects the breaking of time-reversal symmetry exhibited by chaotic systems, possessing a form asymmetric in Λ. This form depends on the exponent ζ of the variance of the coarse-grained expansion rate $\Lambda_n(X_0)$:

$$n^2\left\langle\{\Lambda_n(X) - \Lambda^\infty\}^2\right\rangle = nC_0^\lambda + 2\sum_{t=1}^{n-1}(n-t)C_t^\lambda \tag{6.28}$$

$$\propto n^\zeta \quad (2 > \zeta \geq 0) . \tag{6.29}$$

The quantity $C_t^\lambda \equiv \langle\{\lambda_1(X_t) - \Lambda^\infty\}\{\lambda_1(X_0) - \Lambda^\infty\}\rangle$ here is the *time correlation function* of the orbital expansion rate (6.19). (The derivation of (6.28) is sketched in Sect. A.2 of the Appendix.) In general, C_t^λ is composed of a part that oscillates in time and one that, as a result of mixing, decays in time. Equation (6.29), however, represents an asymptotic ($n \to \infty$) form that is determined by the decaying part alone. In the case of chaos in dissipative systems, this decay is usually exponential in nature, implying $0 \leq \zeta \leq 1$. In the case in which the decaying part remains positive as it decays exponentially to zero, we have $\zeta = 1$, and near Λ^∞ we can write $\psi(\Lambda) \propto (\Lambda - \Lambda^\infty)^2$. Thus in this situation the central limit theorem applies. In a hyperbolic dynamical system, this normal distribution is meaningful, but in a physical system, since tangency structure (Fig. 6.7b) also exists and negative values of λ_1 appear, the normal distribution is not valid for $\Lambda < \Lambda^\infty$. Furthermore, the contribution of $\Lambda_\infty(X_i^*)$ to $\psi(\Lambda)$ is singular near the bifurcation resulting from the collision of the attractor with the saddle X_i^*, and thus, in such a situation, the normal distribution is invalid for $\Lambda > \Lambda^\infty$ as well.

In the chaotic sea of a conservative system, there exist islands of tori whose presence leads to a long-time correlation $C_t^\lambda \propto t^{-(\beta-1)}$ $(1 < \beta < 2)$. For this reason, here we have $1 < \zeta = 3 - \beta < 2$, and in the interval $(0, \Lambda^\infty)$, $\psi(\Lambda) = 0$. In this interval, from (6.24), the probability distribution becomes

$$P(\Lambda; n) = C(\Lambda) n^{-\phi(\Lambda)} \quad (\phi(\Lambda) > 0) . \tag{6.30}$$

We investigate the form of this anomalous distribution in Sect. 10.2.

A chaotic system is characterized by the mixing it displays. The Liapunov number Λ^∞ provides a measure of the coarse-grained mixing rate, and, as it reveals, the nature of a system's mixing behavior changes dramatically in the neighborhood of the point at which chaos first appears. At the emergence point of chaos, $\Lambda^\infty = 0$, and there is no mixing. Here, neutral aperiodic orbits such as those on critical 2^∞ attractors and critical golden tori exist. For these orbits, memory of the initial time remains forever, and time series consisting of self-similar inverted nested structure appears. The treatment of such a situation is completely different from that outlined above. This treatment is given in Sect. 9.3.

6.4.3 The Statistical Structure of Chaos

As discussed in this chapter, there are clear statistical laws governing unstable dynamical systems. The probability density $P(\Lambda; n)$ provides a statistical description of the dynamics of chaotic orbits and the unstable manifolds on which they reside, as well as a characterization of the mixing undergone by such orbits. However, chaos is a multifaceted phenomenon, and to capture its many manifestations, a variety of observable quantities must be considered. The role of statistical physics, as depicted in Fig. 6.8, is based on the geometrical description of the phase structure provided by the theory of dynamical systems. This role is one of unifying the statistical and geometrical viewpoints regarding chaos and elucidating the reproducible statistical structure and physical processes corresponding to chaotic orbits. As mentioned above, the physical processes we consider include such transport phenomena as the mixing and diffusion of chaotic orbits and fluid particles and energy dissipation. The statistical structure of chaos is embodied by the form and structure of the probability distributions ($\psi(\Lambda)$, $f(\alpha)$, etc.) of the coarse-grained local quantities characterizing chaos. The phase structure of chaos is imparted by the huge variety of its orbits. This variety implies the existence of widely varying forms of local structure, including many types of invariant manifolds and other invariant sets. The multitude of bifurcations exhibited by a chaotic system bring about qualitative changes in its phase structure. The investigation of the anomalous statistical structure and universality associated with these bifurcations is a particularly important subject in this field.

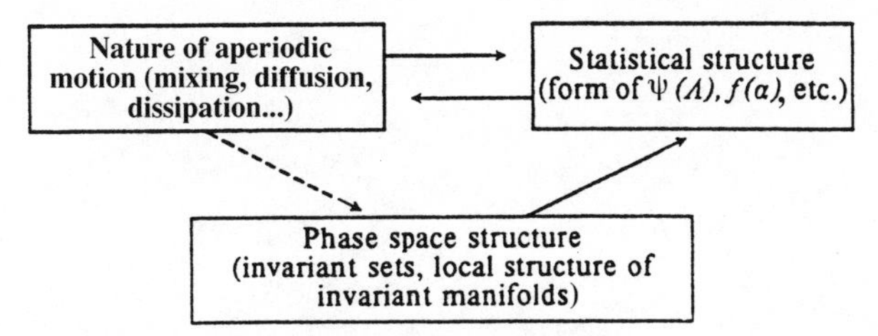

Fig. 6.8. The physical perspective of our description of chaos

7. Bifurcation Phenomena of Dissipative Dynamical Systems

The form and structure of an attractor are characterized by the types of unstable periodic points (saddle points) that it contains. Changing the value of a bifurcation parameter can cause various saddle points to enter and leave attractors. Chaotic bifurcations result from the collision of an attractor with such points and their resultant inclusion into the attractor.

In this chapter, we first consider the geometric structure of chaotic bifurcations resulting from collisions of this nature using the two-dimensional Hénon map. We then derive several one- and two-dimensional maps from a particular ordinary differential equation and outline the general nature of universal bifurcations using well-known one-dimensional maps (a quadratic map and a circle map). We investigate the similarity laws and renormalization transformation corresponding to bifurcation cascades of period doubling, band splitting, and attractor merging phenomena, and also study the structure of chaos arising through the period-doubling and quasi-periodicity routes.

7.1 Band Chaos of the Hénon Map

Let us consider the Hénon map for the two-dimensional vector $X_t = (x_t, y_t)$,

$$\begin{pmatrix} x_{t+1} \\ y_{t+1} \end{pmatrix} = F(X_t) = \begin{pmatrix} 1 - ax_t^2 + by_t \\ x_t \end{pmatrix} \qquad (t = 0, \pm 1, \pm 2, \cdots) . \tag{7.1}$$

The Jacobian of this map is $J = -b$, and in the following we assume $b = 0.3$. Figure 7.1 displays the unstable manifold W^{u} (solid curve) and stable manifold W^{s} (broken curve) of the saddle $X^* = F(X^*)$ for the case $a = 1.0809$. The components of X^* are given by

$$x^* = y^* = \left[-1 + b + \left\{(1-b)^2 + 4a\right\}^{1/2}\right] / 2a ,$$

and the attractor of this system, A, is on the closure of W^{u}. Figure 7.2a describes such an attractor. The attractor A shown in this figure is composed of two bands, and a given chaotic orbit X_t visits these bands alternatingly; that is, if $X_{2t} = F^{2t}(X_0)$ is on the left band (band 1), then $X_{2t+1} = F^{2t}(X_1)$ is on the right band (band 2), while the motion within a given one of these bands is random. Orbits belonging to different bands at a given time will always belong to different bands, and the time correlation function C_t of an orbit X_t contains the period 2 oscillatory part $\cos(\pi t)$. The power spectrum $S(\omega)$ consists of a line spectrum corresponding to the frequency $\omega = \pi$ superimposed on a continuous spectrum. This type of phenomenon is referred to as *two-band chaos.* In such a situation, mixing that extends over the entire attractor does not exist, but there is mixing corresponding to each of the two time series X_{2t} and X_{2t+1} $(t = 0, 1, 2\dots)$. Thus there is mixing within each band, and within each band there is memory loss.

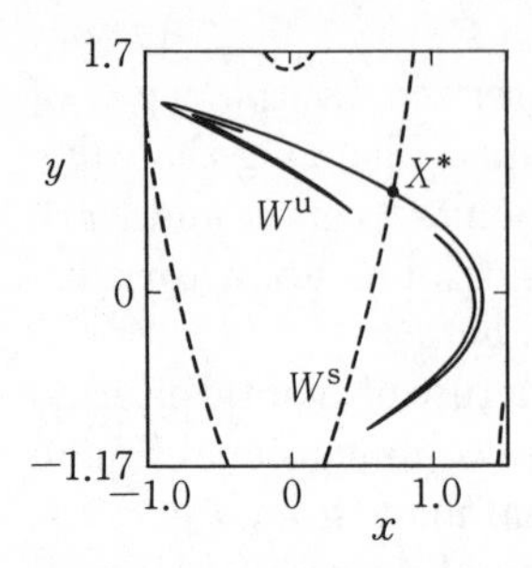

Fig. 7.1. The manifolds W^u (*solid curve*) and W^s (*broken curve*) of the saddle X^* $(a = 1.0809)$. For $a < a_1 = 1.15357\dots$, W^s and W^u do not intersect (except at X^*), but for $a > a_1$ they intersect at an infinite number of points (see Fig. 7.2)

As the value of a is increased, A becomes stretched, and, as shown in Fig. 7.2, the two bands become connected at the saddle X^* at the value $a = a_1 \equiv 1.15357\dots$ (this phenomenon is called *band merging*). For all larger values of a, these two bands form a single band, and both series X_{2t} and X_{2t+1}

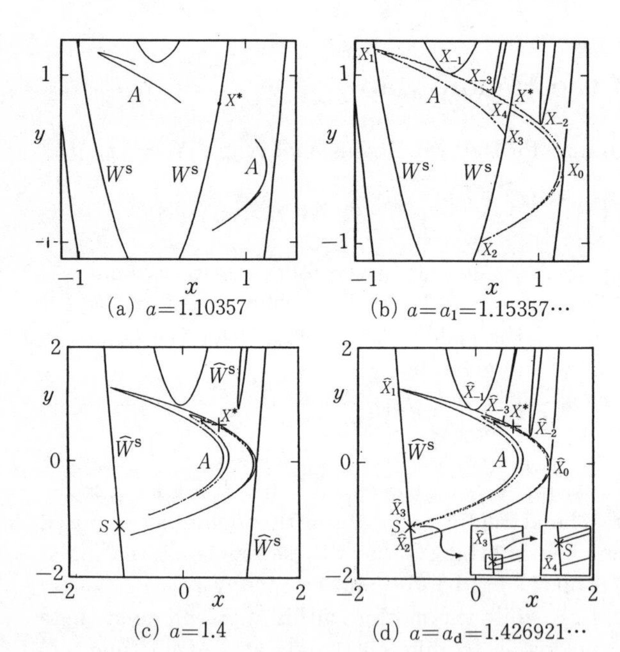

Fig. 7.2. The attractor A and the stable manifolds $W^s(X^*)$ and $\hat{W}^s(S)$ of the saddles X^* and S. In (**b**), W^s and A are tangent, and in (**d**), $\hat{W}^s$ and A are tangent. The inserts in (**d**) describe the self-similar nested structure of $\hat{X}_j$ $(j = 2, 3, 4\dots)$ appearing in the neighborhood of S

pass through every neighborhood of every point on the attractor. In this case C_t does not possess the period 2 oscillatory part, and $S(\omega)$ is continuous. In this manner, the statistical structure of the system undergoes a qualitative change at $a = a_1$. For $a > a_1$, A coincides with the closure of the unstable manifold W^{u} of X^*. As exhibited in Fig. 7.3, the invariant manifolds of X^*, $W^{\mathrm{u}}(X^*)$ and $W^{\mathrm{s}}(X^*)$, cross each other repeatedly on A, forming an infinite number of homoclinic points. This is the so-called infinitely fine net discussed in Sect. 6.3.

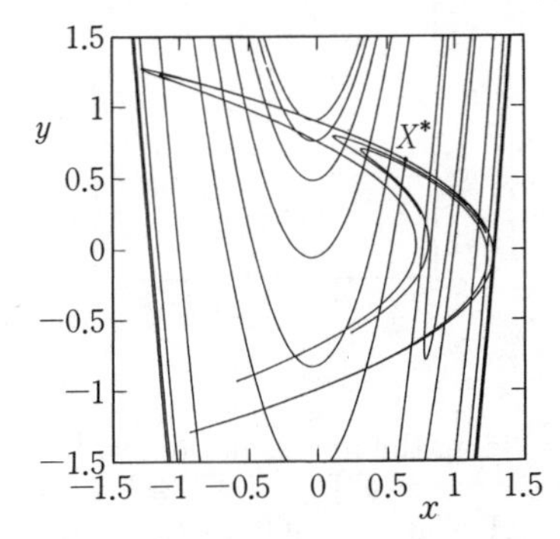

Fig. 7.3. The intersection of $W^{\mathrm{s}}(X^*)$ and $W^{\mathrm{u}}(X^*)$ for $a = 1.4 > a_1$. The intersection points are distributed densely on A. In the largest curved part C of A and its images $F^{\pm j}(C)$, there exist tangency points [Grassberger et al. (1988)]

In the system described above, there exist two saddle points (one in either band) X_1^* and X_2^* satisfying $X_i^* = F^2(X_i^*)$ and for $a < a_1$, the ith band is identical to the closure of $W^{\mathrm{u}}(X_i^*)$. More generally, for $2Q$-band chaos (consisting of chaotic motion on each of $2Q$ bands together with $2Q$-periodic inter-band motion), for values of a satisfying $a_{2Q} < a < a_Q \leq a_1$, each of the $2Q$ bands coincides with the closure of the unstable manifold $W^{\mathrm{u}}(X_i^*)$ corresponding to the saddle $X_i^* = F^{2Q}(X_i^*)$ in that band. Here, $W^{\mathrm{u}}(X_i^*)$ and $W^{\mathrm{s}}(X_i^*)$ intersect repeatedly, forming an infinitely fine net, just as $W^{\mathrm{u}}(X^*)$ and $W^{\mathrm{s}}(X^*)$ for the case of one-band chaos.

The band merging discussed here is a typical chaotic bifurcation. We now discuss the special characteristics of this bifurcation. At $a = a_1$, the points of tangency between W^{s} and A, $X_{\pm j} (j = 0, 1, 2, \ldots)$, are obtained from the tangency point situated at X_0 according to $X_{\pm j} = F^{\pm j}(X_0)$, as shown in Fig. 7.2b. These tangency points accumulate at X^* as $j \to \infty$; the nature of this accumulation is depicted in Fig. 7.4. These tangency points, while appearing to constitute the endpoints of curves terminating at $W^{\mathrm{s}}(X^*)$ in Figs.7.4a and b, in fact lie on continuous curves that bend sharply at W^{s}, as seen in Fig. 7.4c. The manner in which these tangency points accumulate at X^* characterizes the band merging bifurcation point $a = a_1$. The neighborhood of a tangency point X_j is mapped as shown in Fig. 6.7b, with a negative expansion rate $\lambda_1(X_j)$, while the neighborhood of X^* maps as shown in Fig. 6.7a, with the positive expansion rate $\Lambda_\infty(X^*) = \lambda_1(X^*) \approx 0.550$, a value approximately twice the Liapunov number, $\Lambda^\infty \approx 0.315$. The coarse-grained orbital expansion rate for a length n orbit (n finite), $\Lambda_n(X_i)$, will assume a wide range of values, as determined by the initial point X_i. Those orbits which pass near X^* will possess anomalously large values. The large Liapunov number $\Lambda_\infty(X^*) \approx 0.550$ enters the spectrum $\psi(\Lambda)$ at $a = a_1$, as this saddle point enters the attractor. This results in anomalous statistical structure.

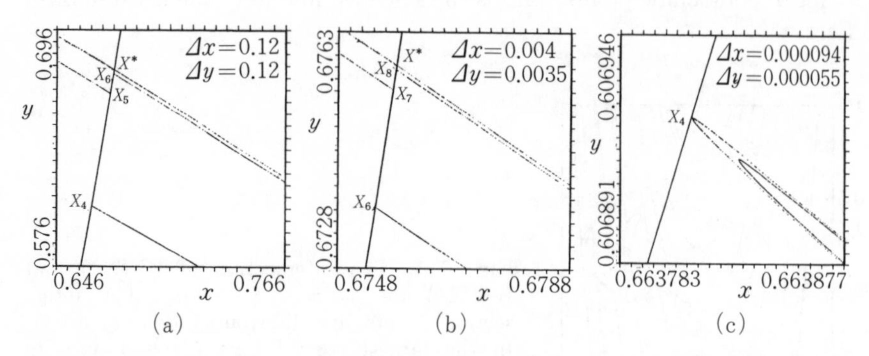

Fig. 7.4. The self-similar nested structure characterizing the accumulation at X^* of tangent points X_j along $W^s(X^*)$ in the case depicted in Fig. 7.2b. **(c)** The continuous nature of $W^s(S)$ in the neighborhood of a given point of tangency [Hata et al. (1988)]

Let us gather together the pieces from the above discussion. We consider the case of $2Q$-band chaos merging into Q-band chaos. The main features of this band merging are as follows:

(1) It occurs when $2Q$ bands constituting the original attractor merge in pairs at Q saddles of period Q, $X_i^* = F^Q(X_i^*)$, thereby forming an attractor of Q bands.
(2) At the bifurcation point $a = a_Q$ corresponding to this merging, the attractor is tangent to the stable manifolds $W^s(X_i^*)$ of the saddles X_i^* at the points $X_{i,\pm j}$ ($j = 0, 1, 2, \ldots$). Each X_i^* is the accumulation point for the tangency points $X_{i,\pm j}$.
(3) The minimum value of the local expansion rate is given by $\Lambda_{\min} = \Lambda_\infty(X_{i,0}) < 0$, and a length n chaotic orbit X_t that passes through the neighborhood of a tangency point $X_{i,0}$ will possess a negative expansion rate satisfying $0 > \Lambda_n(X_t) > \Lambda_{\min}$. (Of course, since $\Lambda^\infty > 0$, the relation $\lim_{n\to\infty} \Lambda_n(X_t) > 0$ holds for almost all X_t, and this is the case for orbits X_t that pass through the neighborhood of a tangency point $X_{i,0}$ as well. However, for an orbit X_t that passes close to some $X_{i,0}$, $\Lambda_n(X_t)$ will remain negative for some finite time, and this time diverges as we consider orbits that pass closer and closer to $X_{i,0}$.) The appearance of such tangency points creates anomalous statistical structure.
(4) The maximum value of the local expansion rate is given by $\Lambda_{\max} = \Lambda_\infty(X_i^*) > \Lambda^\infty$, and chaotic orbits X_t (of sufficiently short length, in

the sense discussed in (3) above) that pass through the neighborhood of a saddle X_i^* possess large positive expansion rates satisfying $\Lambda^\infty < \Lambda_n(X_t) < \Lambda_{\max}$. The existence of such orbits also implies anomalous statistical structure.

We will formulate the anomalous nature of the statistical structure alluded to in (3) and (4) in the next chapter.

In the neighborhood of a tangency point $X_{i,j}$ there exists a multitude of chaotic orbits displaying negative expansion rates. In fact, this is true not only near the bifurcation point. For nonhyperbolic dynamical systems whose attractors possess curved regions (corresponding to the type I tangency structure of Fig. 6.7b), the stable manifold $W^{\mathrm{s}}(X_i^*)$ of the saddle point X_i^* intersects A tangentially, and there always exists an orbit consisting of tangency points, $X_{i,j} = F^j(X_{i,0})$ (see Fig. 7.3). However, only at the bifurcation point do these points accumulate at X_i^*.

The saddle point S shown in Figs. 7.2c and d is created, together with X^*, at the point $a = -(1-b)^2/4$. Its coordinates are given by $x^{\mathrm{s}} = y^{\mathrm{s}} = \{-1 + b - [(1-b)^2 + 4a]^{1/2}\}/2a$. The corresponding stable manifold $\hat{W}^{\mathrm{s}}(S)$ defines the boundary of the basin containing X^*. As the value of a is increased, A becomes elongated, and at the value $a = a_{\mathrm{d}} \equiv 1.42692\ldots$ (depicted in Fig. 7.2d), A collides with $\hat{W}^{\mathrm{s}}(S)$. For $a > a_{\mathrm{d}}$, the unstable manifold $W^{\mathrm{u}}(X^*)$ on which A resides crosses $\hat{W}^{\mathrm{s}}(S)$, creating an infinite number of heteroclinic points. The chaotic orbits moving along $W^{\mathrm{u}}(X^*)$ thus move outside of $\hat{W}^{\mathrm{s}}(S)$ and as a result the attractor A is destroyed. This phenomenon is referred to as an *attractor destruction crisis.* The distinguishing feature of the bifurcation point $a = a_{\mathrm{d}}$ is that, here, the points of tangency $\hat{X}_j = F^j(\hat{X}_0)$ accumulate at S as $j \to \infty$, while the tangency points $\hat{X}_{-j} = F^{-j}(\hat{X}_0)$ accumulate at X^*. In this case, as in the case of band merging, the *tangency orbit* $\hat{X}_j = F^j(\hat{X}_0)$ $(j = 0, 1, 2, \ldots \infty)$ plays an important role, and the probability that the coarse-grained orbital expansion rate $\Lambda_n(X_t)$ will realize its maximum value, $\Lambda_\infty(S) = \lambda_1(S) \approx 1.188$, becomes anomalously high. Here, $\Lambda_\infty(X^*) \approx 0.665$ and $\Lambda^\infty \approx 0.495$. As we will see in Sect. 8.1.4, this large value of $\Lambda_\infty(S)$ induces an anomaly in the statistical structure of the system. In Sect. 9.1, we see that the accumulation of heteroclinic tangency points at the saddle S is a universal phenomenon, appearing also in the cases of discontinuous band merging and attractor-merging crises.

7.2 The Derivation of Several Low-Dimensional Maps

The Hénon map (7.1) can be thought of as a local model able to describe systems that possess attractors with curved regions like C shown in Fig. 6.1b, as can be seen by comparing this figure with Fig. 7.2c. However, forced pendulums also display global dynamical forms, such as tori, which are very different in nature from such local forms. To understand the types of maps that describe such systems, let us consider the Poincaré map of the differential equation

$$\dot{x} = z \tag{7.2a}$$

$$\dot{z} = -\gamma z + kg(x) \sum_{i=-\infty}^{\infty} \delta(t - iT) . \tag{7.2b}$$

This equation is similar to (6.2), with γ acting as the coefficient of friction. Here, however, the forcing term applies a delta-function impulse (a 'kick') at times $t = iT$, for integer i. The quantity k represents the amplitude of the kick, and $g(x)$ is an arbitrary function of x (independent of t). When x represents an angle ϕ, this system is referred to as a *kicked rotor.*

Because of the periodically applied impulse, the 'speed' z changes discontinuously, but let us assume that the 'position' x is a continuous function of time. If we denote times immediately before and immediately after the ith kick by $t = iT \mp 0$ and define $z_i \equiv z(iT - 0)$, $x_i \equiv x(iT \mp 0)$, we obtain the following equation, which is valid for any time t in the range $iT - 0 \le t < (i+1)T - 0$:

$$z(t) = \mathrm{e}^{-\gamma t} \left(z_i \mathrm{e}^{\gamma i T} + kg(x_i) \int_{iT-0}^{t} \mathrm{d}s\, \mathrm{e}^{\gamma s} \delta(s - iT) \right) .$$

Then, for any time satisfying $iT + 0 < t < (i+1)T - 0$,

$$z(t) = \{z_i + kg(x_i)\} \mathrm{e}^{-\gamma(t - iT)} . \tag{7.3}$$

Thus, integrating (7.2a) from $iT + 0$ to $(i+1)T - 0$ yields a two-dimensional map describing the evolution of $X_i \equiv (x_i, z_i)$,

$$\begin{pmatrix} x_{i+1} \\ z_{i+1} \end{pmatrix} = F(X_i) = \begin{pmatrix} x_i + \tau z_{i+1} \\ J\{z_i + kg(x_i)\} \end{pmatrix} \qquad (i = 0, \pm 1, \pm 2, \cdots) , \qquad \begin{matrix} (7.4\mathrm{a}) \\ (7.4\mathrm{b}) \end{matrix}$$

where $J \equiv \mathrm{e}^{-\gamma T} \le 1$ and $\tau \equiv (1 - J)/\gamma J$. This is the Poincaré map [see (6.3)] of this system. The Jacobian of this map, J, satisfies (6.4). The quantity τ represents the effective fundamental period of the map, and when $\gamma T \ll 1$ (i.e., $J \approx 1$), we have $\tau \approx T$.

Now, multiplying (7.4b) by τ and using the resulting equation, together with (7.4a), we can obtain the equation

$$x_{i+1} = (1 + J)x_i + \tau J k g(x_i) - J x_{i-1} , \tag{7.5}$$

free of z. The energy dissipation rate of this system is $W_{\mathrm{f}}(t) = \gamma z^2$. Then, using (7.3) we find that, with $t_i = iT - 0$, the average of this value over a time τ [defined by (6.5b)] is

$$w(X_i) = \Gamma z_{i+1}^2 = \Gamma (x_{i+1} - x_i)^2 / \tau^2 , \tag{7.6}$$

where $\Gamma \equiv (1 - J^2)/2TJ^2$.

In what follows, we will use (7.4) and (7.5) to derive a number of different types of two-dimensional maps. These maps represent the fundamental models describing the local structures that characterize a variety of geometric forms displayed by chaotic systems. As we will see, these maps possess a much wider range of applicability than the rotor model (7.2).

7.2.1 The Hénon Map

If we fix the function $g(x)$ in (7.5) as

$$\tau J k g(x) = -(1+J)x + (1 - ax^2) , \tag{7.7}$$

defining $y_i \equiv x_{i-1}$, we have

$$x_{i+1} = 1 - ax_i^2 - Jy_i, \qquad y_{i+1} = x_i . \tag{7.8}$$

With $J = -b$, this becomes the Hénon map (7.1). In 1976, Hénon proposed this as a model to describe the stretching and bending phenomena exhibited by chaotic systems. It represents a general model of the curved regions existing in attractors.

7.2.2 The Annulus Map

Let us consider the situation in which, in three-dimensional phase space, a periodic orbit γ_1 of frequency ω_1 destabilizes, and there emerges a bi-periodic orbit exhibiting periodic motion of frequency ω_2 ($< \omega_1$) around the path defined by the original orbit γ_1. Then, considering the Poincaré section perpendicular to the path defined by γ_1 (acting as the axis of the slower rotation), if the *rotation number* $\rho = \omega_2/\omega_1$ is irrational, the points at which the bi-periodic orbit intersects this Poincaré section will, on average, rotate by an angle $2\pi\rho$ during each period $T = 2\pi/\omega_1$, describing some closed curve C. This curve is the cross-section of an invariant torus. In the most simple case, C assumes the form of a circle, but, in general, it is an arbitrarily shaped closed curve. Figure 7.5 displays data obtained in an experimental study of Bénard convection which describes such a cross-section. In order to investigate the structure of a torus cross-section C, we use a map in which the intersection points X_i are represented by polar coordinates: $X_i \equiv (R + r_i, 2\pi\theta_i)$. As a representative such model, we use (7.5) with $\theta_i \equiv x_i$ and the function $g(\theta_i)$ defined by

$$\tau J k g(\theta) = (1-J)\Omega - (K/2\pi)\sin(2\pi\theta) . \tag{7.9}$$

Then, with $r_{i+1} \equiv \theta_{i+1} - \theta_i - \Omega$, we obtain the *annulus map*,

$$\begin{cases} \theta_{i+1} = \theta_i + \Omega + r_{i+1} \pmod 1 \\ r_{i+1} = Jr_i - (K/2\pi)\sin(2\pi\theta_i) . \end{cases} \tag{7.10}$$

For $0 \le \Omega \le 1$, $0 \le \theta_i < 1$ and $-2 \le r_i \le 1$, this map corresponds to an annulus of width 3 with circular axis of radius R.

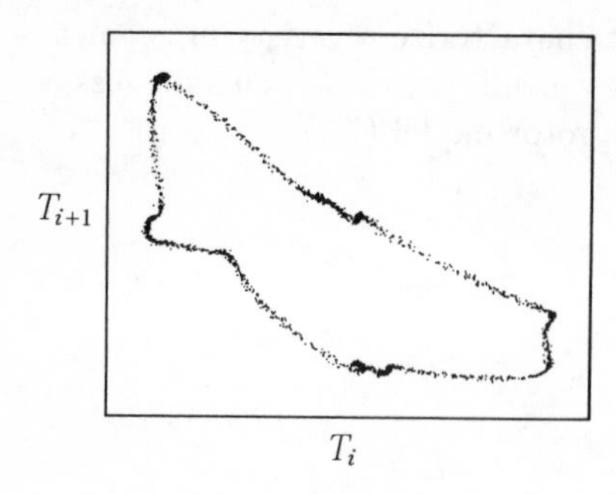

Fig. 7.5. The cross-section of a critical golden torus. This graph represents the plot of temperatures measured at the center of the bottom wall of a container (aspect ratio 2) in a system undergoing Bénard convection at the emergence point of chaos. A time sequence of the temperature T_i ($i = 0, 1, 2 \ldots$) was generated, and the values for successive times i and $i+1$ for 2500 values of i are plotted here [Jensen et al. (1985)]

For a given orbit, the rotation number on the cross-section C is given by

$$\rho(X_0) \equiv \lim_{n\to\infty} (\theta_n - \theta_0)/n \,, \tag{7.11}$$

where in this case we obviously do not take the mod 1 value of θ_n. When the value of the parameter K determining the size of the nonlinear term is set to 0, we have

$$\begin{aligned} R_n &\equiv R + r_n = R + J^n r_0 \\ \theta_n &= \theta_0 + n\Omega + (1 - J^n) J r_0/(1 - J) \,. \end{aligned} \tag{7.12}$$

When $J < 1$, we have $\rho(X_0) = \Omega$, and thus ρ is irrational for almost all values of Ω. In this case, as $n \to \infty$, the points X_n define a torus described about the circle of radius R.

With $J = 0.5$ and $\Omega = \Omega_\infty \equiv 0.61175390\ldots$, increasing the strength K of the nonlinear term causes the attractor in question gradually to assume an increasingly distorted shape, and when K exceeds the value $K_\infty = 0.97883778\ldots$ it becomes a strange attractor. At the point $K = K_\infty$, marking the emergence of chaos, the rotation number $\rho(X_0)$ that characterizes the attractor assumes the value

$$\rho(X_0) = \rho_{\mathrm{G}} \equiv (\sqrt{5} - 1)/2 = 0.61803 \cdots \,, \tag{7.13}$$

the inverse of the golden ratio. The corresponding attractor is called a *golden torus*. This is a type of 'critical' attractor. The value ρ_{G} is the positive root of the equation $\rho^2 + \rho = 1$. The Liapunov number of this *critical golden torus* is 0, and it thus exhibits no mixing behavior, but it is a strange attractor, possessing fractal structure. In fact, the torus shown in Fig. 7.5 is such a critical golden torus. So-called critical orbits, which exist on such critical attractors, occurring at the point marking the emergence of chaos, possess a variety of interesting features. In Sect. 9.3, we will study these in several contexts.

7.2.3 The Standard Map ($J = 1$)

If we set $\gamma = 0$, we have $J = 1$ and $\tau = T$. In this case, defining $y_i \equiv Tz_i$ in (7.4), we obtain

$$\begin{pmatrix} x_{i+1} \\ y_{i+1} \end{pmatrix} = F(X_i) = \begin{pmatrix} x_i + y_{i+1} \\ y_i + Tkg(x_i) \end{pmatrix} . \tag{7.14}$$

This is a conservative map, for which $J = 1$. The special feature of this map is that it possesses time-reversal symmetry; with $\gamma = 0$, (7.2) is unchanged under the simultaneous transformations $t \to -t$, $x \to x$ and $z \to -z$. Correspondingly, the inverse of the map (7.14),

$$\begin{pmatrix} x_i \\ y_i \end{pmatrix} = F^{-1}(X_{i+1}) = \begin{pmatrix} x_{i+1} - y_{i+1} \\ y_{i+1} - Tkg(x_i) \end{pmatrix} , \tag{7.15}$$

can be obtained directly by applying time reversal to X: $X = (x, y) \to \tilde{X} = (x, -y)$. This time-reversal symmetry plays an important role in the study of the statistical structure of critical orbits and chaotic orbits.

If we make the definitions $\theta_i \equiv x_i$ and $J_i \equiv y_i$ and fix the function $g(\theta)$ by

$$Tkg(\theta) = -(K/2\pi)\sin(2\pi\theta) , \tag{7.16}$$

(7.14) becomes

$$\begin{cases} \theta_{i+1} = \theta_i + J_{i+1} \pmod 1 \\ J_{i+1} = J_i - (K/2\pi)\sin(2\pi\theta_i) . \end{cases} \tag{7.17}$$

This so-called *standard map* has been the subject of a great deal of study. Here, θ_i satisfies $0 \le \theta_i < 1$ or $-0.5 \le \theta_i < 0.5$. This map can also be obtained from (7.10) by setting $J = 1$ and defining $J_i \equiv r_i + \Omega$. It is invariant with respect to both the 'shift' $(\theta, J) \to (\theta, J + j)$ (j integer) and the inversion $(\theta, J) \to (-\theta, -J)$. In the following discussion, we refer to the region defined by $(0 \le \theta < 1, j < J < j + 1)$ as the jth cell.

When $K = 0$, we have $J_n = J_0$ and $\theta_n = \theta_0 + nJ_0$, and the phase space (θ, J) is filled with invariant tori for which the rotation number $\rho(X_0) = J_0$ assumes irrational values. In the context of an integrable dynamical system with two degrees of freedom, we can think of these tori as representing the cross-sections of the tori determined by the two action integrals J^α ($\alpha = 1, 2$) and corresponding angular variables $\theta^\alpha = \omega_\alpha t + \theta_0^\alpha$. When the value of the nonlinearity parameter K is increased, as shown in Fig. 7.6, such tori are gradually destroyed, and a phase space consisting of a chaotic sea and islands of tori results. However, for $K \le K_c = 0.971635406\ldots$, there remain tori connecting the two edges of the phase space, corresponding to $\theta = -0.5$ and $\theta = 0.5$. These are the *KAM tori*. Each chaotic region is bounded by two such tori, and it is thus not possible for a chaotic orbit to move between cells. However, for $K > K_c$, all KAM tori are destroyed, and a sea of *widespread chaos*, extending from $J = -\infty$ to $J = +\infty$ appears. Chaotic orbits in such a system diffuse from cell to cell. As depicted in Fig. 7.6b, in such a chaotic sea, a variety of islands of tori exist. As we will discuss in Chap. 10, the mixing and diffusion of chaotic orbits in this case are very similar to the mixing and diffusion undergone by fluid particles in the Bénard convection system considered in Sect. 6.2. In Chap. 10, we will consider a phenomenological model describing this behavior.

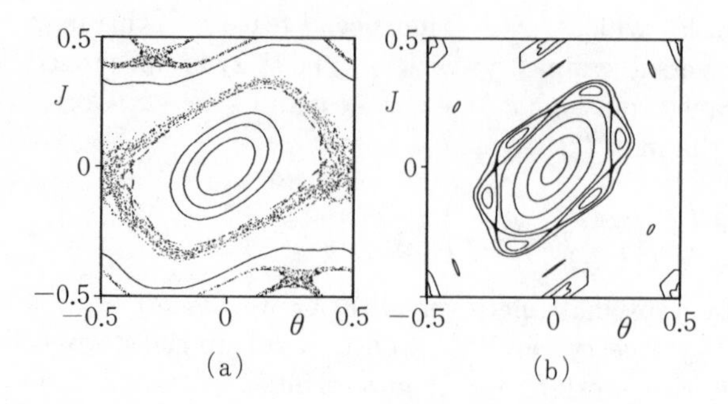

Fig. 7.6. The phase structure of the standard map (7.17). **(a)** For the case $K = 0.97$, two chaotic orbits, two KAM tori, and three tori encircling the elliptic point $(0, 0)$ are shown. **(b)** Islands existing in the chaotic sea of widespread chaos for $K = 1.44$. Here, in addition to the large island centered at the elliptic point $(0, 0)$, smaller, period 2 and period 3 islands can be seen

The KAM torus most resistant to destruction – that which exists until K exceeds the value at which widespread chaos first appears, K_c – is that with rotation number $\rho(x_0)$ equal to the inverse golden ratio, (7.13). This golden torus is also referred to as the *last KAM torus* [Greene (1979)]. The orbits on this golden KAM torus possess a number of interesting properties. For example, their time series display self-similarity and can be expressed in terms of the Fibonacci series (see Chap. 10).

7.2.4 One-Dimensional Maps ($J = 0$)

In the $\gamma \to \infty$ limit, $J \to 0$ and $\tau \to 1/\gamma J$, and with the quantity $\eta \equiv \tau J k = k/\gamma$ considered fixed and finite while taking the $k = \eta\gamma \to \infty$ limit, (7.5) reduces to the *one-dimensional map*

$$x_{i+1} = f(x_i) \equiv x_i + \eta g(x_i) \, . \tag{7.18}$$

This map corresponds to the limit of strong friction for (7.2), in which the inertial term is ignored, leaving

$$\dot{x} = \eta g(x) \sum_{i=-\infty}^{\infty} \delta(t - iT) \, . \tag{7.19}$$

This can be demonstrated directly by integrating this equation from $iT-0$ to $(i+1)T-0$; the result is (7.18). In the limit as $\gamma \to \infty$, the energy dissipation rate (6.5) for this system diverges. Therefore, we take the amplitude of the driving force as $k = \eta\gamma \to \infty$ so that the ratio $\overline{W}_{\rm d}/\overline{W}_{\rm f} \propto k/\gamma = \eta$ of the energy dissipation rate to the energy supply rate can be maintained at a constant value.

If we set $J = -b = 0$ in the Hénon map (7.8), we obtain the *quadratic map*,

$$x_{i+1} = 1 - ax_i^2 \, . \tag{7.20}$$

Also, setting $J = 0$ in the annulus map (7.10) gives the *circle map*,

$$\theta_{i+1} = \theta_i + \Omega - (K/2\pi)\sin(2\pi\theta_i) \pmod 1 \, . \tag{7.21}$$

In this case, $K_\infty = 1$.

The disadvantage of one-dimensional maps is that they lack the two-dimensional fractality displayed in Fig. 7.2. We now turn to a discussion of their advantages.

7.3 Bifurcations of the One-Dimensional Quadratic Map

7.3.1 2^n-Bifurcations and 2^n-Band Bifurcations

Replacing ax in the quadratic map (7.20) by x yields

$$x_{t+1} = f(x_t) = a - x_t^2 \qquad (-1/4 < a \leq 2) \, . \tag{7.22}$$

This equation has been studied in the field of mathematical ecology as a difference model describing the fluctuations in the population of a species from generation to generation. It is also referred to as a *logistic map*. The function $f(x)$ here possesses two fixed points, $x^* = \{-1 + (1+4a)^{1/2}\}/2$ and $s = \{-1 - (1+4a)^{1/2}\}/2$. As shown in Fig. 7.7, these are the intersection points of $f(x)$ and the 45° line (*thick broken line*), satisfying $x = f(x)$. When $a < -1/4$, there is no such solution, and thus we stipulate here that $a > -1/4$. The value $a = -1/4$ corresponds to a saddle-node bifurcation (also known as a *tangent bifurcation*). The eigenvalue corresponding to x^* is $f'(x^*) = -2x^*$, and thus for $a < \hat{a}_1 = 3/4$, $|f'(x^*)| < 1$, and x^* is stable. However, for $a > \hat{a}_1$, $|f'(x^*)| > 1$, and x^* becomes unstable. Contrastingly, the eigenvalue for s is $f'(s) = -2s > 1$, and thus this point is always an unstable saddle point. The functions $f(x)$ and $f^2(x) \equiv f(f(x))$ are illustrated in Fig. 7.7 for the case $a = 1$. The square indicated by the dotted lines in Fig. 7.7a represents a *two-point cycle*. This cycle alternatingly visits the points $x = 0$ and $x = 1$, corresponding to the extrema of $f^2(x)$, as shown in Fig. 7.7b. This two-point cycle first appears when $a = \hat{a}_1$, at which point x^* becomes unstable and $f^2(x)$ and the 45° line first intersect at two points lying on either side of x^*. The appearance of this form represents a period-doubling phenomenon.

In order to understand the mechanism involved in period doubling and to obtain a physical picture of its corresponding 'cascade', let us study Fig. 7.8.

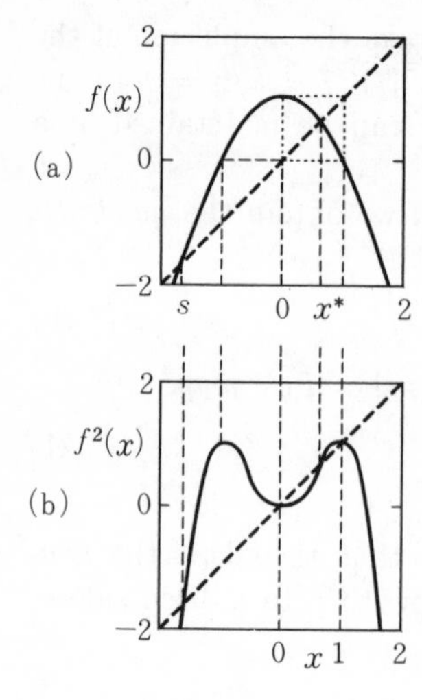

Fig. 7.7. The two-point cycle of $f(x)$ and the arc formed in the interval $0 < x < 1$ defined by $f^2(x)$ (here $a = 1$). The fixed points of $f(x)$, i.e. x^* and s, satisfy $f(x) = x$. The domain of attraction of the attractor is the region satisfying $s < x < f^{-1}(s)$. The slope of $f^2(x)$, $Df^2(x) = f'(f(x)) \cdot f'(x)$, is 0 at $x = 0, \pm 1$

We define $Q \equiv 2^n$ $(n = 1, 2, 3 \ldots)$ and suppose that a assumes a value near $\hat{a}_n$, so that $f(x)$ possesses a period $Q/2$ limit cycle. In this case, the map $\phi(x) \equiv f^{Q/2}(x)$ has $Q/2$ stable fixed points. Note that if n is odd, the central extremal point of ϕ is a maximum. Let us assume this to be the case and consider the fixed point $x_i^* = \phi(x_i^*)$ situated nearest this central peak. The behavior of ϕ at a value of a slightly smaller than $\hat{a}_n$, the value at which this fixed point becomes unstable, is described by Fig. 7.8a. The upper and lower graphs in this figure correspond to $\phi(x)$ and $\phi^2(x)$, respectively. Note that in the case in which n is an even number, and the central extremal point of $\phi(x)$ is a minimum, rotating the graphs for $\phi(x)$ and $\phi^2(x)$ by 180° produces graphs qualitatively the same as those shown in Fig. 7.8a. Note also that all of the discussion given below for arbitrary n holds also in the case $n = 1$, with the fixed point x^* and the map $f(x)$.

If we increase a, the slope of the curve in the neighborhood of x_i^* becomes steeper, as understood from Fig. 7.8b. If a is increased beyond $\hat{a}_n$, x_i^* becomes unstable (i.e., $|\phi'(x_i^*)| > 1$ and $D\phi^2(x_i^*) = \{\phi'(x_i^*)\}^2 > 1$), and an undulation of $\phi^2(x)$ about the 45° line appears, creating the intersection points y_1^* and y_2^* on either side of x_i^*. (For such values of a, x_i^* marks the point at which the curvature of ϕ^2 changes sign.) In this case, the values $x = y_j^*$ $(j = 1, 2)$ represent fixed points of $\phi^2(x)$. When $D\phi^2(y_j^*) = \phi'(y_1^*)\phi'(y_2^*) < 1$, these fixed points are stable. Returning to the original map, $\phi(x)$, for such values of a, we find y_1^* and y_2^* constitute a two-point cycle – if the system starts at one of these points, two applications of $\phi(x)$ bring it back to the same point. This two-point cycle corresponds to the square in Fig. 7.8b. The destabilization of a fixed point of ϕ in response to the increase of a and the subsequent appearance of two new fixed points of ϕ^2 constitutes the mechanism of period doubling.

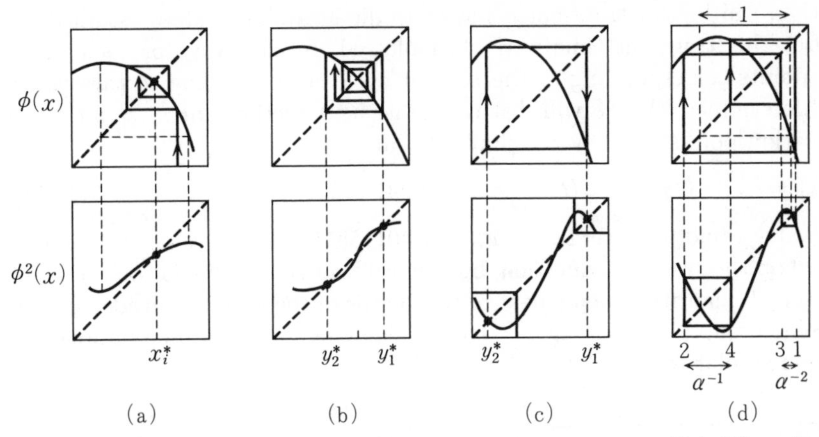

Fig. 7.8. Bifurcation of the central fixed point x_i^* of the map $\phi(x)$. The value of a increases from (**a**) to (**d**), where $a = \hat{a}_n - 0$ for (a), $\hat{a}_n + 0$ for (**b**), $\hat{a}_{n+1} - 0$ for (**c**), and $\hat{a}_{n+2} - 0$ for (**d**). At (**d**), defining the distance between y_1^* and y_2^* here to be 1, the lengths of the intervals corresponding to the two two-point cycles of $\phi^2(x)$ shown here are α^{-1} and α^{-2}

If we increase a further, the distance between y_1^* and y_2^* increases (see Fig. 7.8c), and the maximum of ϕ comes to be situated between these two values. Then, let us consider the boxes containing the local maximum and minimum of ϕ^2. The behavior within either of these boxes is very similar to that within the entire graph for $\phi(x)$ in case (a). Then, as one may expect, further increasing a, upon reaching some value $\hat{a}_{n+1}$, the fixed points of ϕ^2, y_1^* and y_2^*, destabilize (just as the fixed point of $\phi(x)$, x_i^*, destabilizes at $\hat{a}_n$), and in the neighborhood of each fixed point, a two-point cycle of $\phi^2(x)$ appears. This implies the appearance of a four-point cycle of $\phi(x)$. The form assumed by this four-point cycle $\{1, 2, 3, 4\}$ for $a = \hat{a}_{n+2} - 0$ is shown in Fig. 7.8d. Note that for such values of a, the distances between periodic points are characterized by the quantity α (see Fig. 7.11).

As described above, increasing a causes a series of bifurcations in which the destabilization of the fixed point x_i^* of $f^{Q/2}(x)$ results in the appearance of the fixed points y_1^* and y_2^* of $f^Q(x)$ on either side of x_i. These bifurcations correspond to the successive destabilization of period $Q/2 = 2^{n-1}$ ($n = 1, 2, \ldots \infty$) cycles and the appearance of period $Q = 2^n$ cycles of the original map, $f(x)$. Thus the period 2^{n-1} cycle becomes unstable at $a = \hat{a}_n$, and a stable period 2^n cycle appears for $a > \hat{a}_n$. The bifurcation points $\hat{a}_n$ accumulate at some value $\hat{a}_\infty$. The infinite period 2^∞ attractor characterized by $a = \hat{a}_\infty$ is referred to as the *critical* 2^∞ *attractor*. The orbits on this attractor are nonclosed, neutral, and aperiodic. For $a > \hat{a}_\infty$, the aperiodic orbits of band chaos appear.

Figure 7.9 is the bifurcation diagram of the quadratic map. This figure was obtained by fixing a at the desired value, giving x some initial value, and then waiting for transient behavior to die away. The values assumed by x_t after this transient time were then plotted. In this way, for each value of a, points corresponding to the n-point cycle corresponding to that value were determined. As we will discuss in the next section, for $n \gg 1$ the 2^n bifurcation points $\hat{a}_n$ obey the similarity law

$$\hat{a}_n = \hat{a}_\infty - C\delta^{-n} \qquad (\delta = 4.6692016091\cdots) . \tag{7.23}$$

This value rapidly converges to $\hat{a}_\infty = 1.40115518\ldots$. The values $\hat{a}_\infty$ and C depend on the system in question, but the value of the accumulation ratio δ is a universal quantity, independent of the details of the system considered. For the Hénon map (7.1), setting $b = 0.3$ and changing a, we obtain a bifurcation diagram similar to Fig. 7.9.

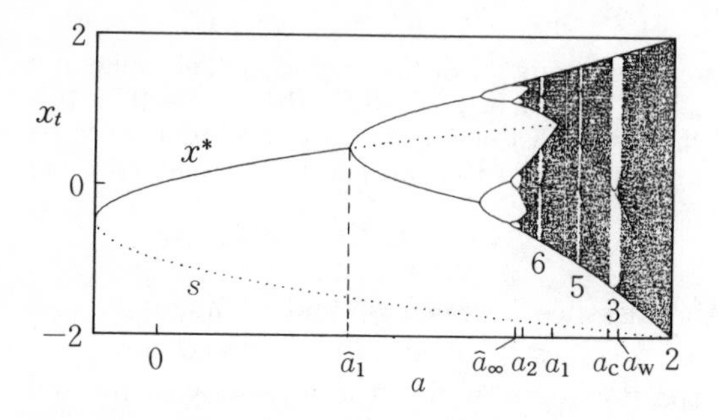

Fig. 7.9. Bifurcation diagram of the quadratic map (7.22). At $a = a_1$, x^* collides with the two-band attractor, while at $a = a_d = 2$, s collides with the one-band attractor. The white bands labeled 3, 5, and 6 are the periodic windows corresponding to these periods. These three bands constitute the most easily discerned periodic windows

For $a > \hat{a}_\infty$, the values $a_1 = 1.5436890\ldots$ and $a_d = 2$ in the case of Fig. 7.9 correspond to a_1 and a_d in Fig. 7.2, describing the Hénon map. In other words, for values of a between $a_2 \approx 1.43036$ and a_1 (except in windows), the attractor exhibits two-band chaos. However, at $a = a_1$ these bands collide with the saddle x^*, and for $a > a_1$ the attractor exists as a single band. Then, at a_d this band collides with the saddle s, and the attractor is destroyed. As discussed in Sect. 7.1, these bifurcation points are characterized by the vertex orbit $y_j = f^j(0)$, emanating from the vertex $x_0 = 0$ of the curved region. At this vertex, we have $f'(0) = 0$ and $\Lambda_{\min} = \lambda_1(0) = -\infty$. However, at $a = a_1$, where $x^* = f^3(0)$, this orbit is absorbed into the saddle x^* in three time steps, and at $a = a_d$, where $s = f^2(0)$, it is absorbed into the saddle s in two time steps. Thus in these cases, the infinite accumulation depicted in Fig. 7.4 does not occur. This is due to the fact that for the $J = 0$ quadratic map the chaotic attractor does not possess two-dimensional fractality. However, as in the case of the Hénon map, as a becomes larger and the attractor collides with x^* and then s, causing $\Lambda_n(x^*)$ and then $\Lambda_n(s)$ to enter the spectrum

$\psi(\Lambda)$, $\Lambda_{\max}$ comes to assume values larger than those assumed before these collisions, and anomalies in the statistical structure result.

As depicted in Fig. 7.9, decreasing a from 2 to $\hat{a}_\infty$, at each value $a = a_n$ $(n = 1, 2, \ldots; a_\infty = \hat{a}_\infty)$ a 2^n-band band-splitting bifurcation occurs. For values of a satisfying $a_n > a > a_{n+1}$ (except in windows), the attractor is composed of 2^n separate bands of period $Q = 2^n$. Furthermore, for $n \gg 1$, we have the similarity law

$$a_n = a_\infty + C'\delta^{-n} \qquad (a_\infty = \hat{a}_\infty)\,, \tag{7.24}$$

where δ is the same constant as that appearing in (7.23). The value of the nth bifurcation point a_n quickly converges to $a_\infty = \hat{a}_\infty$.

Actually, the chaotic region existing for $a > a_\infty$ contains a multitude of small 'windows' in which nonchaotic behavior is displayed. In a period p window, the 2^n-band chaos existing on either side of this window disappears, and in its place, period $q = p \times 2^m$ $(m = 0, 1, 2 \ldots)$ q-point cycles and q-band attractors appear. Figure 7.10 displays a $p = 3$ window spanning the interval $a_c = 1.75 < a < a_w \approx 1.79032$. Each band contained within this window itself contains a myriad of even smaller windows, and the bifurcation diagram corresponding to each such branch is similar to the entire diagram shown in Fig. 7.9. The self-similar hierarchical structure of the bifurcation diagrams thus becomes clear. As described by Fig. 7.10, as a consequence of the tangency bifurcation of $f^3(x)$, the three-point cycles $x_i^* = f^3(x_i^*)$ and $s_i = f^3(s_i)$ $(i = 1, 2, 3)$ appear, and, for this reason, when a is increased beyond a_c, band chaos disappears. Then, as a is increased further, owing to the period doubling experienced by $f^3(x)$, from the three-point cycle x_i^*, period 3×2^m (m increasing with a) cycles are successively generated. This continues to a value $a \approx 1.78$, corresponding to $\hat{a}_\infty$ in Fig. 7.9, after which point a band attractor structure appears. When this point is reached, an inverse cascade resulting from band merging causes period 3×2^m (m decreasing with a) band attractors to appear successively. Then, as a reaches the value a_w, the last of these attractors, a period 3, three-band attractor, collides with the three-point saddle s_i, merging discontinuously into a one-band attractor. Such behavior is referred to as a *band-merging crisis.*

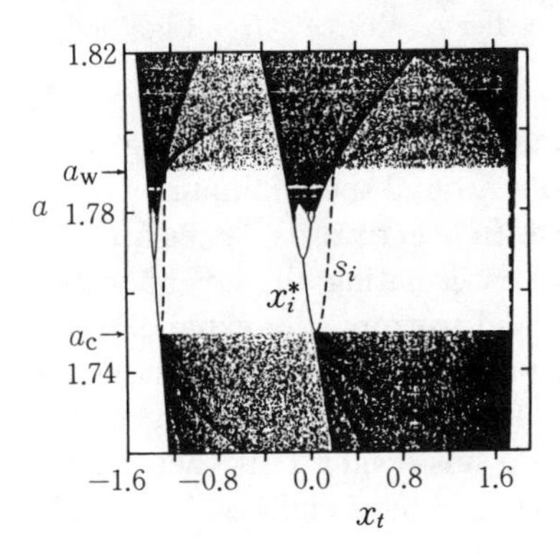

Fig. 7.10. Magnification of the period 3 window that exists in the range $a_c < a < a_w$. At a_c, x_i^* and s_i are created, and at a_w, the attractor collides with s_i [Grebogi et al. (1983)]

7.3.2 The Self-Similarity and Renormalization Transformation of 2^n-Bifurcations

Figure 7.11 is a simplification of the bifurcation diagram given in Fig. 7.9, characterizing the 2^n-bifurcation cascade resulting from period doubling. Here, x_i^* represents the central fixed point of the map $\phi(x) \equiv f^{Q/2}(x)$ ($Q = 2^n \gg 1$). This point becomes unstable at $a = \hat{a}_n$, and from it is generated the two-point cycle $\{y_1^*, y_2^*\}$. We define the distance between y_1^* and y_2^* at $a = \hat{a}_{n+1} - 0$ to be 1. Now, at $a = \hat{a}_{n+1}$, these two cycles themselves become unstable, each producing a new two-point cycle of the map $\phi^2(x)$. Then we define the distance between these two-point cycles at $a = \hat{a}_{n+2} - 0$ as α^{-1} and α^{-2}. The situation at such a value of a is represented in Fig. 7.8d. There, the interval $[2, 4]$ containing the maximum is mapped by $\phi(x)$ to the interval $[3, 1]$, and in the process, the size of the interval is shortened by a factor of α^{-1}. The distance between the points of the two-point cycle $\{y_1^*, y_2^*\}$ at this value of a (now an *unstable* two-point cycle) changes little from that at $a = \hat{a}_{n+1} - 0$, and we can approximate it as 1. Repeating this procedure, and at each step applying a similar approximation, we obtain the universal constant representing the size ratio α^{-1},

$$\alpha = \alpha_{\mathrm{PD}} \equiv 2.502907875 \cdots \tag{7.25}$$

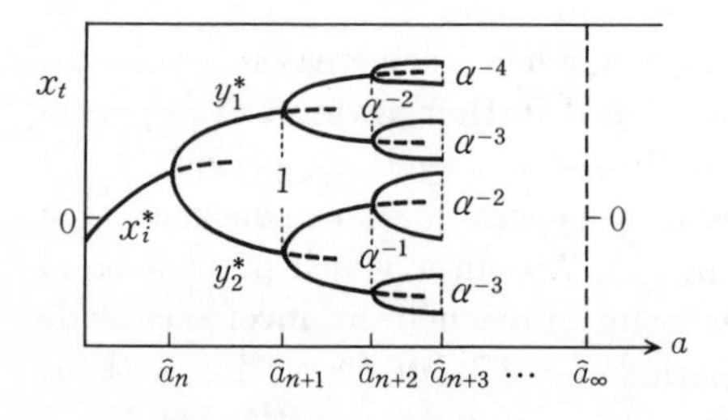

Fig. 7.11. The self-similarity of the 2^n-bifurcation cascade. The distance along the a-axis between bifurcations decreases as a increases toward $\hat{a}_\infty$, with the ratio of consecutive distances given by the value δ^{-1}. Also, the distance along the x_t-axis between a pair of 'sibling' branches in this diagram is smaller than that between their 'parent' branch and sibling by a factor of either α^{-1} or α^{-2} (see Fig. 7.8d)

In order to make explicit the kind of similarity represented by Fig. 7.11, let us first consider the similarity of the regions around the minimum of $\phi^2(x)$ and the maximum of $\phi(x)$ shown in Fig. 7.8d. In other words, let us attempt to obtain a graph identical to that for $\phi(x)$ by isolating the neighborhood of the minimum of $\phi^2(x)$, rotating by 180°, and isotropically expanding the resulting graph. For this purpose, we have plotted the graphs of these two neighborhoods in Fig. 7.12, where we have used a normalization such that the maximum of $\phi(x)$ is located at $x = 0$ and has height 1. It then follows that the minimum of $\phi^2(x)$ is located at $x = 0$ and has height $\alpha^{-1} \equiv -\phi(1)$.

Then, let us introduce the transformation $\mathcal{T}$ mapping this minimum to the maximum of $\phi(x)$:

$$\mathcal{T}\phi(x) \equiv -\alpha\phi^2(-x/\alpha) = -\alpha\phi(\phi(-x/\alpha))\,. \tag{7.26}$$

Here,

$$\mathcal{T}\phi(0) = \phi(0) = 1, \qquad \alpha = -1/\phi(1)\,. \tag{7.27}$$

We can think of $\mathcal{T}$ as the composition $\mathcal{T}_2\mathcal{T}_1$, where $\mathcal{T}_1$ is the operator producing ϕ^2 from ϕ, and $\mathcal{T}_2$ is the operator which rotates the coordinate system by 180° about the origin and isotropically expands it by the factor α. The transformation $\mathcal{T}$ is called a *renormalization transformation*. The curve obtained when this transformation is applied to $\phi(x)$ should itself be similar to $\phi(x)$. In fact it is easily imagined that for an appropriate choice of a (which we later find is in fact $\hat{a}_\infty$), $\mathcal{T}\phi(x)$ becomes identical to $\phi(x)$ in the limit as $n \to \infty$. Thus we may expect that there exists a fixed point $g(x)$ of $\mathcal{T}$ represented by such a limit. In fact, such a map $g(x)$ does exist, and it satisfies

$$g(x) = -\alpha g(g(-x/\alpha)) \tag{7.28}$$

$$g(0) = 1, \quad \alpha = -1/g(1)\,. \tag{7.29}$$

The relation between $g(x)$ and the quadratic map with $a = \hat{a}_\infty$ is made clear below.

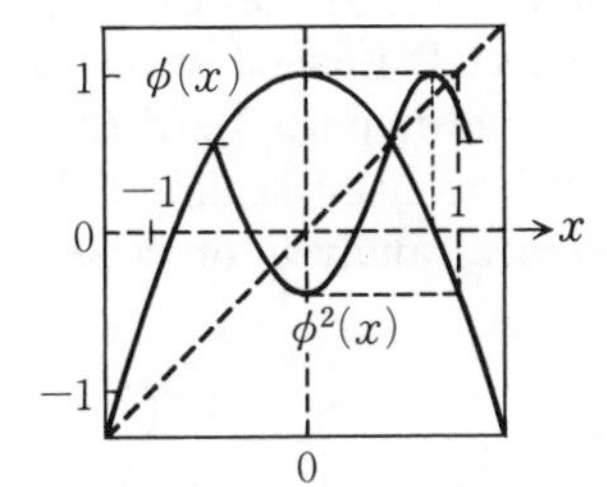

Fig. 7.12. The similarity of the maximum of $\phi(x)$ and the minimum of $\phi^2(x)$. Here $\phi(0) = 1$, and $\phi^2(0) = \phi(1) = -\alpha^{-1}$

Note that $g(x)$ in (7.28) is a function of x^2. With this in mind, using the polynomial approximation

$$g(x) = 1 + g_1x^2 + g_2x^4 + \cdots \tag{7.30}$$

and Newton's method, Feigenbaum (1979) numerically obtained the values

$$g_1 = 1.52763\cdots, \qquad g_2 = 0.104815\cdots \tag{7.31}$$

(and higher order coefficients) together with the value for α given in (7.25).

In order to construct a geometric description of the renormalization transformation (7.26) and to derive the similarity law (7.23), let us consider the function space V consisting of all one-dimensional maps containing either exactly one maximum or one minimum. Each point in this space represents one such map. Thus, for example, the family of maps corresponding to the quadratic map (7.22) defined by some continuous set of values of a is represented in V by a continuous curve C. The transformation $\mathcal{T}$ transforms a given point in V to another, and $g(x)$ represents a fixed point of $\mathcal{T}$ in this space.

To understand the flow under $\mathcal{T}$ in the neighborhood of $g(x)$, let us consider the infinitesimal shift $e(x)$ from $g(x)$ and the linearization of its transformation under $\mathcal{T}$:

$$\mathcal{T}\{g(x) + e(x)\} - \mathcal{T}g(x) \approx Le(x) . \tag{7.32}$$

Defining the eigenvectors of the linear operator L as $E_l(x)$ and the corresponding eigenvalues as δ_l, we can write

$$LE_l(x) = \delta_l E_l(x) \qquad (l = 1, 2, \cdots) \tag{7.33}$$

$$e(x) = \sum_l t_l E_l(x) . \tag{7.34}$$

We refer to the amplitude t_l corresponding to the eigenvector $E_l(x)$ as a *scaling field*. If the eigenvalue δ_l satisfies $|\delta_l| > 1$, the corresponding scaling field is termed *relevant*, while if $|\delta_l| < 1$, it is termed *irrelevant*. Relevant scaling fields are unstable, and their values grow exponentially under the renormalization transformation. For this reason, they determine the asymptotic nature of the flow in V under $\mathcal{T}$ corresponding to maps situated in the neighborhood of the fixed point g, and also for maps in the neighborhood of the quadratic map (7.22) at the 2^n-bifurcation accumulation point $a = \hat{a}_\infty$. Irrelevant scaling fields, on the other hand, are stable, and their effect is washed away by the renormalization transformation. In 1979, Feigenbaum hypothesized that in the case of 2^n-bifurcations there exists just one such relevant scaling field δ_1 and confirmed this hypothesis through the numerical analysis of (7.28), obtaining the value

$$\delta = \delta_1 \equiv 4.6692016091 \cdots \tag{7.35}$$

The quantities δ and α are known as the *Feigenbaum universal constants.*

The fixed point g is a saddle possessing a single unstable eigenvector E_1. Figure 7.13 depicts the geometric structure of the function space V in the neighborhood of this point. Here, the one-dimensional curve $W^{\mathrm{u}}(g)$ is the unstable invariant manifold tangent to E_1 at g, and Σ represents the codimension 1 stable invariant manifold cutting through E_1 transversely at g. The curve C represents the one-parameter family of maps expressed by (7.22), where the position of a given map on this curve is determined by its value of a. This curve crosses the codimension 1 surface Σ at some point P. This point corresponds to the value $a = \hat{a}_\infty$ and represents the map $f_\infty(x) \equiv \hat{a}_\infty - x^2$. This map is not the fixed point function $g(x)$, but it lies on the stable manifold Σ of $g(x)$, and therefore under the repeated action of the transformation $\mathcal{T}$, it converges to g, as indicated by the arrow in the figure. However, if a differs even slightly from $\hat{a}_\infty$, under such mapping, it first approaches $W^{\mathrm{u}}(g)$ and then moves away from g, following $W^{\mathrm{u}}(g)$ in one direction or the other. In any case, all one-dimensional maps in the neighborhood of g will asymptotically approach $W^{\mathrm{u}}(g)$ under $\mathcal{T}$ and come to reside on the closure of this manifold. In this sense, the one-parameter family of functions corresponding to $W^{\mathrm{u}}(g)$ represents a set of universal maps. [Here, "universal" simply implies that the map in question has zero component along an 'irrelevant direction', or, in other words, that the map is on $W^{\mathrm{u}}(g)$.]

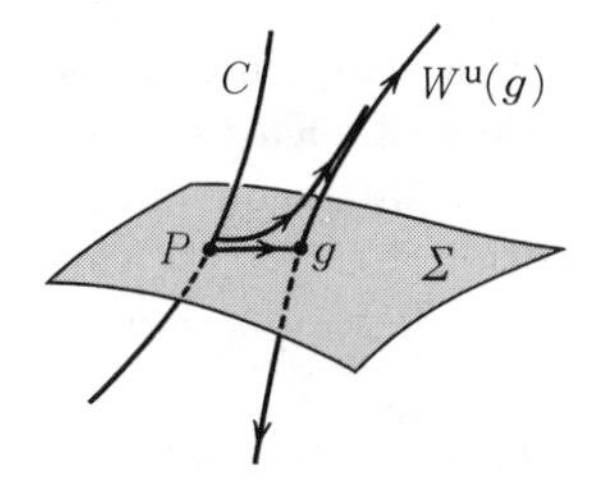

Fig. 7.13. The structure near the fixed point g of the renormalization transformation $\mathcal{T}$. The arrows denote the direction of flow under $\mathcal{T}$

The similarity law (7.23) for $n \gg 1$ is equivalent to the expression

$$\hat{a}_{n+1} = \hat{a}_\infty + \delta^{-1}(\hat{a}_n - \hat{a}_\infty) , \tag{7.36}$$

relating $\hat{a}_n$ and $\hat{a}_{n+1}$. Also, the similarity with regard to the 2^n-bifurcations expressed by Fig. 7.11 in fact takes the form of a relation between the cycle existing at $a = \hat{a}_{n+1} - 0$ and $a = \hat{a}_{n+2} - 0$. We can apply the renormalization transformation (7.26) to the investigation of such relationships between neighboring bifurcation points. This results from the similarity exhibited by the 'box' containing the minimum of $\phi^2(x)$ shown in Fig. 7.8c and that containing the maximum of $\phi(x)$ in Fig. 7.8a.

The similarity described above has the meaning

$$\mathcal{T}\phi(x, \hat{a}_{n+1}) = -\alpha\phi^2(-x/\alpha, \hat{a}_{n+1}) \sim \phi(x, \hat{a}_n) , \tag{7.37a}$$

where we have expressed the a dependence of the map explicitly for clarity. (For direct comparison with Fig. 7.8, $\hat{a}_n$ and $\hat{a}_{n+1}$ here should be replaced by $\hat{a}_n - 0$ and $\hat{a}_{n+1} - 0$.) In fact, we can construct families of maps for which this similarity holds exactly. Consider an arbitrary map $\psi_n(x)$ situated at the nth period-doubling bifurcation point. Then, we define members of the family of maps $g_i(x)$ [$i = 1, 2 \ldots$; $g_n(x) \equiv \psi_n(x)$] to satisfy

$$\mathcal{T} g_{i+1}(x) = -\alpha g_{i+1}^2(-x/\alpha) = g_i(x). \tag{7.37b}$$

There are several points to note here. First, for any given map $\psi_n(x)$ of this type (i.e., a one-dimensional map at the nth period-doubling point) with finite n, there exists such a family of maps to which it belongs. Then, for any such family, we have $\lim_{i\to\infty} g_i(x) = g(x)$. In this limit, the irrelevant fields quickly become washed away, and $g_i(x)$ approaches g along $W^{\mathrm{u}}(g)$.

Now, in the space V, let us consider the set of all maps $\psi_n(x)$ situated at the nth period-doubling bifurcation point. This set forms a surface (of codimension 1) in V. Let us call this set Σ_n. Note that the map (7.22) with $a = \hat{a}_n$ is on Σ_n. Then since each map for any finite n belongs to a family of the type described by (7.37), we see that, through these families, for any such n, $\mathcal{T}$ maps Σ_{n+1} to Σ_n. Now, while these families $g_i(x)$ have been specially defined in order to satisfy (7.37b), we wish to consider more 'natural' families of maps and, in particular, one-parameter families that can be defined in terms of a single mapping rule, such as (7.22). Such a one-parameter family of maps defines a curve in V intersecting each of the surfaces Σ_i at a single point representing the element of that family situated at the ith period-doubling bifurcation point. Clearly, because of the variety of such families, the values assumed by the parameter that characterizes a particular family at each of these surfaces depends on the family in question, but despite this dependence, in the $i \to \infty$ limit, these parameter values display a type of universal behavior. Let us consider this point.

In the neighborhood of g, such a surface Σ_n is approximately parallel to the stable manifold Σ. The intersection of Σ_n with $W^{\mathrm{u}}(g)$ represents the universal map corresponding to the value n. Now, let us designate the distance between Σ_n and Σ along $W^{\mathrm{u}}(g)$ by d_n. Then, consider a particular one-parameter family of maps in V of the type discussed above. Suppose that the element of this family on Σ_n is characterized by the parameter value $\hat{\mu}_n$. Then, writing the parameter value corresponding to the element of this family on Σ as $\hat{\mu}_\infty$ and the distance between Σ_n and Σ along this family as l_n, we can write l_n as an expansion in $\hat{\mu}_\infty - \hat{\mu}_n$. Thus, if n is sufficiently large, we should be able to approximate this by the lowest-order term, and write $l_n \propto \hat{\mu}_\infty - \hat{\mu}_n$. Hence, since one application of $\mathcal{T}$ to a map on $W^{\mathrm{u}}(g)$ causes the distance from this map to g to be increased by a multiplicative factor of δ, we have $d_n/d_{n+1} = \delta$. Therefore, if the map corresponding to $\hat{\mu}_\infty$ is sufficiently close to g (so that we can treat the surfaces as parallel here and assume that $l_n \approx d_n$), for large n it should be the case that $(\hat{\mu}_\infty - \hat{\mu}_n)/(\hat{\mu}_\infty - \hat{\mu}_{n+1}) \approx \delta$. In the neighborhood of g, then we have

$$\delta = \lim_{n\to\infty} \frac{\hat{\mu}_n - \hat{\mu}_{n-1}}{\hat{\mu}_{n+1} - \hat{\mu}_n} . \tag{7.38}$$

Both this similarity relation and that represented by Fig. 7.11 have been verified experimentally in a variety of systems.

For the purpose of simplicity, we have to this point concentrated on the similarity displayed at the bifurcation points $\mu = \hat{\mu}_n$. However, this discussion can be generalized to the case of arbitrary μ. In fact, replacing $\hat{\mu}_n$ ($n = 1, 2, \ldots$) in the above discussion by $\mu = \hat{\mu}_\infty - (\hat{\mu}_\infty - \hat{\mu}_n)/\delta^k$ ($n = 1, 2, \ldots$), for some arbitrary k satisfying $0 < k < 1$, the above discussion holds unchanged.

7.3.3 The Similarity of 2^n-Band Bifurcations

As seen in the bifurcation diagram given in Fig. 7.9, the 2^n-band (splitting) bifurcations and cyclic 2^n-bifurcations lie symmetrically on either side of $a =$

$\hat{a}_\infty = a_\infty$. For this reason, a similarity relation akin to (7.24), obtained for the case of cyclic 2^n-bifurcations, also exists for 2^n-band bifurcations. In addition, in this case there is a similarity relation concerning the internal structure of the bands. For values of a satisfying $a_n > a > a_{n+1}$, the attractor consists of period $Q = 2^n$ bands (excluding values of a corresponding to 'windows'). In order to investigate the similarity exhibited by the corresponding cascade, it is useful to consider the continuous decrease of a toward a_∞. As this is done, the band merging bifurcations discussed above become band-splitting bifurcations. Figure 7.14 depicts such a process. This figure represents the graph near the fixed point x_i^* of the map $\phi(x) = f^{Q/2}(x)$. The behavior displayed by this map in this region is characterized by the vertex orbit $y_j = \phi^j(0)$. For the case $a > a_n$, shown in Fig. 7.14c, a chaotic orbit x_t passes through every open sub-interval of the interval $[y_2, y_1]$ as $t \to \infty$, implying the existence of a single band. When $a = a_n$ (Fig. 7.14b), $y_3 = x_i^*$, and the orbit emanating from the vertex terminates at this point. When $a < a_n$ (Fig. 7.14a), the gap $[y_4, y_3]$ forms, and the attractor is split into two bands, $[y_2, y_4]$ and $[y_3, y_1]$. In particular, when $a = a_{n+1} + 0$, defining the distance between the centers of these two bands to be 1, their respective widths become α^{-1} and α^{-2}, just as in the case of the component intervals of the four-point cycle displayed in Fig. 7.8d.

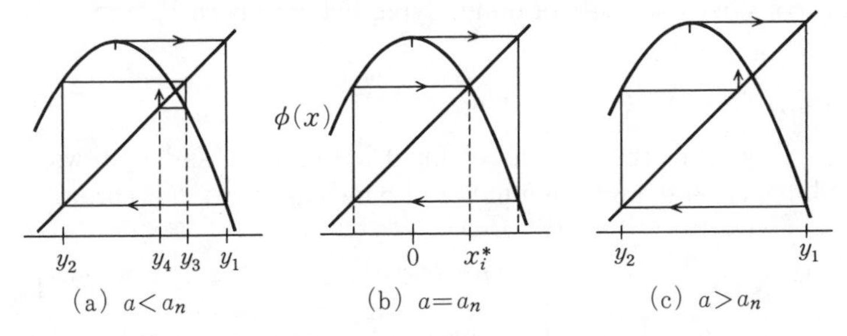

Fig. 7.14. The graph of $\phi(x) \equiv f^{Q/2}(x)$ on either side of a_n and the vertex orbit $y_j = \phi^j(0)$

As a decreases, the type of band splitting described here occurs at each a_n. For values of a satisfying $a_n > a > a_{n+1}$, there exist 2^n bands. Then, for some such value of a, let us designate the smallest width among this set of bands as l_w and the shortest distance between the centers of two neighboring bands as l_d ($> l_\mathrm{w}$). The dependence of these quantities on a is shown in Fig. 7.15 (note that this is a log–log scale). We see that if a decreases, l_w and l_d decrease discontinuously at each bifurcation point a_n. The graphs of l_w and l_d consist of disconnected pieces with the following characteristics:

(α) The length of each piece is $\ln \delta$, independent of n.
(β) The shapes of all pieces are similar.

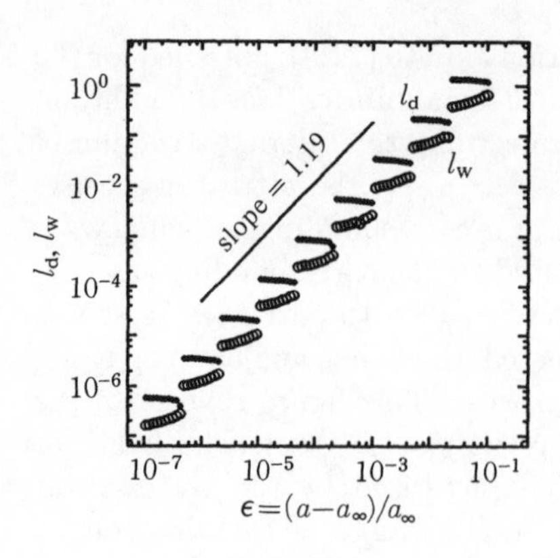

Fig. 7.15. The similarity displayed by the ϵ dependence of l_d and l_w seen in 2^n-band bifurcations. Here a assumes values ranging from $a_1 = 1.54368\ldots$ ($\epsilon \approx 1.02 \times 10^{-1}$) to $a_{10} = 1.40115\ldots$ ($\epsilon \approx 0.93 \times 10^{-7}$) [Tominaga and Mori (1991)]

The similarity displayed in this figure provides a clear description of the similarity characterizing 2^n-band bifurcations.

The quantity

$$k \equiv \ln\{(a - a_\infty)/(a_n - a_\infty)\}/\ln \delta^{-1} \tag{7.39}$$

can be used to designate any point on the segment corresponding to values of a satisfying $a_n \geq a > a_{n+1}$. Note $0 \leq k < 1$. For some such arbitrary value of k, let us now consider the series of points lying between each a_n and a_{n+1} defined by

$$\epsilon_k \equiv (a - a_\infty)/a_\infty = \delta^{-k}(a_n - a_\infty)/a_\infty \propto \delta^{-n} \tag{7.40}$$

and consider moving from the value so defined between a_1 and a_2 toward a_∞. As this is done, $\epsilon_k \to 0$, and the number of bands Q grows in accordance with

$$Q = 2^n = \delta^{\kappa n} \propto \epsilon_k^{-\kappa}\,, \tag{7.41}$$

where $\kappa \equiv \ln 2/\ln \delta = 0.44980\ldots$ When 2^n-band splitting takes place, each band splits into two bands which are smaller than their parent band by factors of α^{-1} and α^{-2}. Thus we can write

$$l_d, l_w \propto \alpha^{-2n} \propto \epsilon_k^{\nu}\,, \tag{7.42}$$

where $\nu \equiv \ln \alpha^2/\ln \delta = 1.190732\ldots$. This value is in agreement with the slope 1.19 of the line drawn in Fig. 7.15, obtained through numerical computation. Similarly, the largest band width is $\alpha^n l_w$, and the scaling exponent of this quantity is $\nu' = \nu/2$. We note that these similarity relations also hold for the Hénon map (7.1), and they provide a framework within which to understand the similarity concerning the two-dimensional band chaos appearing in the neighborhood of the emergence point of chaos for that system.

It is natural to expect that behind the similarities discussed here there exists an 'invariant set similarity'. This connection can be understood by considering the Šarkovskii order [Šarkovskii (1964)],

$$
\begin{aligned}
&3 \vdash 5 \vdash 7 \vdash \cdots \vdash 2 \times 3 \vdash 2 \times 5 \vdash 2 \times 7 \vdash \cdots \\
&\cdots \vdash 2^n \times 3 \vdash 2^n \times 5 \vdash 2^n \times 7 \vdash \cdots (n \to \infty) \cdots \\
&\cdots \vdash 2^n \vdash \cdots \vdash 2^3 \vdash 2^2 \vdash 2 \vdash 1 ,
\end{aligned}
\tag{7.43}
$$

corresponding to periodic orbits of unimodal maps (maps that possess unique maxima). Here, $i \vdash j$ expresses the meaning that if a period i orbit exists, then a period j orbit exists in the same basin. The bottom line in (7.43) constitutes the 2^n series corresponding to the 2^n-bifurcations that exist for $a < a_\infty$. The series on the left side of the top line is that of all odd numbers greater than 3. This is followed by series consisting of this series multiplied by 2, 2^2, 2^3, ... The characteristic feature of the $a > a_\infty$ chaotic region is the appearance of these $2^n\times$(odd series)-periodic orbits. For the quadratic map (7.22) these orbits appear in the reverse order of (7.43), resulting from tangent bifurcations, as a is increased from a_∞. For $a > a_c$, as shown in Fig. 7.10, a period 3 orbit exists, and for this reason, as seen from (7.43), periodic orbits of all (integer) periods exist. In the range $a_{n+1} < a < a_n$, the chaotic attractor consists of 2^n bands and includes $2^n\times$(odd series)-periodic orbits. The "(odd series)" here corresponds to odd-period orbits contained within each band. These orbits determine the internal structure of the bands. This series of odd-period orbits appears, as depicted in the bifurcation diagrams shown in Figs. 7.9 and 7.10, as windows in each band. Decreasing a from a_n to a_{n+1}, the orbits corresponding to the odd series disappear in the order $3, 5, 6 \ldots$, all vanishing as a approaches a_{n+1}. Next, 2^n-periodic orbits become separated from the band, and band splitting occurs. Then for $a < a_{n+1}$, a similar process takes place for the new 2^{n+1} bands, and this is repeated again and again as $a \to a_\infty$. Thus, fixing the value of k in (7.39), each band possesses the same odd-series periodic orbits, corresponding to the series of values of a given by (7.40). This guarantees that all 'pieces' appearing in Fig. 7.15 possess similar forms.

From the above discussion, we see that there are two types of similarity suggested by Fig. 7.15. First, there is the *external similarity* – that corresponding to the spatial distribution of the many small bands which appear. In addition, there is the *internal symmetry* – that corresponding to the internal band structure. As can be understood from (7.40) and (7.41), the external symmetry gives rise to a self-similarity in the distribution of orbital points in the critical 2^∞ attractor of the 2^n-bifurcations in the $\epsilon_k \to 0$ limit. This point is treated in Sects. 9.3.1 and 9.3.2. The internal similarity characterizes each 'piece' in Fig. 7.15 and is described by the Šarkovskii odd series. However, in the quadratic map (7.22), because J has been set to 0, multiplex string-like fractal structure depicted in Figs. 7.3 and 7.4 does not exist here. This two-dimensional fractality is an important feature of chaotic behavior, and, as will be seen in Sects. 9.3.3 and 9.3.4, the similarity that characterizes intra-band chaos in more general settings than that considered in this section includes such two-dimensional fractality. This can be understood easily from the form of the spectrum $\psi(\Lambda)$.

7.4 Bifurcations of the One-Dimensional Circle Map

Let us consider the map

$$\theta_{t+1} = f(\theta_t) = \theta_t + \Omega - (K/2\pi)\sin(2\pi\theta_t) \tag{7.44}$$

with $0 < \Omega < 1$, $K > 0$, and $0 \le \theta_t < 1$ (mod 1). This is the angle $2\pi\theta$ circle map (7.21), defined on the (circular) cross-section of a torus. Figure 7.16 gives the graphs of this map for the cases $K = 0.8$ and $K = 1.3$. When $K < 1$, $f(\theta)$ represents a one-to-one transformation, and its inverse map exists, while for $K > 1$ the inverse map does not exist. In addition, because in the latter case both a maximum and a minimum exist [i.e., $f'(\theta) = 0$ is satisfied for two values of θ], there exist two attractors. For $K < 1$ and any $\theta_1 < \theta_2$, we have $f(\theta_1) < f(\theta_2)$, but when $K > 1$ this is not true for all such θ_1, θ_2. The reversal of order of a series of points under f is due to the *folding* which takes place at the extremal points. This behavior suggests the existence of chaos, and thus we refer to the $K > 1$ region as the 'chaotic region'. Actually, the $K > 1$ region consists of intervals in which the attractor exhibits chaotic behavior, punctuated by windows in which it is characterized by cyclic motion.

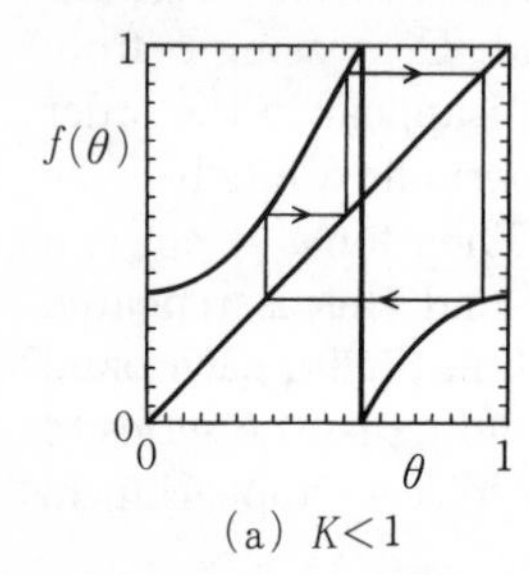

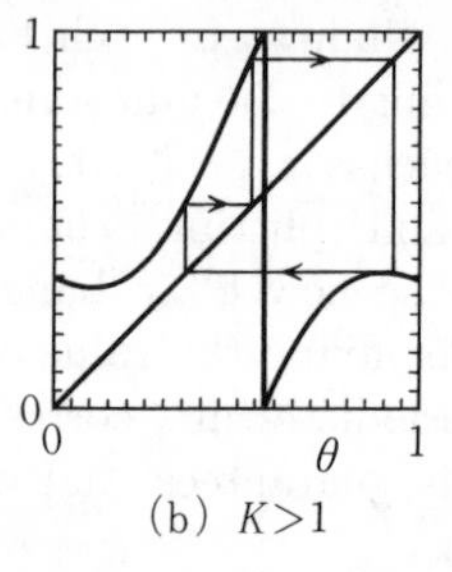

Fig. 7.16. The graph of the circle map (7.44) and a three-point cycle. Here $\Omega = 0.34$, while $K = 0.8$ and 1.3 in **(a)** and **(b)**, respectively

7.4.1 Phase-Locked Band Chaos

The rotation number of an arbitrary orbit defined on a circle is given [from (7.11)] by

$$\rho(\theta_0) \equiv \lim_{n\to\infty} \rho_n(\theta_0), \qquad \rho_n(\theta_0) \equiv (\theta_n - \theta_0)/n\,. \tag{7.45}$$

For (7.44), in the case $K = 0$ we have $\rho(\theta_0) = \Omega$. When $K > 0$, $\rho(\theta_0)$ becomes locked to some fixed rational number p/q, remaining independent of Ω over a finite interval. This phenomenon is known as *phase locking* or *phase synchronization.* In the case of a q-point cycle defined on a circle, or in the case of q-band chaos, when $\rho(\theta_0) = p/q$, the orbit visits these points or bands in such a way that there exist $p - 1$ such points or bands between successive θ_t, and it returns to the original periodic point or band after p revolutions. This kind of phase locking occurs in the inverted triangular regions shown in Fig. 7.17, which intersect the Ω-axis at points $(p/q, 0)$. Such regions are referred to as p/q *Arnold tongues* and are designated by $A_{p/q}$. In the case of forced pendulums, such an Arnold tongue represents the situation in which the ratio of the average angular velocity $\langle\dot{\theta}\rangle$ and the frequency of the driving force ω becomes locked to some p/q over a finite range of values of the amplitude of the driving force, Ω (see Sect. 9.2.2). The size of this range of values $\Delta\Omega_{p/q}(K)$ is a monotonically increasing function of K. Emanating from each rational number along the Ω-axis there exists such a triangular region. For $K \leq 1$, these regions do not overlap, but for $K > 1$ overlap occurs. The sum over all rational numbers (in the interval $[0, 1]$) of the corresponding interval lengths $\Delta\Omega_{p/q}(k)$ is exactly 1 at $K = 1$, and thus for this value of K the Lebesgue measure (the sum over the lengths of all intervals in this set) of values of Ω yielding tori for which $\rho(\theta_0)$ is irrational is 0. However, for $K > 1$ (as discussed in the following subsection), the Lebesgue measure of this set is nonzero, owing to the existence of chaotic regions (U *regions*) in which phase locking is not exhibited.

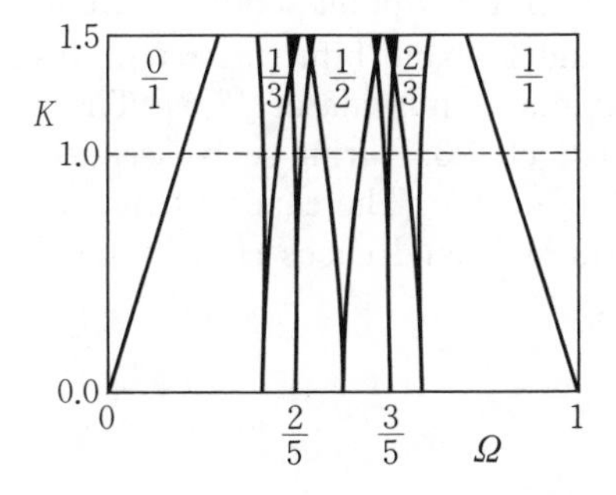

Fig. 7.17. Examples of Arnold tongues $A_{p/q}$. Here, the tongues corresponding to $p/q = 0/1, 1/3, 2/5, 1/2, 3/5, 2/3$, and $1/1$ are shown. For $K > 1$ there exists overlap among the tongues, as can be seen

The form of an Arnold tongue $A_{p/q}$ is exhibited in Fig. 7.18. We designate the left and right boundaries of this region by $A^{\mp}_{p/q}$ and the upper left and right boundaries by $C^{\mp}_{p/q}$. Then, $C^{\mp}_{p/q}$ exist in the $K > 1$ region. Also, crossing $A^{\mp}_{p/q}$ from the external to the internal region, q-point cycles θ_i^* and q-point saddles s_i with $\rho(\theta_0) = p/q$ appear as the result of tangent bifurcations of the map $f^q(\theta)$. The graph of $f^q(\theta)$ on $A^{+}_{p/q}$ shown in the figure is that between the two tangency points, where $f^q(\theta) = \theta$. This is indicated in the lower small graph in the figure. However, an attractor existing within $A_{p/q}$ which arises from bifurcation out of a q-point cycle θ_i^* is destroyed at $C^{\pm}_{p/q}$, as a result of collision with the q-point saddle s_i. As indicated in the upper small graph in the figure, on the *crisis curve* $C^{+}_{p/q}$, the vertex orbit passing through the value of θ situated at the local minimum of $f^q(\theta)$ is absorbed by the fixed point s_i in the upper right corner of the graph.

In the shaded region $A_{p/q}$ lying between $A^{\pm}_{p/q}$ and $C^{\pm}_{p/q}$, all orbits on the attractor are phase-locked, with $\rho(\theta_0) = p/q$. Just inside $A^{\pm}_{p/q}$, this attractor

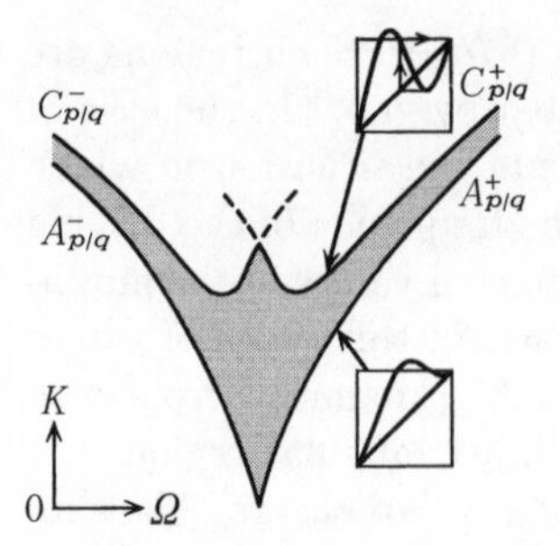

Fig. 7.18. The form of $A_{p/q}$ (*shaded region*). In this region, the rotation number $\rho(\theta_0)$ is locked to the value p/q

consists of q-point cycles, but moving further inside this curve, as shown in Fig. 7.19, as a consequence of period doubling, a cascade of $2^n \times q$-point cycles ($n = 1, 2, \ldots, \infty$) occurs, and above the corresponding accumulation line $2^\infty q$, band chaos is realized. In this way, the shaded region inside the crisis curve $C^{\pm}_{p/q}$ is characterized by $2^n \times q$ bands and a 2^n-band bifurcation cascade. In particular, just inside $C^{\pm}_{p/q}$, q-band chaos is displayed. This is then destroyed at $C^{\pm}_{p/q}$, where these bands collide with the q-point saddle s_i. Here, the cascades resulting from 2^n-bifurcations and 2^n-band bifurcations are locally brought about by either the local maxima or minima of $f^q(\theta)$. These cascades obey the similarity laws discussed in Sect. 7.3. During such cascades, the rotation number remains fixed at some value p/q. The chaotic behavior displayed within $A_{p/q}$ is in all cases phase-locked band chaos characterized by the value p/q.

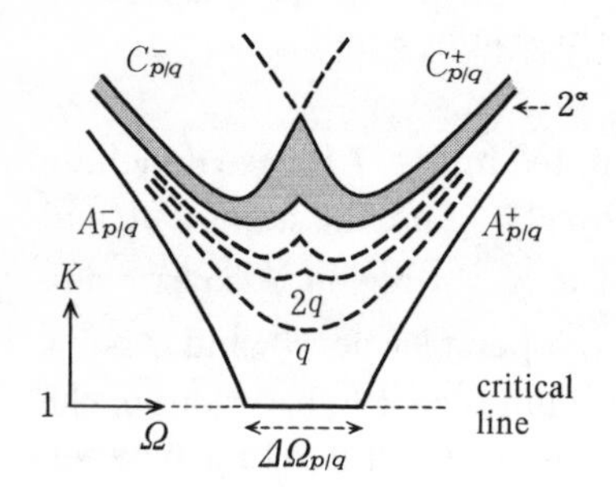

Fig. 7.19. The structure of $A_{p/q}$. Increasing K beyond 1, the system exhibits 2^n-bifurcations of q-point cycles, and then (in the *shaded region*) 2^n-band bifurcations of q-band chaos

The quantities defining the size of the region described in Fig. 7.19, the height $\Delta K_{p/q}$ along the K-axis of the central peak at which $C^-_{p/q}$ and $C^+_{p/q}$ are joined and the distance $\Delta\Omega_{p/q}(1)$ between $A^-_{p/q}$ and $A^+_{p/q}$ along the critical ($K = 1$) line, are decreasing functions of the period q. Furthermore, as discussed in the next subsection, there is a similarity relation satisfied by the sizes of these attractors.

Suppose that we consider some $K \approx 0$ and some Ω such that $\rho(\theta_0)$ is irrational. If we then increase K, since the region of irrational $\rho(\theta_0)$ gradually becomes smaller as K approaches 1 (where its Lebesgue measure becomes 0) for almost all such values of Ω, the system will eventually enter the neighboring Arnold tongue before crossing the critical curve. For this reason, the normal route to the appearance of chaos for the circle map (7.44) is: torus $\to$ phase locking $\to$ 2^n-bifurcations $\to$ phase-locked band chaos. However, in the (Ω, K) parameter space, the following route also exists: golden torus $\to$ critical golden torus $\to$ phase-unlocked chaos (chaos extending over the entire circle). This is known as the *quasi-periodicity route*. We next investigate this route, following Horita et al. (1990).

7.4.2 Phase-Unlocked Fully Extended Chaos

If we define ρ_+ and ρ_- by

$$\rho_+ \equiv \lim_{n\to\infty} \sup_{\theta_0\in[0,1]} \rho_n(\theta_0), \qquad \rho_- \equiv \lim_{n\to\infty} \inf_{\theta_0\in[0,1]} \rho_n(\theta_0) , \tag{7.46}$$

then in the two wings of an Arnold tongue $A_{p/q}$, we have $\rho_+ \geq p/q \geq \rho_-$, while in the convex central region, $\rho_\pm = p/q$. The boundaries between these regions are represented by the broken lines in Fig. 7.20. The two broken lines labeled "$\rho_+ = p/q$" and "$\rho_- = p/q$" coincide with the crisis curve $C^\pm_{p/q}$ just above the convex central region. In the region of coexistence of $A_{p/q}$ and $A_{p'/q'}$ (indicated by the dark shading in Fig. 7.20), we have $\rho_- \leq p/q < p'/q' \leq \rho_+$. Here, two attractors (of rotation numbers p/q and p'/q') coexist. However, above this region there is a U region in which there exists no Arnold's tongue. In the figure, this is the lightly shaded region bounded by the solid lines representing $C^+_{p/q}$ and $C^-_{p'/q'}$ and the broken lines representing $\rho_- = p/q$ and $\rho_+ = p'/q'$. Following the route defined by the arrows, entering the U region from the coexistence region, the two attractors merge into a single attractor. We next study the type of chaos exhibited in the U region.

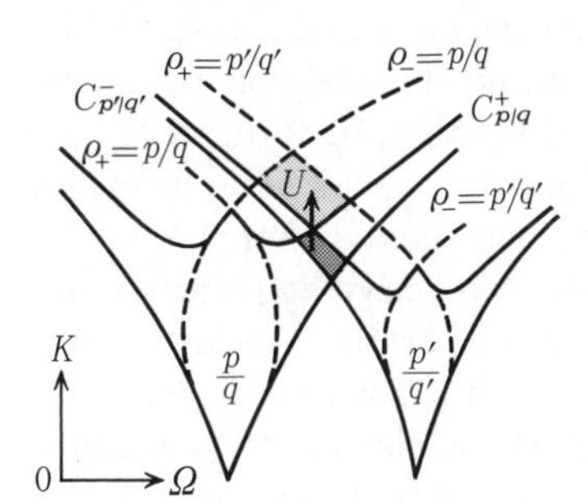

Fig. 7.20. The coexistence region (*darkly shaded region*) of the attractors corresponding to rotation numbers p/q and p'/q', and the U region (*lightly shaded region*)

In the darkly shaded region shown in Fig. 7.21, attractors of rotation numbers 0/1 and 1/2 coexist. Here, the point of intersection of the two crisis curves $C^+_{0/1}$ and $C^-_{1/2}$ is designated by (Ω_c, K_c). In the present case, this point,

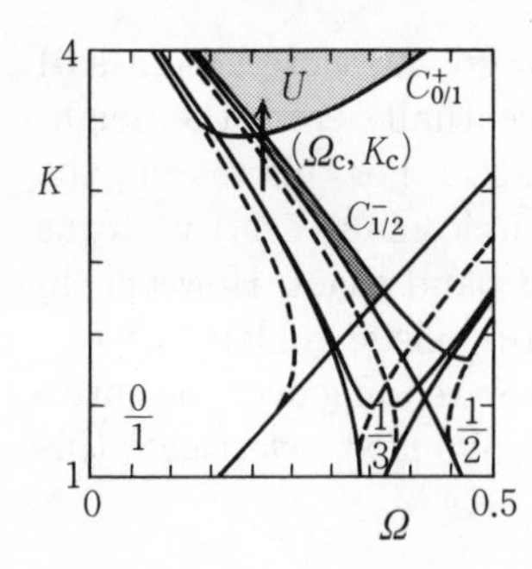

Fig. 7.21. The region of coexistence of $A_{0/1}$ and $A_{1/2}$ (*darkly shaded region*), the intersection point (Ω_c, K_c) of $C^+_{0/1}$ and $C^-_{1/2}$, and the U region of phase-unlocked, fully extended chaos with $K > K_c$ (*lightly shaded region*)

representing the tip of the coexistence region of these two attractors, is given by $(\Omega_c, K_c) \approx (0.21388, 3.40573)$. The bifurcation diagram corresponding to the case in which the system passes through this point in moving from the coexistence region into the U region is given in Fig. 7.22. Here, $\Omega = \Omega_c$ and $3 < K < 3.5$. Figure 7.22b is a plot of the rotation number $\rho(\theta_0)$ as a function of K, and, as seen there, for $3.33 < K < K_c$ the rotation numbers $\rho = 0/1$ and $\rho = 1/2$ coexist. In increasing K from 3, with internal structure as that shown in Fig. 7.19, $A_{0/1}$ begins exhibiting two-band chaos, which later becomes one-band chaos, and at $K = K_c$ this attractor is destroyed as it collides with the single saddle point. The attractor $A_{1/2}$ appears together with a two-point saddle at $K \approx 3.33$, undergoes a 2^n-bifurcation cascade and then a 2^n-band inverse cascade, and finally becomes a two-band chaotic attractor just below K_c, at which point it collides with the two-point saddle and collapses. We thus see that for $K > K_c$ neither of these attractors exhibiting band chaos exist, and in the U region the system is released from its phase-locked state. Here, chaotic behavior extending over the entire circle $(0 \leq \theta_t < 1)$ results. In the sense that this behavior consists of rotation around the circle in which any given (open) interval is eventually visited, it does not differ from bi-periodic motion on a torus. However, in the present case there is also a folding of intervals on the circle. For the two-dimensional annulus map (7.10), this folding implies fully extended chaos on a 'wrinkled torus' possessing the multiplex string-like structure of a Cantor set along the radial direction. As the value of K approaches K_c, the attractors come to possess periodic orbits of periods mq and mq', respectively, for all positive integers m. For this reason we know that the sets of periodic orbits for the respective attractors must also include such infinite sets of period mq and mq' orbits. We investigate the form and structure of this kind of fully extended chaos in Sect. 9.2.

This kind of phase-unlocked, fully extended chaos also exists just above the $K = 1$ critical line. Its existence affords a quasi-periodicity route to the emergence of chaos. Let us now consider this situation. For $K = 1$ and $\Omega = \Omega_\infty = 0.60666106\ldots$, we have $\rho(\theta_0) = \rho_G = (\sqrt{5}-1)/2$, and a critical golden torus corresponding to Fig. 7.5 results. Using the Fibonacci numbers F_m

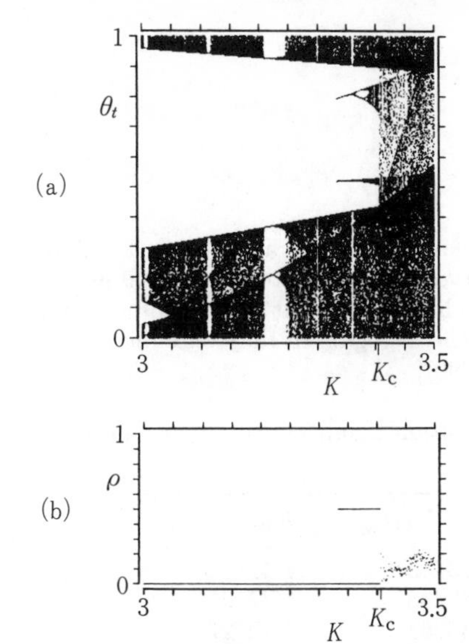

Fig. 7.22. The bifurcation diagram corresponding the path indicated by the arrow in Fig. 7.21. In the U region, where $K > K_c$, the system exhibits phase-unlocked chaos extending over all values of θ_t [Horita et al. (1988)]

($\{F_m\} = 1, 1, 2, 3, 5, 8, \ldots$), defined by $F_0 = F_1 = 1$ and $F_{m+1} = F_{m-1} + F_m$ ($m = 1, 2, \ldots$), the inverse of the golden ratio can be written as

$$\rho_G = \cfrac{1}{1+\cfrac{1}{1+\cfrac{1}{1+\cdots}}} = \lim_{m\to\infty} \frac{F_{m-1}}{F_m} . \tag{7.47}$$

This is the asymptotic expression obtained from continued fractions. We can also write

$$F_m = \{\rho_G^{-(m+1)} - (-\rho_G)^{m+1}\}/(\rho_G^{-1} + \rho_G) , \tag{7.48}$$

and thus for $m \gg 1$, $F_m \propto \rho_G^{-m}$ becomes approximately valid. As we will see in Sect. 8.2.3, this critical golden torus reproduces the fractality exhibited in Fig. 7.5 quite well. Let us now consider the chaotic series that converges to this golden torus when we take the $K \to 1$ limit from the $K > 1$ region. The dark curve in Fig. 7.23 represents the route of such a series. The attractors A_{ρ_m} ($m = 1, 2, \ldots$) with rotation numbers $\rho(\theta_0) = \rho_m \equiv F_{m-1}/F_m$ are represented by $A_{1/1}, A_{1/2}, A_{2/3}, A_{3/5} \ldots$. The curves $C_{1/1}, C_{1/2}, C_{2/3}, \ldots$ are the corresponding crisis curves. The dark curve follows downward along these curves. The corners T_m of this path correspond to the intersections of pairs of crisis curves: $T_m = C_{\rho_m} \cap C_{\rho_{m+1}}$. These corners $T_m = (\Omega_m, K_m)$ represent the points at which two attractors of rotation numbers ρ_m and ρ_{m+1} merge into a single attractor. The coordinates of a number of such points are given in Table 7.1. For example, $T_1 = (\Omega_1, K_1) = (0.786\ldots, 3.405\ldots)$. In the $m \to \infty$ limit, $\Omega_m \to \Omega_\infty$ and $K_m \to K_\infty = 1$, and the corresponding attractor converges to the critical golden torus.

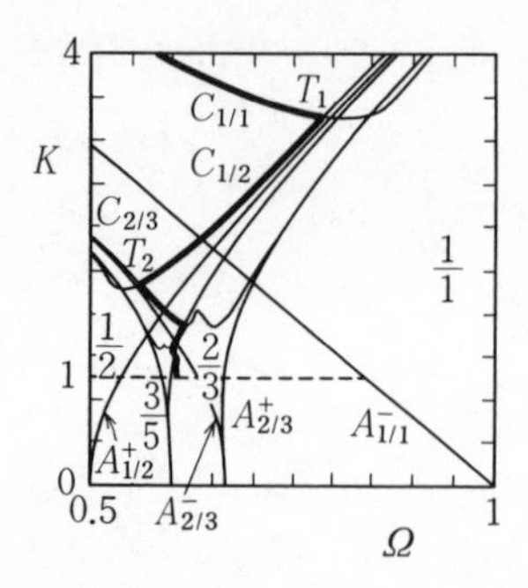

Fig. 7.23. The dark curve indicates the path following the crisis curves C_{ρ_m} down to the critical golden torus. Above each corner point T_m situated along this path exists the U region of fully extended chaos [Horita et al. (1990)]

Table 7.1. A series of attractor-merging bifurcation points converging to a critical golden torus [Horita et al. (1990)]

m	Ω_m	K_m	$F_{m+1}\Lambda_\infty(s^+_{m,i})$	$F_m\Lambda_\infty(s^-_{m,i})$
1	0.786119517337	3.405735089365	1.413443	1.418130
2	0.558100372495	1.873689867485	1.445438	1.471071
3	0.614616583648	1.505301778413	1.436753	1.497271
4	0.597157073317	1.268447545603	1.445470	1.499052
5	0.605744125674	1.156702126804	1.442263	1.504854
6	0.604461101859	1.090765794772	1.444407	1.504013
7	0.605935686595	1.053918954236	1.443443	1.505230
8	0.606024071255	1.032051470891	1.443994	1.504835
9	0.606352103083	1.019198787693	1.443717	1.505117
10	0.606450291104	1.011510501298	1.443865	1.504988
11	0.606543436191	1.006917946440	1.443787	1.505059
12	0.606587157848	1.004159781420	1.443828	1.505021
13	0.606617715411	1.002503309560	1.443806	1.505040
14	0.606634576177	1.001506788007	1.443817	1.505030
15	0.606645256408	1.000907210078	1.443811	1.505035
⋮	⋮	⋮	⋮	⋮
∞	0.606661063470	1.0	1.443813	1.505033

For points in the neighborhood of T_m defined by $T_m^+ \equiv (\Omega_m, K_m + \Delta)$ $(\Delta \to 0)$, as for the region lying just above (Ω_c, F_c) shown in Fig. 7.21, a U region exists, and the system exhibits phase-unlocked, fully extended chaos on a wrinkled torus. The behavior in such a region T_m^+ with $m \gg 1$ represents the chaos exhibited for the case $K = 1 + \Delta$ just beyond the critical golden torus. The set that makes the transition to the golden torus is that consisting of F_m (or F_{m+1}) narrow bands of rotation number ρ_m (or ρ_{m+1}) existing at $T_m^- \equiv (\Omega_m, K_m - \Delta)$. The conditions involved in this transition are similar to those accompanying the transition of the 2^n narrow bands to the critical 2^∞ attractor described in Sect. 7.3.3. Thus the circle map possesses the following *quasi-periodicity route*: golden torus → critical golden torus → phase-unlocked, fully extended chaos.

When $m \gg 1$, the cascade of attractor-merging bifurcation points $T_m = (\Omega_m, K_m)$ is also characterized by a kind of similarity. The attractor A_{ρ_m} possesses structure like that shown in Fig. 7.19. This attractor becomes smaller as m increases. In going from m to $m+1$, the length of the attractor along the K-axis shrinks by a factor of $1/\alpha_{\mathrm{GM}}^2$, and that along the Ω-axis shrinks by a factor of $1/\delta$, where

$$\alpha_{\mathrm{GM}}^2 = 1.66043\cdots, \qquad \delta = -2.83361\cdots \tag{7.49}$$

Here, the minus sign on δ indicates that in going from m to $m+1$, right and left are reversed. Thus along the K-axis, the similarity relation

$$\epsilon_m \equiv K_m - K_\infty \propto \alpha_{\mathrm{GM}}^{-2m} \qquad (m \gg 1) \tag{7.50}$$

holds. Using the relation $F_m \propto \rho_{\mathrm{G}}^{-m}$, which is valid for large m, this becomes

$$F_m \propto \epsilon_m^{-\kappa_{\mathrm{GM}}}, \qquad \epsilon_m \propto F_m^{-\hat{\nu}} \qquad (\hat{\nu} \equiv 1/\kappa_{\mathrm{GM}})\,. \tag{7.51}$$

Here, $\kappa_{\mathrm{GM}} \equiv \ln \rho_{\mathrm{G}}^{-1} / \ln \alpha_{\mathrm{GM}}^2 = 0.9489\ldots$, and $\hat{\nu} = 1.0538\ldots$. The existence of this kind of similarity, as in the case of the 2^n-bifurcations treated in the previous section, is guaranteed by the existence of the fixed point g of the renormalization transformation corresponding to the cascade of points T_m in the function space of the circle map. The two constants in (7.49) represent the eigenvalues of the linear operator L corresponding to this renormalization transformation.

In the present context, the fixed point g possesses two unstable eigenvectors, E_1 corresponding to changes in K, with eigenvalue α_{GM}^2, and E_2 corresponding to changes in Ω, with eigenvalue δ. Therefore, as a generalization of Fig. 7.13, for the present case we have Fig. 7.24 describing the structure of flow in the neighborhood of g. In other words, referring to the unstable manifold tangent to E_l at g as $W_l^{\mathrm{u}}(g)$, the codimension 2 stable manifold as $W^{\mathrm{s}}(g)$, and the two-dimensional surface representing the two-parameter family of maps represented by (7.44) as S, the line $K = 1$ defines the intersection of S and the codimension 1 invariant manifold $\Sigma_1(g)$, which cuts transversely through W_1^{u}. The set of circle maps at $K = K_m$ defines a codimension 1 surface $\Sigma_{1,m}$ which is parallel to Σ_1 in the neighborhood of g, and the distance between $\Sigma_{1,m}$ and Σ_1 is given approximately by d_m, where $d_m/d_{m+1} = \alpha_{\mathrm{GM}}^2$. This yields the relation (7.50). The intersection of the codimension 1 invariant manifold $\Sigma_2(g)$, which cuts transversely through W_2^{u}, and S is the curve $\Omega = \overline{\Omega}(K)$,

$$\bar{\Omega}(K) = \Omega_\infty - 0.017482\epsilon - 0.0005\epsilon^2 + \cdots\,, \tag{7.52}$$

where $\epsilon = K - 1$ [MacKay and Tresser (1986)]. Similarly, the set of circle maps with $\Omega = \Omega_m$ constitutes the codimension 1 surface $\Sigma_{2,m}$. In the neighborhood of g, this surface is approximately parallel to Σ_2, and the distance between these surfaces is d'_m, where $d'_m/d'_{m+1} = |\delta|$. Thus for large m we have the similarity relation

$$\Omega_m - \bar{\Omega}(K_m) \propto \delta^{-m} \propto F_m^{-y}\,, \tag{7.53}$$

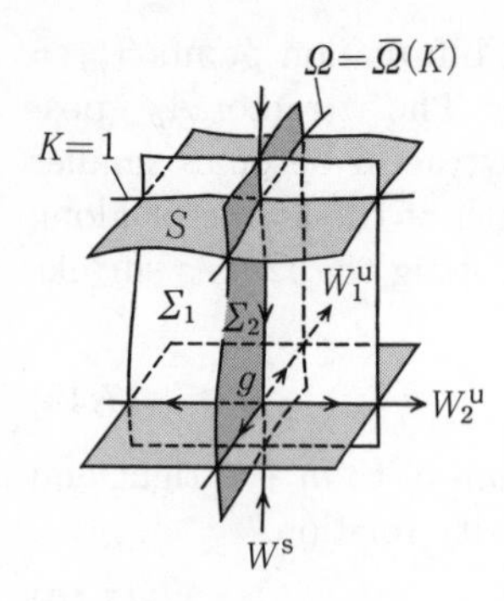

Fig. 7.24. The structure in the region surrounding the fixed point g of the renormalization transformation for the attractor-merging cascade

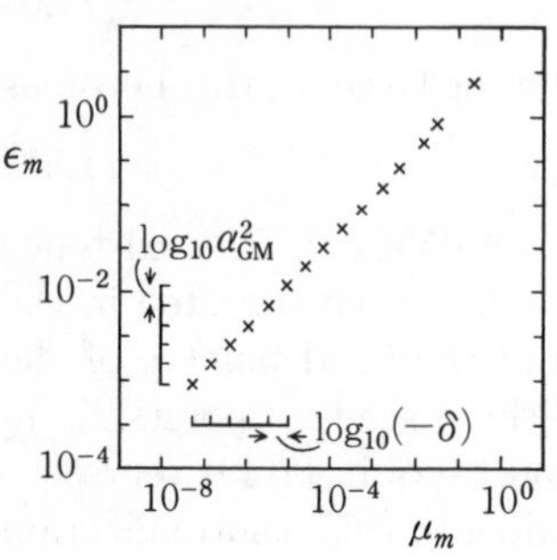

Fig. 7.25. The dependence of $\epsilon_m \equiv K_m - K_\infty$ on $\mu_m \equiv |\Omega_m - \overline{\Omega}(K_m)|$. These data provide numerical confirmation of the relation $\log_{10}\epsilon_m/\log_{10}\mu_m = \log_{10}\alpha_{\mathrm{GM}}^2/\log_{10}|\delta| = 0.4868\ldots$

where $y \equiv \ln|\delta|/\ln\rho_{\mathrm{G}}^{-1} = 2.1644\ldots$. Figure 7.25 displays numerical results which verify the similarity relations (7.50) and (7.53).

At the attractor-merging point T_m, the F_m-band attractor a_m^- and the F_{m+1}-band attractor a_m^+ collide with the F_m-point saddle $s_{m,i}^-$ and F_{m+1}-point saddle $s_{m,i}^+$, respectively. Hence, defining $Q_m^- \equiv F_m$ and $Q_m^+ \equiv F_{m+1}$, at the attractors a_m^+ and a_m^- the orbits passing through the extremal points of $f^{Q_m^+}$ and $f^{Q_m^-}$ become absorbed into the saddles $s_{m,i}^+$ and $s_{m,i}^-$, respectively. At this point, the very large Liapunov numbers $\Lambda_\infty(s_{m,i}^+)$ and $\Lambda_\infty(s_{m,i}^-)$ enter the spectrum $\psi(\Lambda)$. The presence of these saddle points and their large Liapunov numbers thus come to determine the statistical structure of the system. In addition, $\Lambda_\infty(s_{m,i}^+)$ and $\Lambda_\infty(s_{m,i}^-)$ and the Liapunov numbers corresponding to a_m^+ and a_m^-, $\Lambda^\infty(a_m^+)$ and $\Lambda^\infty(a_m^-)$, satisfy the following dynamical similarity relation :

$$\Lambda_\infty(s_{m,i}^\pm) = c_\pm^*/Q_m^\pm \propto \epsilon_m^{\kappa_{\mathrm{GM}}} \tag{7.54a}$$

$$\Lambda^\infty(a_m^\pm) = c_\pm^\infty/Q_m^\pm \propto \epsilon_m^{\kappa_{\mathrm{GM}}} . \tag{7.54b}$$

Here $c_+^* \approx 1.444$, $c_-^* \approx 1.505$, $c_+^\infty \approx 0.693$, and $c_-^\infty \approx 0.692$. The rate of convergence of c_+^* and c_-^* to their asymptotic values can be seen in Table 7.1. As we will see in Sect. 9.3.3, the statistical universality corresponding to the internal structure of the chaotic behavior existing on both sides $T_m^\pm$ of the merging points T_m can be understood in the framework provided by the various similarity relations appearing above.

8. The Statistical Physics of Aperiodic Motion

What physical quantity can we use to capture and describe the multitude of invariant sets and the local structure of unstable manifolds W^{u} that determine the form and structure of chaos? By introducing the expansion rate of neighboring orbits, which expresses the stretching and folding of segments of W^{u}, and the local dimension, which describes the self-similarity of the nested structure of strange attractors, the geometric and statistical descriptions given in terms of chaotic orbits through the fluctuations of these quantities can be unified. We show that the chaotic bifurcations and tangency structure of unstable manifolds can be directly understood in terms of the spectrum $\psi(\Lambda)$, and also that a fixed relationship exists between the spectra of this expansion rate and the local dimension.

As concrete examples, we consider here the chaotic behavior displayed on either side of band-merging, attractor-destruction, and attractor-merging bifurcations, and we also study the critical attractors existing at the emergence point of chaos. Our goal is to make clear the statistical structure and multifractality exhibited in each of these contexts.

8.1 The Statistical Structure Functions of the Coarse-Grained Orbital Expansion Rate

Let us introduce the *partition function*

$$Z_n(q) \equiv \int \mathrm{d}\Lambda \exp[(1-q)n\Lambda]P(\Lambda;n) \tag{8.1a}$$

corresponding to the probability density function (6.23), where q is a real parameter extending from $-\infty$ to ∞. With the definition (6.23), this becomes

$$Z_n(q) = \langle \exp[(1-q)n\Lambda_n(X)] \rangle \ . \tag{8.1b}$$

Let us now consider a ferromagnetic material to which a magnetic field H is applied and for which the magnetization corresponding to a single spin is m. Then, identifying m with Λ and H with $1-q$, we can introduce the following quantities corresponding to the Helmholtz free energy and its derivatives:

$$\Phi_n(q) \equiv -(1/n)\ln[Z_n(q)] \tag{8.2}$$
$$\Lambda_n(q) \equiv \Phi'_n(q) = \langle \Lambda \rangle_{q,n} \tag{8.3}$$
$$\sigma_n(q) \equiv -\Lambda'_n(q) = n\left\langle \{\Lambda - \Lambda_n(q)\}^2 \right\rangle_{q,n} \geq 0\,. \tag{8.4}$$

Here, $\langle \cdots \rangle_{q,n}$ represents an average weighted by $\mathrm{e}^{(1-q)n\Lambda}$:

$$< G >_{q,n} \equiv \frac{1}{Z_n(q)} \int \mathrm{d}\Lambda G(\Lambda) \mathrm{e}^{(1-q)n\Lambda} P(\Lambda; n) \tag{8.5a}$$
$$= \frac{1}{Z_n(q)} \left\langle G(X) \mathrm{e}^{(1-q)n\Lambda_n(X)} \right\rangle\,. \tag{8.5b}$$

Since $\Lambda'_n(q) \leq 0$, $\Lambda_n(q)$ is a nonincreasing function of q, and thus,

$$\Lambda_n(-\infty) = \Lambda_{\text{max}}, \quad \Lambda_n(1) = \Lambda^{\infty}, \quad \Lambda_n(\infty) = \Lambda_{\text{min}}\,, \tag{8.6}$$

where Λ_{max} and Λ_{min} are the maximum and minimum values assumed by Λ. In this way, the parameter q explicitly extracts the various values assumed by $\Lambda_n(X_i)$, as determined by the initial point X_i. Because $\Phi''_n(q) = -\sigma_n(q) \leq 0$, $\Phi_n(q)$ is an upward convex function of q. Also, $\Phi_n(1) = 0$, and thus $\Phi_n(q) = \int_1^q \mathrm{d}q \Lambda_n(q)$. The function $\Phi_n(q)$ is therefore referred to as the q-potential of Λ_n. The quantities $\Lambda_n(q)$ and $\sigma_n(q)$ are called the q-average and q-variance, respectively. In general, all of these functions depend on n, and we wish to consider the form that they assume in the temporal coarse-graining limit, $n \to \infty$.

Let us suppose that the form (6.27) of the probability distribution holds. Then, substituting this into (8.1a), we obtain $Z_n(q) = \int \mathrm{d}\Lambda \exp[-n\{\psi(\Lambda) + (q-1)\Lambda\}]$. Therefore, taking the maximal value of the integrand in the $n \to \infty$ limit, we obtain the variational principle

$$\Phi(q) \equiv \Phi_\infty(q) = \min_{\Lambda}\{\psi(\Lambda) + (q-1)\Lambda\}\,, \tag{8.7}$$

where the minimum of the quantity $\psi(\Lambda) + (q-1)\Lambda$ represents the maximal value of the integrand. If we assume that this minimum is realized at $\Lambda = \Lambda(q)$, we have $\Lambda_\infty(q) = \Lambda(q)$ and $\Phi(q) = \psi[\Lambda(q)] + (q-1)\Lambda(q)$. The function $\Lambda(q)$ is the solution $\Lambda = \Lambda(q)$ of the equation $\psi'(\Lambda) = 1 - q$. Also, $\psi''(\Lambda) = -\mathrm{d}q/\mathrm{d}\Lambda = 1/\sigma_\infty(q) \geq 0$, and thus $\psi(\Lambda)$ is a downward convex function. Furthermore, $\psi(\Lambda^\infty) = 0$ is the minimum of ψ. When it is easier to compute $\Phi(q)$, in order to determine $\psi(\Lambda)$ we can use the function

$$\psi(\Lambda) = \Phi(q(\Lambda)) - \{q(\Lambda) - 1\}\Lambda\,, \tag{8.8}$$

which can be obtained from the relation $\psi(\Lambda) = \max_q\{\Phi(q) - (q-1)\Lambda\}$. The function $q(\Lambda)$ here is the solution $q = q(\Lambda)$ of $\Lambda(q) = \Lambda$.

The expansion rate of neighboring orbits $\Lambda_n(X)$ represents the stretching and folding of segments of the unstable manifold W^{u} on which the chaotic orbits come to reside. The structure functions $\psi(\Lambda)$ and $\Lambda(q)$ characterize the global structure of chaos, as we will see below. Statistical thermodynamic formalism was first introduced to the study of dynamical systems from a mathematical point of view [Ruelle (1978)]. The concept of fluctuations of the expansion rate of orbits has been introduced and discussed in a variety of forms by Fujisaka, Takahashi and Oono, Sano et al. (1986), Eckmann and Procaccia (1986), Fujisaka and Inoue (1987), and others, but these works are for the most part confined to low-dimensional hyperbolic systems. Here, in order to make the physical meaning of this expansion rate clear, we have been considering two-dimensional nonhyperbolic systems, following Morita et al. (1987), Grassberger et al. (1988), and Mori et al. (1989a), and we have focused on the basic process involved in the mixing of chaotic orbits, the stretching and folding of segments of the unstable manifold. The expansion rate $\Lambda_n(X)$, as discussed in Sect. 6.3.3, expresses the coarse-grained mixing rate.

While the fluctuations experienced by molecules due to thermal agitation are characterized by, among other quantities, the Boltzmann entropy and Helmholtz free energy, chaotic fluctuations existing on a macroscopic level are characterized by $\psi(\Lambda)$ and $\Lambda(q)$. In hyperbolic dynamical systems, $\psi(\Lambda)$ represents the distribution of Liapunov numbers of coexisting periodic orbits.

8.1.1 The Baker Transformation

Let us seek the statistical structure functions discussed above for the map

$$F(x,y) = \begin{cases} (a^{-1}x, h_a y) & (0 \le x \le a) \\ (b^{-1}x - b^{-1}a, h_b y + 0.5) & (a < x \le 1) \end{cases} \tag{8.9}$$

$$(0 < a < 0.5 < b = 1 - a, \quad 0 < h_a, h_b < 0.5) .$$

As shown in Fig. 8.1, this defines a map of the unit square to itself. Under this map, the regions shown become stretched by factors of a^{-1} and b^{-1} along the x-axis and contracted by factors of h_a and h_b along the y-axis. After n applications of this map, the image consists of 2^n bands of widths $h_a^r h_b^{n-r}$ $(r = 0, 1, \dots n)$. Thus in the limit as $n \to \infty$ the attractor becomes a Cantor set of length 1 strips. The distance along the x-axis between two neighboring orbits is expanded by a factor of $(a^{-1})^r (b^{-1})^{n-r}$ after the nth iteration of (8.9). For a given orbit with initial point X_0, we thus have

$$\Lambda_n(X_0) = p \ln a^{-1} + (1-p) \ln b^{-1} , \tag{8.10}$$

where $1 \ge p \equiv r/n \ge 0$ for some fixed value of r. The map (8.9) possesses two unstable fixed points, $S = (0,0)$ and $X^* = (1, 0.5/(1-h_b))$, and the maximal and minimal values of $\Lambda_n(X_0)$, Λ_1 and Λ_2, are given by the Liapunov numbers of these points: $\Lambda_1 = \Lambda_\infty(S) = \ln a^{-1}$, and $\Lambda_2 = \Lambda_\infty(X^*) = \ln b^{-1}$. If we make the value assumed by $\Lambda_n(X_0)$, Λ, the independent variable in (8.10), we have $p = (\Lambda - \Lambda_2)/(\Lambda_1 - \Lambda_2)$. Now, if we choose the initial point X_0 of an orbit randomly (with a flat sampling measure) in the unit square, the probability that after n mappings the resultant orbit will be in a given band of width $h_a^r h_b^{n-r}$ is $a^r b^{n-r}$. Then, since there are ${}_nC_r$ bands of this width, the probability that Λ will assume the value $\Lambda_n(X_0)$ in (8.10) (in other words, that $p = r/n$) is $P(\Lambda; n) = {}_nC_r a^r b^{n-r}$. Thus, from (6.27), in the limit as $n \to \infty$, we obtain the expression

$$\psi(\Lambda) = \left(\frac{\Lambda - \Lambda_2}{\Lambda_1 - \Lambda_2}\right) \ln\left(\frac{\Lambda - \Lambda_2}{\Lambda_1 - \Lambda_2}\right) + \left(\frac{\Lambda_1 - \Lambda}{\Lambda_1 - \Lambda_2}\right) \ln\left(\frac{\Lambda_1 - \Lambda}{\Lambda_1 - \Lambda_2}\right) + \Lambda \tag{8.11}$$

for the spectrum $\psi(\Lambda)$. (We have used the Stirling approximation to simplify the form of this equation.) In the attractor in question, there exist infinitely many periodic orbits whose Liapunov numbers are densely distributed in the interval (Λ_2, Λ_1). The spectrum of this distribution coincides with the spectrum (8.11) of the chaotic orbits. The minimum of $\psi(\Lambda)$ is located at $\Lambda = \Lambda^\infty = a \ln a^{-1} + b \ln b^{-1}$. If we expand about this point, retaining only the lowest-order nonzero term, we have $\psi(\Lambda) = (\Lambda - \Lambda^\infty)^2/2\sigma$, where $\sigma = ab(\Lambda_1 - \Lambda_2)^2$. This is, of course, the normal distribution, which can be thought of as resulting from the central limit theorem.

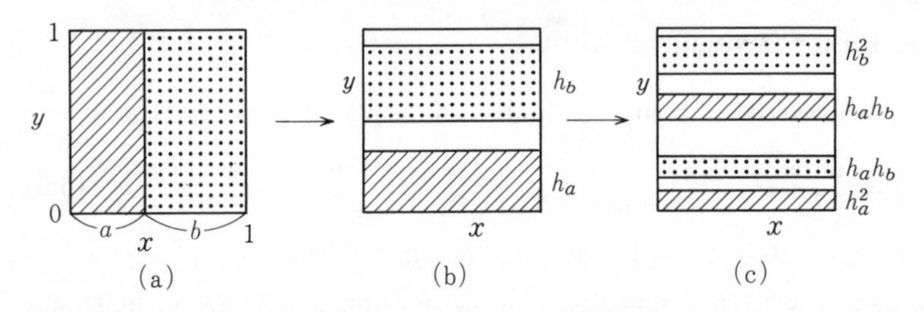

Fig. 8.1. A mapping of the unit square under the baker transformation (8.9). The bands in (c) have widths h_a^2, $h_a h_b$, $h_a h_b$, and h_b^2

The partition function (8.1b) can, with the use of (8.10), be written

$$Z_n(q) = \sum_{r=0}^{n} {}_nC_r a^r b^{n-r} \{a^{-r} b^{-(n-r)}\}^{1-q} , \tag{8.12}$$

which, by the binomial theorem, reduces to $Z_n(q) = (a^q + b^q)^n$. Thus the statistical structure functions become

$$\Phi_n(q) = -\ln(a^q + b^q) \tag{8.13a}$$

$$\Lambda_n(q) = (a^q \Lambda_1 + b^q \Lambda_2)/(a^q + b^q) \tag{8.13b}$$

$$\sigma_n(q) = a^q b^q (\Lambda_1 - \Lambda_2)^2/(a^q + b^q)^2 . \tag{8.13c}$$

Note that these functions do not depend on n because they are determined by the eigenvalues a^{-1} and b^{-1} of the saddle points S and X^*. The baker transformation represents the simplest hyperbolic dynamical system and, in general, for this type of hyperbolic system, the spectrum $\psi(\Lambda)$ is identical to that of the distribution of Liapunov numbers corresponding to the coexisting periodic orbits [see Takahashi and Oono (1984), Morita et al. (1988)].

8.1.2 Attractor Destruction in the Quadratic Map

As the simplest nonhyperbolic system, let us consider the following quadratic map:

$$x_{t+1} = f(x_t) = 2 - x_t^2 \quad (-2 < x_t < 2) . \tag{8.14}$$

This is a particular realization of (7.22). Note that (8.14) represents a system whose attractor has just collided with the saddle s. The special features of this map are that, as shown in Fig. 8.2, $f^2(0) = s = -2$ and $x^* = 1$, and that periodic orbits of all periods coexist here. However, with the exception of that corresponding to the saddle s, all Liapunov numbers for these orbits are equal to $\Lambda_\infty(x^*) = \Lambda^\infty = \ln 2$. For the saddle s, we have $\Lambda_\infty(s) = 2\Lambda^\infty$.

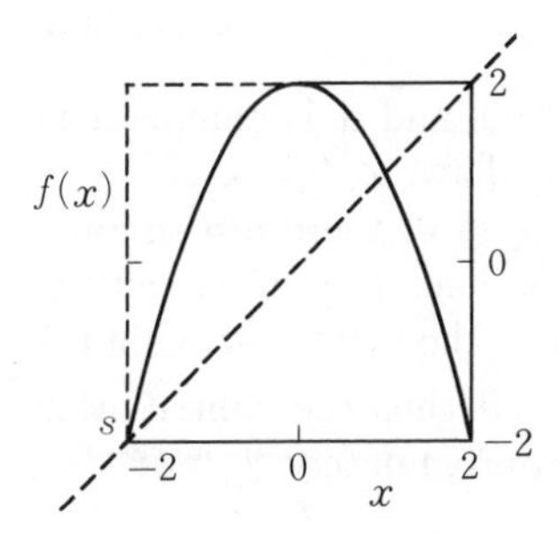

Fig. 8.2. Graph of the quadratic map (8.14) for the case $a = 2$. For $a > 2$, $f^2(0) < s$, orbits in the neighborhood of the apex orbit $f^j(0)$ run away to $-\infty$, and the attractor is destroyed

Under the transformation $x = 2\sin(\pi z/2)$, this map becomes the tent map,

$$z_{t+1} = 1 - 2|z_t| \quad (-1 < z_t < 1) . \tag{8.15}$$

Under this kind of continuously differentiable, one-to-one transformation, chaotic Liapunov numbers and the invariant probability measure are unchanged, and thus the maps (8.14) and (8.15) are topologically conjugate. Because the natural probability density of the tent map (8.15) is uniform [$p(z) = 1/2$], we obtain

$$p(x) = \frac{1}{2}\frac{\mathrm{d}z}{\mathrm{d}x} = \frac{1}{2\pi\sqrt{(1-(x/2)^2)}} . \tag{8.16}$$

Note that this function is singular at $x = \pm 2$.

In general, for one-dimensional maps, as in the case of (6.15), we can write $n\Lambda_n(x) = \ln|\mathrm{d}f^n(x)/\mathrm{d}x|$, and therefore the partition function (8.1b) becomes

$$Z_n(q) = \int \mathrm{d}x |\mathrm{d}f^n(x)/\mathrm{d}x|^{1-q} p(x) . \tag{8.17}$$

Using $|\mathrm{d}z_n/\mathrm{d}z_0| = 2^n$, the derivative of $x_n = f^n(x_0)(x_0 \neq \pm 2)$ is

$$\left|\frac{\mathrm{d}x_n}{\mathrm{d}x_0}\right| = \left|\frac{\mathrm{d}x_n}{\mathrm{d}z_n}\frac{\mathrm{d}z_n}{\mathrm{d}z_0}\frac{\mathrm{d}z_0}{\mathrm{d}x_0}\right| = 2^n \left|\frac{\cos(\pi z_n/2)}{\cos(\pi z_0/2)}\right| . \tag{8.18}$$

Thus the partition function (8.17) with $p(x_0)\mathrm{d}x_0 = (1/2)\mathrm{d}z_0$ can be written

$$Z_n(q) = \frac{1}{2} 2^{n(1-q)} \int_{-1}^{1} \mathrm{d}z_0 \left|\frac{\cos(\pi z_n/2)}{\cos(\pi z_0/2)}\right|^{1-q} . \tag{8.19}$$

In the $n \to \infty$ coarse-grained limit, this integral can be evaluated as follows. When $1 - q > 0$, the integrand yields a singular contribution at the zeroes of $\cos(\pi z_0/2)$. In other words, in the interval $1 > |z_0| \gtrsim 1 - 2^{-n}\epsilon$, if we set $z_j = \pm 1 \mp \epsilon_j$ (with $\epsilon_j = 2^{j-n}\epsilon \ll 1$), we have $\cos(\pi z_j/2) = \sin(\pi\epsilon_j/2) \approx \pi\epsilon_j/2$, and thus the integrand here becomes approximately $2^{n(1-q)}$. Then, multiplying this by the factor 2^{-n} representing the approximate length the interval in question, we have $Z_n(q) \propto 2^{n(1-2q)}$. We thus obtain

$$Z_n(q) \approx 2^{n(1-2q)} c_1 + 2^{n(1-q)} c_0 , \tag{8.20a}$$

where the second term is the nonsingular contribution, and c_0 is independent of n. Thus, in the $n \to \infty$ limit, when $q < 0$, we have $Z_n(q) \propto 2^{n(1-2q)}$, while for $0 < q < 1$, $Z_n(q) \propto 2^{n(1-q)}$. When $1 - q < 0$, there are singular contributions to (8.19) at the zeroes of $\cos(\pi z_n/2)$. Therefore, if we extract the orbits which satisfy $1 > |z_n| \gtrsim 1 - \delta(\delta \to 0)$, after the nth iteration of the map the corresponding integrand becomes δ^{1-q}, and thus the contribution from this interval (of length $\sim 2^{-n}\delta$) becomes approximately $2^{-nq}\delta^{2-q}c_2$, where c_2 is independent of n. We thus obtain

$$Z_n(q) \approx 2^{-nq}\delta^{2-q}c_2 + 2^{n(1-q)}c_0 . \tag{8.20b}$$

Hence, if $q > 2$, taking $\delta \to 0$, we have $Z_n(q) \to \infty$, while if $2 > q > 1$, we obtain $Z_n(q) \propto 2^{n(1-q)}$.

To summarize the above discussion, if $q < 0$, $\Phi(q) = (2q - 1)\ln 2$, if $0 < q < 2$, $\Phi(q) = (q - 1)\ln 2$, and if $q > 2$, $\Phi(q) = -\infty$. Hence, with $\Lambda^\infty = \ln 2$, we have

$$\Lambda(q) = \begin{cases} 2\Lambda^\infty & (q < 0) \\ 2\Lambda^\infty \sim \Lambda^\infty & (q = 0) \\ \Lambda^\infty & (0 < q < 2) \\ \Lambda^\infty \sim -\infty & (q = 2) \\ -\infty & (q > 2) \end{cases} . \tag{8.21}$$

For the corresponding inverse function $q(\Lambda)$, we have $q(\Lambda) = 2$ for $\Lambda < \Lambda^\infty$, $q(\Lambda) = 0$ for $\Lambda^\infty < \Lambda < 2\Lambda^\infty$, and $q(\Lambda) = -\infty$ for $\Lambda > 2\Lambda^\infty$. Then, from (8.8), we have $\psi(\Lambda) = \infty$ for $\Lambda > 2\Lambda^\infty$, $\psi(\Lambda) = \Lambda - \Lambda^\infty$ for $\Lambda^\infty \leq \Lambda \leq 2\Lambda^\infty$, and $\psi(\Lambda) = \Lambda^\infty - \Lambda$ for $\Lambda \leq \Lambda^\infty$. Figure 8.3 graphically displays these functions. The striking feature seen here is the discontinuity of both these functions and their derivatives. For example, $\Lambda(q)$ assumes only three values, $\Lambda_\infty(s) = 2\Lambda^\infty$, $\Lambda_\infty(x^*) = \Lambda^\infty$, and $\lambda_1(0) = \ln|f'(0)| = -\infty$, changing discontinuously between these values. The transition of $\Lambda(q)$ between Λ^∞ and $2\Lambda^\infty$ at $q = 0$ is similar to the discontinuous phase transition undergone by a ferromagnet at $H = 0$. This transition is thus referred to as a *q-phase transition*. As shown in Fig. 8.3a, $\psi(\Lambda)$ consists of two linear segments of slopes ± 1. These segments represent the states of two-phase coexistence present at the two q-phase transition points $q = 0, 2$ of $\Lambda(q)$.

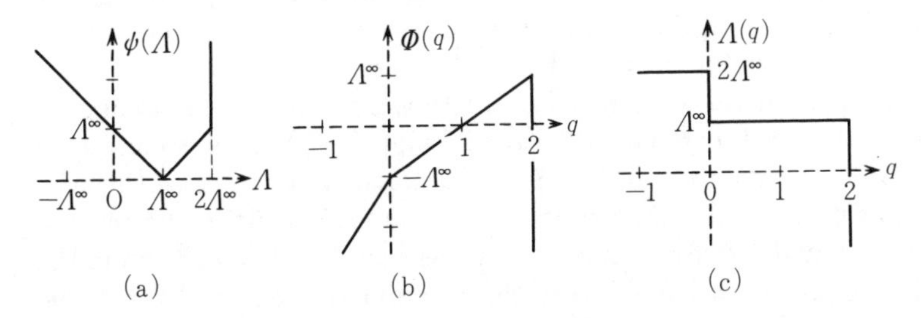

Fig. 8.3. The statistical structure functions of the quadratic map (8.14). The function $\psi(\Lambda)$ has a sharp minimum at $\Lambda^\infty = \ln 2 = 0.693\ldots$. The linear segments on either side of this point have slopes $s_\alpha = -1$ and $s_\beta = 1$

The Liapunov numbers of the coexisting periodic orbits are equal to either Λ^∞ or $2\Lambda^\infty$, but $\psi(\Lambda)$ consists of a line of slope -1 extending from $(\Lambda^\infty, 0)$ to $\Lambda = -\infty$, and a line of slope $+1$ connecting $\psi(\Lambda^\infty)$ and $\psi(2\Lambda^\infty - 0)$. The existence of the slope -1 line and the corresponding appearance of negative values of Λ are due to the chaotic orbits that pass through the neighborhood of the apex of the map located at $x = 0$, while the slope $+1$ linear segment results from the chaotic orbits that pass through the neighborhood of the saddle s. In this way, the spectrum $\psi(\Lambda)$ clearly reveals the influence of the apex and reflects the collision with the saddle s and the resulting bifurcation.

8.1.3 Attractor Merging in the Circle Map

We now investigate the kind of phase structure and statistical structure that exist in the neighborhood of the bifurcation point at which phase-unlocked fully extended chaos appears as a result of the destruction of phase-locked band chaos on a torus with rational rotation number p/q. The nature of this structure was first investigated by Horita et al. (1988a) for the circle map and by Tomita et al. (1988) for the annulus map. In the (Ω, K) parameter space (Fig. 7.21) of the circle map, this type of bifurcation occurs, for example, at the attractor-merging point (Ω_c, K_c). The bifurcation diagram corresponding to the situation in which this bifurcation is encountered when moving from the phase-locked region into the phase-unlocked region is depicted in Fig. 7.22. In the region just below the bifurcation point (Ω_c, K_c), a single-band attractor with rotation number $\rho = 0/1$ and a two-band attractor with rotation number $\rho = 1/2$ coexist, but upon encountering the bifurcation, these attractors are both destroyed, and phase-unlocked chaos extending over the entire circle results. At this bifurcation point, these two attractors can be represented graphically as the single square in Fig. 8.4a [corresponding to $f(\theta)$], and the two squares in Fig. 8.4b, corresponding to $f^2(\theta)$. The former represents the $\rho = 0/1$ one-band attractor $\tilde{A}_{0/1}$ that has just collided with the one-point saddle s_0. The orbit passing through the minimum of $f(\theta)$ lying on the bottom side of this square is absorbed by s_0, creating a situation similar to that depicted in Fig. 8.2, and the statistical structure functions in the present case are thus expected to be similar to those shown in Fig. 8.3, corresponding to the map (8.14). This is confirmed by the numerical results given in Fig. 8.5a. (Unfortunately, because the type of analysis given in the previous subsection is quite difficult in the present case, we have no analytic results here.) As we see, the spectrum shown in the parts labeled (i) in Fig. 8.5 has a very sharp minimum at $\Lambda = \Lambda^{\infty} \approx 0.690$ and possesses a form approximated by

$$\psi(\Lambda) = \begin{cases} \infty & (\Lambda > \Lambda_1) \\ s_\beta(\Lambda - \Lambda^{\infty}) & (\Lambda_1 \geq \Lambda \geq \Lambda^{\infty}) \\ s_\alpha(\Lambda - \Lambda^{\infty}) & (\Lambda^{\infty} \geq \Lambda > \Lambda_2) \end{cases} \tag{8.22}$$

with $s_\alpha \approx -1.00$, $s_\beta \approx 0.99$, $\Lambda_1 \approx 1.28$, and $\Lambda_2 \approx -0.29$. With these numerical results, let us assume that over some interval $\Lambda_- < \Lambda < \Lambda_+$, the spectrum $\psi(\Lambda)$ assumes the form of a line of slope s_k. Then, with $\psi(\Lambda) = s_k\Lambda +$ constant, the variational principle (8.7) specifies the value of Λ for which $(q-q_k)\Lambda$ is a minimum, where here $q_k \equiv 1 - s_k$. For $q = q_k \mp \epsilon(\epsilon \to +0)$, this minimum is realized when $\mp\epsilon\Lambda$ is minimized. Then, since $\Lambda_+ > \Lambda > \Lambda_-$, we see that the minimum of the quantity that appears in (8.7) is realized for such q at $\Lambda(q) = \Lambda_{\pm}$. Therefore, when increasing q, $\Lambda(q)$ changes discontinuously at $q = q_k \equiv 1 - s_k$ from Λ_+ to Λ_-, signaling a q phase transition.

Figure 8.5a(i) is a plot of $\psi_n(\Lambda)$ with $n = 10$, obtained from (6.26) by using $N = 5 \times 10^5$ in the average (6.22) and computing $P(\Lambda; n)$ defined in (6.23) numerically. This result was obtained by taking the group average represented by (6.22) with $N = 5 \times 10^5$. The continuous curves in (ii) and (iii) here are the results for $\Lambda_n(q)$ and $\sigma_n(q)$ obtained from (8.3) and (8.4) by computing $Z_n(q)$ in (8.1b) numerically. The discontinuous function described by the data points in (ii) represents the result of using the variational principle (8.7) and an approximate expression for $\Lambda_n(q)$ in which only the largest term of the integrand of (8.1a) is kept. This computational method provides an approximate description of the q-phase transitions that occur in the coarse-graining, $n \to \infty$ ($n \ll N$), limit and, as we see, the q-variance $\sigma(q)$ displays sharp peaks at the corresponding points, $q = q_\alpha \equiv 1 - s_\alpha \approx 2.00$ and $q = q_\beta \equiv 1 - s_\beta \approx 0.01$. On theoretical grounds, we expect that at these points we

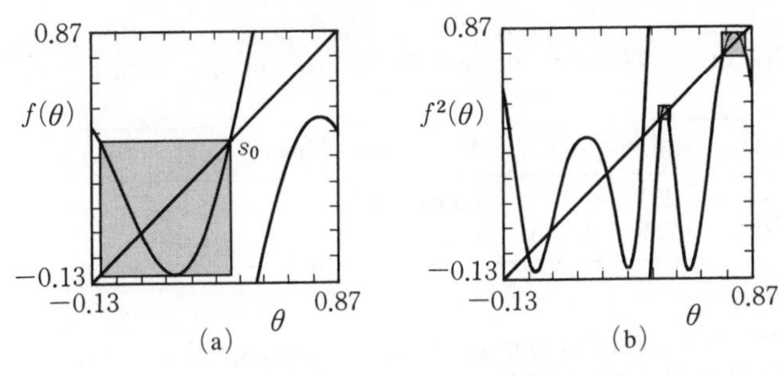

Fig. 8.4. The graph of the circle map (7.44) corresponding to the attractor-merging bifurcation point (Ω_c, K_c). The square shown in (**a**) represents the one-band attractor $\tilde{A}_{0/1}$ characterized by $\rho = 0/1$. The two squares in (**b**) represent the $\rho = 1/2$, two-band attractor $\tilde{A}_{1/2}$

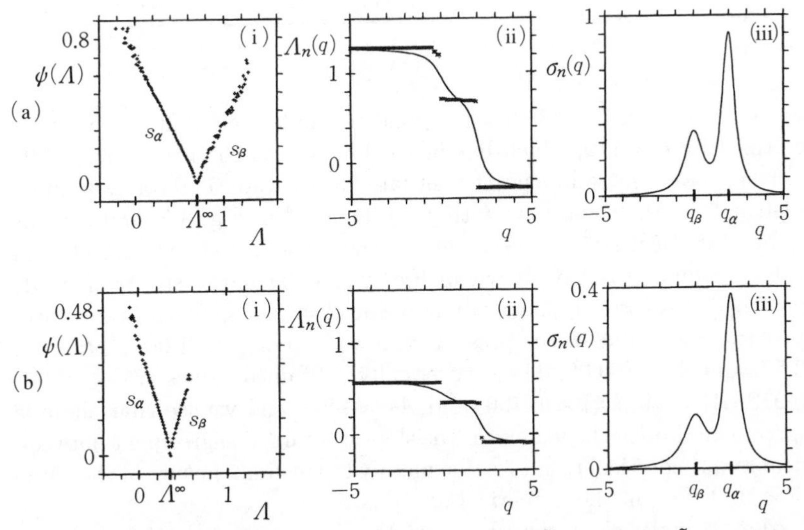

Fig. 8.5. The statistical structure functions of (**a**) $\tilde{A}_{0/1}$ and (**b**) $\tilde{A}_{1/2}$. In (**a**) $n = 10$, and in (**b**) $n = 20$, while in both cases $N = 5 \times 10^5$. Note the sharp minimum of $\psi(\Lambda)$ that occurs in each case at Λ^∞ with linear segments of slopes s_α and s_β on either side

will have $\Lambda_1 = \Lambda_\infty(s_0) \approx 1.418$ and $\Lambda_2 = \ln|f'(\theta^0)| = -\infty$, where θ^0 is the minimum point of the map f. The difference between these values and the numerical results given above is due chiefly to the finiteness of the number of orbits $N = 5 \times 10^5$ used for the average (6.22) in the numerical evaluation, and the resulting situation in which the number of orbits taken into account that pass near the saddle s_0 and the minimum θ^0 is not sufficiently large. However, we note that the slopes obtained numerically are in good agreement with those given by the theory described in Sect. 8.3, $s_\alpha = -1$ and $s_\beta \approx 0.974$, displayed in Table 8.1.

Table 8.1a. Theoretical values of $\Lambda_\infty(S)$, $\alpha_1(S)$, s_β, and related quantities, along with the numerical value of s_β. These results are for the Hénon map with $b = 0.3$ and $R \approx 1.204$

	$\Lambda_\infty(S)$	$\alpha_1(S)$	$\alpha(S)$	$h(S)$	Λ^{o}	s_β (Theor.)	s_β (Num.)
$a = a_{\mathrm{d}}$	1.188	0.856	1.281	1.199	0.530	1.85	1.90
$a = a_1$	0.550	0.728	0.956	1.136	0.325	2.02	1.80

Table 8.1b. Similar to Table 8.1a above, but for the quadratic map with $a = a_{\mathrm{d}} = 2$ and the circle map with $\Omega = \Omega_{\mathrm{c}}$ and $K = K_{\mathrm{c}}$

	$\Lambda_\infty(s)$	$\alpha_1(s) = \alpha(s)$	$h(s)$	Λ^{o}	s_β (Theor.)	s_β (Num.)
$a = a_{\mathrm{d}}$	$2\ln 2$	0.5	1	$\ln 2$	1	1.0
$\rho = 0/1$	1.418	0.5	1	0.690	0.974	0.99
$\rho = 1/2$	0.706	0.5	1	0.343	0.972	1.03

The two squares in Fig. 8.4b correspond to the two-band attractor $\tilde{A}_{1/2}$, characterized by $\rho = 1/2$, which has just collided with the two-point saddle s_1, s_2. In each case, the saddle point is situated at the value of θ corresponding to the left side of the square, as with $f(x)$ depicted in Fig. 8.2, and for this reason the statistical structure existing in the present case, as exhibited in Fig. 8.5b, is similar to that shown in Fig. 8.3. In fact, the spectrum $\psi(\Lambda)$ in Fig. 8.5b(i) possesses a sharp minimum at $\Lambda = \Lambda^\infty \approx 0.343$ and a form described by (8.22), but in the present case we have $s_\alpha \approx -1.00$, $s_\beta \approx 1.03$, $\Lambda_1 \approx 0.57$, and $\Lambda_2 \approx -0.09$. The theoretically[1] obtained values are $s_\alpha = -1$, $s_\beta \approx 0.972$, $\Lambda_1 = \Lambda_\infty(s_i) \approx 0.706$, and $\Lambda_2 = -\infty$, and we see that there is good agreement among the values for the slopes. Again, disagreement between numerical and theoretical results is due mainly to the finite nature of the value N $(= 5 \times 10^5)$ used in the numerical analysis.

In order to make clear the nature of the fully extended chaos existing in the U region beyond the bifurcation, it is necessary to consider its two-dimensional fractality. For this reason, the statistical structure in this region is treated in Sect. 9.2 in the context of the attractor-merging bifurcations of annulus maps and forced oscillators.

8.1.4 Bifurcations of the Hénon Map

Let us consider the band merging and attractor destruction exhibited by the Hénon map (7.1) discussed in Sect. 7.1. The geometric structure of these bifurcations is described by Figs. 7.2b and d.

[1] In keeping with the current trend in the study of chaos and related phenomena, we wish to distinguish three forms of investigatin of such systems: experimental, theoretical, and numerical

Numerically generated statistical structure functions at the band merging bifurcation point $a = a_1$ are given in Fig. 8.6. Here, $n = 40$ and $N = 10^8$. The spectrum $\psi(\Lambda)$ shown here possesses a somewhat rounded minimum at $\Lambda = \Lambda^\infty \approx 0.315$. The linear segments located on either side of this minimum have slopes $s_\alpha \approx -0.95$ and $s_\beta \approx 1.80$. As can be seen from Fig. 8.6b, the value of $\Lambda_{\max}$ is approximately 0.550. The variance $\sigma_n(q)$ possesses sharp peaks at $q_\alpha = 1 - s_\alpha \approx 1.95$ and $q_\beta = 1 - s_\beta \approx -0.80$, reflecting the fact that $\Lambda(q) = \Lambda_\infty(q)$ undergoes a q-phase transition at each of these points. These values agree well with those obtained theoretically, $s_\alpha = -1$, $s_\beta \approx 2.02$, and $\Lambda_{\max} = \Lambda_\infty(X^*) \approx 0.550$ (see Table 8.1 in Sect. 8.3).

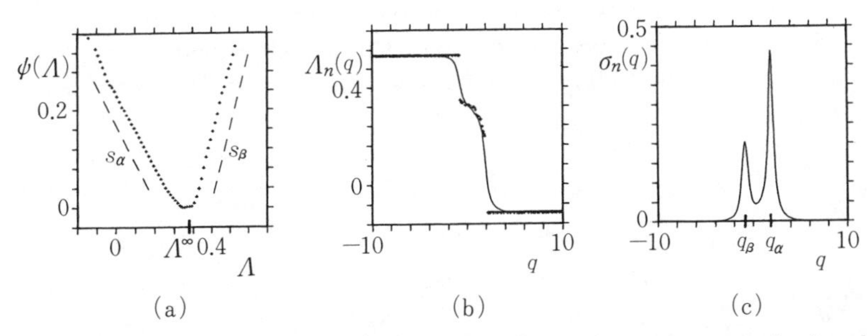

Fig. 8.6. Statistical structure functions of the Hénon map at the band-merging bifurcation point $a = a_1$ (here $n = 40$ and $N = 10^8$). The spectrum $\psi(\Lambda)$ possesses a sharp minimum situated between linear segments of slopes $s_\alpha \approx -0.95$ and $s_\beta \approx 1.80$ [Hata et al. (1988)]

The statistical structure functions at the attractor-destruction bifurcation point $a = a_d$ possess forms similar to those shown in Fig. 8.6 for the band-merging bifurcation. In the case of the former, numerical analysis with $n = 21$ and $N = 10^9$ shows that $\psi(\Lambda)$ possesses a similarly rounded minimum at $\Lambda = \Lambda^\infty \approx 0.495$ and linear segments on either side with slopes $s_\alpha \approx -1.00$ and $s_\beta \approx 1.90$. In this case, $\Lambda_{\max} \approx 0.97$. Here, $\sigma_n(q)$ possesses sharp peaks at $q_\alpha \approx 2.00$ and $q_\beta \approx -0.90$, reflecting the q-phase transitions taking place at these points. Again, these values show good agreement with those obtained theoretically, $s_\alpha = -1$, $s_\beta \approx 1.85$, and $\Lambda_{\max} = \Lambda_\infty(S) \approx 1.188$.

8.1.5 The Slopes s_α and s_β of $\psi(\Lambda)$

As in the case depicted in Figs. 7.2a and c, when the system is sufficiently far from any bifurcation point, in general, the statistical structure functions have forms similar to those shown in Fig. 8.7. As seen there, the distribution of Liapunov numbers $\Lambda_\infty(X_i^*)$ corresponding to the coexisting unstable periodic orbits possesses a smooth spectrum $g(\Lambda)$ existing in the interval $(\Lambda_g, \Lambda_{\max})$ of positive Λ. Contrastingly, the local expansion rate of chaotic orbits can assume negative values in the interval $\Lambda_{\min} < \Lambda < 0$, reflecting the folding of the unstable manifold corresponding to type I tangency structure (Fig. 6.7). The spectrum of these chaotic orbits, $\psi(\Lambda)$, contains a linear segment of slope $s_\alpha = -1$ defined by

$$\psi(\Lambda) = s_\alpha(\Lambda - \Lambda_\alpha) + g(\Lambda_\alpha), \quad (\Lambda_\alpha \geq \Lambda \geq \Lambda_{\min}) , \tag{8.23}$$

which is tangent to the spectrum $g(\Lambda)$ of periodic orbits. Here, $0 < \Lambda_g \leq \Lambda_\alpha \leq \Lambda^\infty$, and Λ_α is the point of tangency between this segment and $g(\Lambda)$. Designating the tangency point at which the largest curved part of the attractor contacts the stable manifold by X_T, we have $\Lambda_{\min} = \Lambda_\infty(X_T) < 0$. The q-average of Λ, $\Lambda(q) = \Lambda_\infty(q)$ given by (8.3), undergoes a q-phase transition at $q = q_\alpha \equiv 1 - s_\alpha = 2$, but corresponding to values satisfying $\Lambda_{\min} < \Lambda(q) < \Lambda_\alpha$, characterized by two-phase coexistence, there exists a multitude of chaotic orbits passing through the neighborhood of X_T. The linear segment defined by (8.23) is due to the presence of such chaotic orbits with negative Λ.

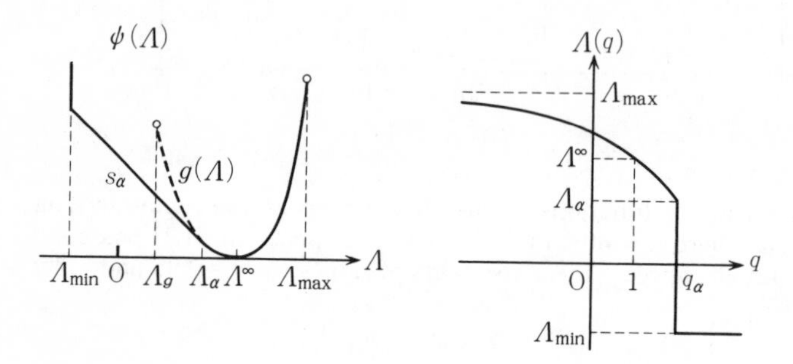

Fig. 8.7. The q-phase transition of $\psi(\Lambda)$ and $\Lambda(q)$ occurring in a nonhyperbolic system. The spectrum $\psi(\Lambda)$ displays a linear segment of slope $s_\alpha = -1$ to the left of its minimum. The curve $g(\Lambda)$ ($\Lambda_g < \Lambda < \Lambda_{\max}$) represents the spectrum of the distribution of Liapunov numbers $\Lambda_\infty(X_i^*)$ corresponding to the set of coexisting periodic orbits. The variance $\sigma_\infty(q)$ diverges at $q = q_\alpha$

Figure 8.8 represents a summary of the important features of the bifurcations described in Figs. 8.5 and 8.6. In the situation depicted in this figure, $\Lambda_{\max} = \Lambda_\infty(S_i)$, where S_i represents the saddle with which the attractor has just collided. In addition to the linear segment defined by (8.23), $\psi(\Lambda)$ possesses a linear segment to the right of $\Lambda = \Lambda^\infty$ of slope $s_\beta > 0$:

$$\psi(\Lambda) = s_\beta(\Lambda - \Lambda_\beta) + g(\Lambda_\beta) \quad (\Lambda_\beta \leq \Lambda \leq \Lambda_{\max}) . \tag{8.24}$$

Here, $\Lambda^\infty \leq \Lambda_\beta < \Lambda_{\max} = \Lambda_\infty(S_i)$, where Λ_β is the tangency point of the segment in question to $g(\Lambda)$. In order for a linear segment so defined to exist (i.e., in order for Λ_β to be a tangency point rather than an intersection point), it is necessary that the value of $\psi(\Lambda)$ induced by the existence of the saddle S_i, $\psi(\Lambda_\infty(S_i))$, be to the right of $g(\Lambda)$. While $\Lambda(q)$ displays a q-phase transition at $q = q_\beta \equiv 1 - s_\beta$, this linear segment corresponds to the many values of $\Lambda(q)$

in the interval $\Lambda_\beta < \Lambda(q) < \Lambda_{\max}$ characterized by the coexistence of these two phases. This linear segment is created as a result of the influence of the many chaotic orbits passing through the neighborhood of the saddle S_i. As in the situation described by Figs. 8.3 and 8.5, when the Liapunov numbers of all periodic orbits other than S_i are equal to Λ^∞, we have $\Lambda_\alpha = \Lambda_\beta = \Lambda^\infty$, and the minimum of $\psi(\Lambda)$ becomes very sharp.

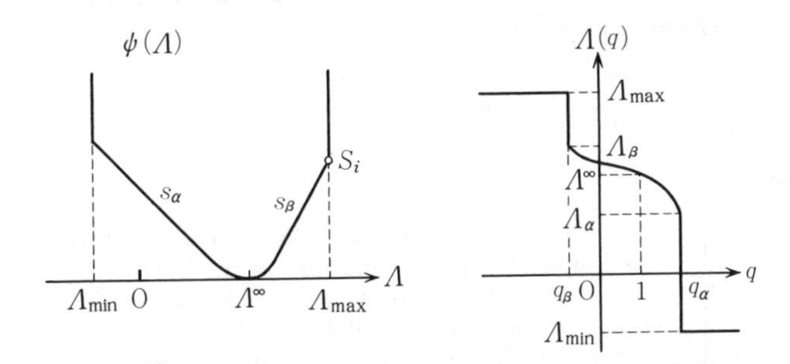

Fig. 8.8. The q-phase transition of $\psi(\Lambda)$ and $\Lambda(q)$ at the bifurcation point corresponding to collision with the saddle S_i. As seen, $\psi(\Lambda)$ possesses slope s_α and s_β linear segments on either side of its minimum, while $\sigma_\infty(q)$ diverges at $q = q_\alpha$ and q_β

There are cases in which $\Lambda_\infty(S_i) < \Lambda^\infty$ and in which $\psi(\Lambda_\infty(S_i))$ appears in the left-hand portion of $\psi(\Lambda)$ plotted in Fig. 8.7. In this case, the linear segment of $\psi(\Lambda)$ with slope s_β $(-1 < s_\beta < 0)$ defines a line of tangency to $g(\Lambda)$, and this segment extends from the point $[\Lambda_\infty(S_i),\ \psi(\Lambda_\infty(S_i))]$ to $[\Lambda, \psi(\Lambda)]$ with Λ given by some value near Λ^∞. In this case, the linear segment of slope s_α then appears to the left of this [in the $\Lambda < \Lambda_\infty(S_i)$ region].

8.2 The Singularity Spectrum $f(\alpha)$

On a strange attractor, the natural probability density $p(X)$ possesses a large number of singularities. Let us now study the manner in which we can understand this situation. Our presentation here is based on the study by Halsey et al. (1986) and the body of work which followed it.

Let us consider a small box $B_x(l)$ of linear extent l centered at some point X in the attractor of the system in question. Then, if we designate the number of points X_t on some aperiodic orbit $\{X_t\}$ $(t = 1, 2, \ldots N)$ which enter this box by $N(B)$, the *natural probability* $\mu_X(l)$ of this box is defined by $\mu_X(l) \equiv \lim_{N\to\infty} N(B)/N$. Now, let us define the exponent $\alpha(X, l)$ by

$$\mu_X(l) \propto l^{\alpha(X,l)} . \tag{8.25}$$

Then, let us assume that the *local dimension* $\alpha(X) \equiv \lim_{l\to 0} \alpha(X, l)$ exists. Such a function $\alpha(X)$ exists when the neighborhood of X possesses self-similar nested structure like that in the neighborhood of S in Fig. 7.2d and X^* in Fig. 7.4. Figure 8.9 presents a comparison of forms of $\mu_X(l)$ in the neighborhood of the saddle S from Fig. 7.2d obtained from numerical and theoretical studies. (The value $\alpha(S) \approx 1.281$ obtained theoretically appears as one of the entries in Table 8.1.) These results indicate that a function $\alpha(X)$ of this nature indeed exists. The undulations exhibited by the numerically generated form of $\mu_S(l)$ result from the discrete nature of the multiplex string-like structure existing in the neighborhood of S.

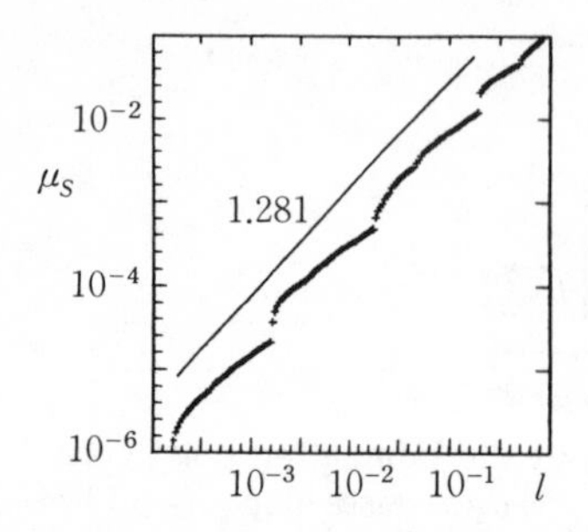

Fig. 8.9. A comparison between the theoretical and numerical determinations of the l dependence of the probability measure $\mu_S(l)$ in the neighborhood of the saddle S of Fig. 7.2d. The line representing the theoretical results corresponds to $\alpha(S) \approx 1.281$. The numerical data were generated using $N = 10^8$

In general, the upper limit of the scale on which the system displays self-similar structure in the neighborhood of a given point X depends on the position of this point. In addition, the local dimension $\alpha(X)$ also depends on X and will assume a range of values between some $\alpha_{\min}$ and $\alpha_{\max}$. If we define the probability $P(\alpha; l)\mathrm{d}\alpha$ that the value of $\alpha(X, l)$ will be found in the interval $(\alpha, \alpha + \mathrm{d}\alpha)$, taking $l \to 0$ we can write

$$P(\alpha; l) \equiv \langle \delta(\alpha(X, l) - \alpha) \rangle = l^{\alpha - f(\alpha)} P(\bar{\alpha}; l) , \tag{8.26}$$

where the angle brackets represent the long-time average defined by (6.22). This average is identical to the average with respect to the measure $\mu_X(l)$. Also here, $\overline{\alpha} \equiv \langle \alpha(X) \rangle$, and f is some function satisfying the relations $f(\overline{\alpha}) = \overline{\alpha}$ and $\alpha \geq f(\alpha) \geq 0$. As we will see later, $f(\alpha)$ is the Hausdorff dimension of the set of points $\{X\}$ at which $\alpha(X) = \alpha$.

With the partition function corresponding to the measure $\mu_X(l)$,

$$\chi(q; l) \equiv \left\langle \{\mu_X(l)\}^{q-1} \right\rangle \propto \int \mathrm{d}\alpha \, l^{(q-1)\alpha} P(\alpha; l) , \tag{8.27}$$

let us introduce the functions

$$\tau(q) \equiv (1/\ln l) \ln[\chi(q; l)] \tag{8.28}$$

$$\alpha(q) \equiv \tau'(q) = \langle \alpha \rangle_{q,l} \tag{8.29}$$

$$\sigma^\alpha(q) \equiv -\alpha'(q) = |\ln l| \left\langle \{\alpha - \alpha(q)\}^2 \right\rangle_{q,l} \geq 0 \tag{8.30}$$

corresponding to (8.2)–(8.4). Here, $\langle \cdots \rangle_{q,l}$ represents an average with respect to the weight $l^{(q-1)\alpha} P(\alpha; l)$. Since $\chi(q; l) = 1$ for $q = 1$, (8.29) leads to

$\tau(q) = \int_1^q dq\alpha(q)$. It is natural to refer to this quantity as the q-potential of α. Note that $\alpha'(q) \leq 0$, and therefore $\alpha(q)$ is a nonincreasing function of q, and $\alpha(-\infty) = \alpha_{\max}$, $\alpha(1) = \overline{\alpha}$, and $\alpha(\infty) = \alpha_{\min}$. Next, substituting (8.26) into (8.27), we obtain $\tau(q) \sim (1/\ln l) \ln \left[\int d\alpha l^{\alpha q - f(\alpha)}\right]$. Then taking the maximal value of the integrand in the $l \to 0$ limit, we obtain $\tau(q) = \min_\alpha \{q\alpha - f(\alpha)\}$. From this we have $f'(\alpha) = q$ and $f''(\alpha) = -1/\sigma^\alpha(q) \leq 0$. Hence we see that $f(\alpha)$ is an upward convex function. Furthermore, as shown in Fig. 8.10, $f(\alpha)$ is tangent at $\alpha = \alpha(1)$ to the 45° line defined by $f = \alpha$ and assumes its maximal value at $\alpha = \alpha(0)$. Also note that at the points $\alpha(\pm\infty)$ we have $f'(\alpha) = \pm\infty$. When $\tau(q)$ can be easily determined, the variational principle

$$f(\alpha) = \min_q \{q\alpha - \tau(q)\} = \alpha q(\alpha) - \tau(q(\alpha)) \tag{8.31}$$

can be used to find $f(\alpha)$, where here, $q(\alpha)$ is the solution of the equation $\tau'(q) = \alpha$.

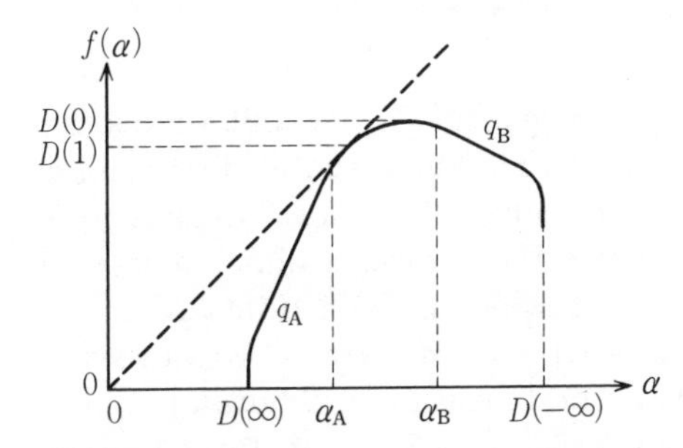

Fig. 8.10. The form of $f(\alpha)$ for a chaotic attractor. At $\alpha = D(1) = \alpha(1)$, $f(\alpha)$ is tangent to the 45° line $f = \alpha$ (broken line). It assumes its maximum value of $D(0)$ at $\alpha = \alpha(0)$. At the endpoints $\alpha = \alpha(\pm\infty) = D(\pm\infty)$, we have $f'(\alpha) = \pm\infty$. At the chaotic bifurcation points, $f(\alpha)$ displays linear segments of slopes q_A (>2) and q_B (<0)

The upper limit on the scale L_X at which self-similarity appears in multifractal structure in general depends, as $\alpha(X)$, on X (see Sect. 8.2.3). In the case in which L_X does not depend on X, the quantity in (8.26) is independent of l, and thus a constant local dimension $D \equiv \alpha = f(\alpha)$ exists. The q-potential of α then becomes $\tau(q) = (q-1)D$.

In the band attractor existing just beyond the emergence point of chaos, in the neighborhood of each point X there are two types of self-similarity, one that exists on a lengthscale smaller than some characteristic value l_w and one that exists on a length scale larger than some second characteristic value l_d ($> l_w$). (This will be discussed in Sect. 9.3.4.) The types of self-similarity existing on these two levels differ, and for this reason here it is necessary to take the $l \to 0$ limit mentioned above twice, once imposing the condition $l_w > l$ and once the condition $l_d < l$. In this way, two different functional forms of $f(\alpha)$ are generated, reflecting the two different types of self-similarity.

8.2.1 The Multifractal Dimension $D(q)$

In a uniform set, if we divide the sides of a small box of linear extent l into m equal pieces, we will obtain $N = m^D = r^{-D}$ boxes each of length $l' \equiv l/m = rl$. Here, D is the *fractal dimension.* Note that this quantity can be written $D = \ln N/\ln r^{-1}$. For example, for the Cantor ternary set shown in Fig. 8.11, we have $D = \ln 2/\ln 3 \approx 0.63$.

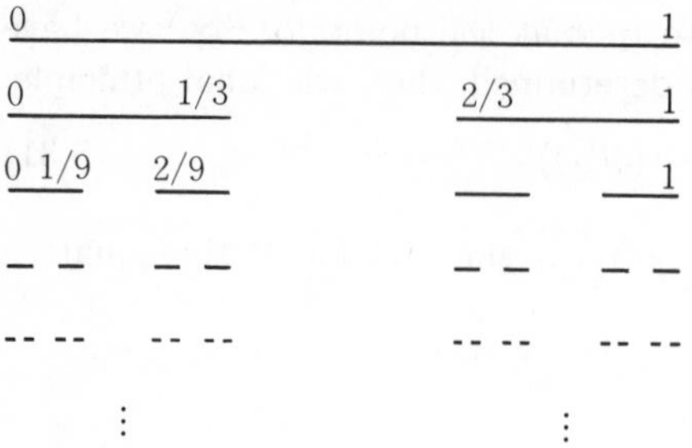

0 1

0 1/3 2/3 1

0 1/9 2/9 1

Fig. 8.11. The construction of a Cantor ternary set. We begin by breaking the segment $[0, 1]$ into three equal length pieces and removing the center piece. Repeating this procedure n times yields a set of 2^n segments and 2^{n+1} endpoints. The $n \to \infty$ limit of the set of endpoints defined in this way is a Cantor set

However, in the case in which the set in question is not uniform, as, for example, the critical golden torus shown in Fig. 7.5, the local dimension is a function of X. The distribution of such values of the local dimension is given by $f(\alpha)$. Let us now explicitly extract the various dimensions contained therein. First, let us consider a set S which is divided into N boxes, $S_1, S_2, \ldots S_N$, each of length l. Assume that no two of these boxes have a nonempty intersection. Then, with the natural measure of S_i given by $\mu_i(l)$, the partition function (8.27) can be written $\chi(q;l) = \sum_{i=1}^{N} \mu_i^q$. We define the multifractal dimension here in the $l \to 0$ limit as

$$D(q) \equiv \frac{\tau(q)}{q-1} = \frac{1}{q-1}\left[q\alpha(q) - f(\alpha(q))\right] . \tag{8.32}$$

The quantity $D(0) = \ln N/\ln l^{-1}$ is the *Hausdorff dimension* of the support for the measure in question. Note that $D'(q) = \{f(\alpha) - \alpha(q)\}/(q-1)^2 \leq 0$, and thus $D(q)$ is a nonincreasing function of q.

When $q \to 1$, we obtain $\mu_i^q = \mu_i\{1+(q-1)\ln\mu_i\}$, and thus, using $\chi(1;l) = 1$, we have $D(1) = -\sum_i \mu_i \ln\mu_i/\ln l^{-1}$. The numerator of this quantity is known as the information entropy. This value increases logarithmically with the resolution index (the inverse of l). Its rate of increase is given by $D(1)$, which is referred to as the *information dimension.* Note that we can also obtain the relations between $D(q)$ and $f(\alpha)$ appearing in Fig. 8.10, $D(1) = \tau'(1) = f(\alpha(1)) = \alpha(1)$, $D(0) = -\tau(0) = f(\alpha(0))$, and $D(\pm\infty) = \alpha(\pm\infty)$.

8.2.2 Partial Local Dimensions $\alpha_1(X)$ and $\alpha_2(X)$

A chaotic attractor for a system of dimension ≥ 2 possesses a multiplex string-like structure similar to that shown in Figs. 6.1b and 7.4. The structure in the neighborhood of every point X in such an attractor is anisotropic. In order to capture this anisotropy, let us split $\alpha(X)$, writing $\alpha(X) = \alpha_1(X)+\alpha_2(X)$, where $\alpha_1(X)$ is the dimension at X along the direction of $u_1(X)$, and $\alpha_2(X)$ is that along $u_2(X)$. Here, $u_1(X)$ is the unit vector tangent to W^{u} at X, and $u_2(X)$ is that transverse to $u_1(X)$; that is, we choose a box centered at X of length l_1 along $u_1(X)$ and length l_2 along $u_2(X)$. Then, expanding (8.25) and (8.28) and taking the $l_1, l_2 \to 0$ limit, we obtain

$$\mu_X(l_1, l_2) \propto l_1^{\alpha_1(X,l_1)} l_2^{\alpha_2(X,l_2)} \tag{8.33}$$

$$\chi(q; l_1, l_2) \equiv \langle \{\mu_X(l_1, l_2)\}^{q-1} \rangle \propto l_1^{\tau_1(q)} l_2^{\tau_2(q)} , \tag{8.34}$$

where $\alpha_k(X) = \lim_{l\to 0} \alpha_k(X, l)$ and $\tau(q) = \tau_1(q) + \tau_2(q)$. Also, we have

$$\alpha_k(q) = \tau_k'(q), \qquad \alpha(q) = \alpha_1(q) + \alpha_2(q) \tag{8.35}$$

$$D_k(q) = \tau_k(q)/(q-1), \quad D(q) = D_1(q) + D_2(q) . \tag{8.36}$$

The quantities $D_1(q), D_2(q), \alpha_1(q)$ and $\alpha_2(q)$ are called *partial dimensions.*

Now, if we use the solution $q = q(\alpha_k)$ of the equation $\tau_k'(q) = \alpha_k$, we can write

$$f_k(\alpha_k) \equiv \alpha_k q(\alpha_k) - \tau_k(q), \quad f(\alpha) = f_1(\alpha_1) + f_2(\alpha_2) . \tag{8.37}$$

With these identifications, $\alpha_k(q)$ is the solution of $f_k'(\alpha_k) = q$, and the function $f_k(\alpha_k)$ describes a curve tangent to the 45° line $f_k = \alpha_k$ at $\alpha_k = \alpha_k(1) = D_k(1)$. At $\alpha_k = \alpha_k(0)$ this function assumes its maximum value, $f_k(\alpha_k) = D_k(0)$.

In a two-dimensional hyperbolic system, such as the baker transformation (8.9), the chaotic attractor consists of a uniform string-like Cantor set with $D_1(q) = 1$ and $D_2(q) < 1$. This implies that $\tau_1(q) = q-1, \alpha_1(q) = f_1(\alpha_1) = 1$, and $\alpha_2(q) < 1$.

8.2.3 $f(\alpha)$ Spectra of Critical Attractors

In order to generalize the definition (8.28) of the q-potential $\tau(q)$, let us consider a set S embedded in a bounded region in a d-dimensional phase space. Then, we divide S into N nonintersecting pieces $S_1, S_2 \dots S_N$. We designate by μ_i the natural probability measure corresponding to S_i and assume that S_i is contained within a d-dimensional sphere of radius l_i. Then, with $l_i < l$, we define the *generalized partition function*

$$\Gamma(q, \tau) \equiv \lim_{l\to 0} \sum_{i=1}^{N} \frac{\mu_i^q}{l_i^\tau} , \tag{8.38}$$

where we choose $\{S_i\}$ so as to maximize the sum (8.38) when $q > 1$ and $\tau > 0$ [i.e. $D(q) > 0$] and minimize this sum when $q < 1$ and $\tau < 0$. In this case, we can define the function $\tau = \tau(q)$ as the solution of the relation

$$\Gamma(q, \tau) = 1 . \tag{8.39}$$

Then, for $\tau < \tau(q)$, $\Gamma(q,\tau) = 0$ and for $\tau > \tau(q)$, $\Gamma(q,\tau) = \infty$. In the case $q = 0$ this is consistent with the mathematical definition of the Hausdorff dimension $D(0) = -\tau(0)$. Also, if $l_i = l$ for each i, (8.39) reduces to (8.28).

Using this $\tau(q)$, we can obtain $D(q)$ from (8.32), and from (8.29) and (8.31) we can obtain $\alpha(q)$ and $f(\alpha)$. In what follows we consider a coherent decomposition $\{S_i\}$ for which all μ_i are equal.

(1) $f(\alpha)$ for a Critical 2^∞ Attractor. 2^n-bifurcations accumulate at $\hat{a}_\infty$ according to the similarity law expressed in Fig. 7.11. Let us find the spectrum $f(\alpha)$ for the attractor at the accumulation point $a = \hat{a}_\infty$.

We focus on the apex orbit $y_j = f^j(0)$ $(j = 1, 2, \ldots 2^n)$ for the quadratic map (7.22). Figure 8.12 presents a plot of the orbits $\{y_j\}$ for the cases $n = 2, 3, 4, \ldots$. The line segments appearing here represent intervals within which elements of $\{y_j\}$ exist, and we consider the process depicted in the figure in which this set of intervals is continually redefined to exclude a successively greater number of 'empty' intervals in which no $\{y_j\}$ are found. Each row in this figure corresponds to some fixed value of n, and the endpoints of the intervals there represent the points in the set $\{y_j\}$ for that value of n. Taking the $n \to \infty$ limit, this set of points defines a Cantor set. When n is increased by 1, each component, 'nonempty' interval is split into two such intervals separated by an empty interval. The two nonempty intervals thus obtained are shorter than the original by factors of $\alpha_{\mathrm{PD}}^{-1}$ and $\alpha_{\mathrm{PD}}^{-2}$, as in Fig. 7.11. Thus, using the average contraction ratio $r = (\alpha_{\mathrm{PD}}^{-1} + \alpha_{\mathrm{PD}}^{-2})/2$ with $N = 2$, we obtain the fractal dimension $D = \ln 2 / \ln r^{-1} \approx 0.544$.

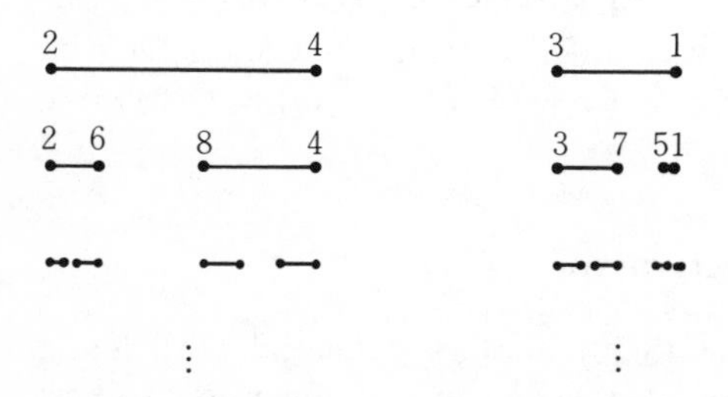

Fig. 8.12. The construction of the critical 2^∞ attractor. The numbers here designate points on the apex orbit $y_j = f^j(0)$ $(j = 1, 2, \ldots, 2^n)$. The line segments shown here represent the component intervals corresponding to their respective endpoints. In the limit as $n \to \infty$, as in Fig. 8.11, the endpoints of these segments come to define a Cantor set

Now, if we consider the decomposition $\{S_i\}$ used in (8.38) as being comprised of 2^{n-1} component intervals of the above type, then l_i represents the length of the ith such interval. Now, note that for a given n an order can be defined for the corresponding component intervals according to the action of the map, under which each interval is mapped to its successor in this order. For this reason, all 2^{n-1} component intervals have equal measure:

$\mu_i = 1/2^{n-1}$ for all i. With $n = 11$ ($2^n = 2048$), Halsey et al. (1986) solved (8.39) numerically, determining $\tau(q)$ and the curves $D(q)$ and $f(\alpha)$ shown in Fig. 8.13.

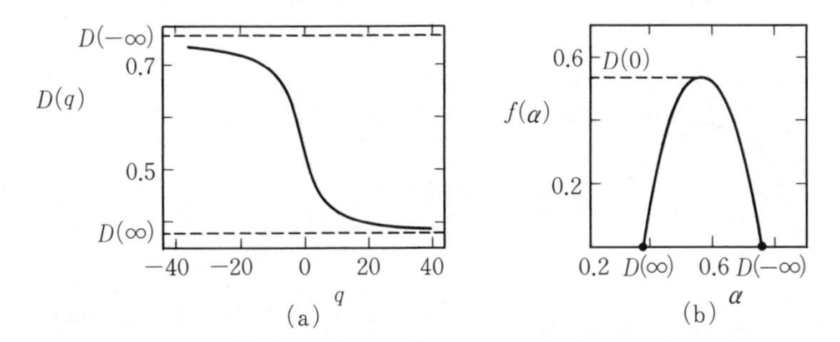

Fig. 8.13. The functions $D(q)$ and $f(\alpha)$ for the critical 2^∞ attractor as obtained numerically with 2^{11} orbital points. The broken lines designating $D(\pm\infty)$ represent theoretically obtained values [Halsey et al. (1986)]

In order to understand the function $f(\alpha)$ appearing in Fig. 8.13, let us determine the endpoints $\alpha_{\max} = D(-\infty)$ and $\alpha_{\min} = D(\infty)$. The value $D(-\infty)$ comes from the sparsest region of the set $\{S_i\}$, and its natural scale is thus $l_{-\infty} \propto \alpha_{\mathrm{PD}}^{-n}$, while $D(\infty)$ comes from the most dense region, and thus its natural scale is $l_\infty \propto \alpha_{\mathrm{PD}}^{-2n}$. The measure of a single interval S_i satisfies $\mu_i \propto 2^{-n}$, and thus using the relation $D(\mp\infty) = \ln \mu_i / \ln l_{\mp\infty}$, which can be obtained from (8.25), we find

$$D(-\infty) = \frac{\ln 2}{\ln \alpha_{\mathrm{PD}}} \approx 0.755, \quad D(\infty) = \frac{\ln 2}{\ln \alpha_{\mathrm{PD}}^2} \approx 0.377\,.$$

These values are in good agreement with the endpoints of $f(\alpha)$ appearing in Fig. 8.13. The maximal value of $f(\alpha)$ is equal to $D(0) \approx 0.537$. This is in close agreement with the analytical value obtained from (7.28) for the fixed point of the renormalization equation. However, the curve of $D(q)$ shown in Fig. 8.13 is far from $D(\mp\infty)$ even at $q = \mp 40$. Thus we can say that numerical calculation of $f(\alpha)$ produces results superior to those obtained in the same way for $D(q)$.

(2) $f(\alpha)$ of a Critical Golden Torus. As discussed in Sect. 7.4.2, the circle map (7.44) possesses a quasi-periodicity route to chaos. We now find $f(\alpha)$ for the golden torus of rotation number ρ_{G} at the critical point $K = 1, \Omega = \Omega_\infty$.

When $K = 1$, the circle map $f(\theta)$ has a cubic inflection point at $\theta = 0$. Let us consider the orbit $\theta_i = f^i(0)$ ($i = 1, 2, \ldots F_n$) emanating from this point. Here, F_n represents the nth order Fibonacci number, and we assume $n \gg 1$. Then, considering (8.38), we define the F_n intervals $l_i = |\theta_{i+F_{n-1}} - \theta_i|$ (mod 1) as the decomposition $\{S_i\}$. Note that since $\rho_{\mathrm{G}} \approx F_{n-2}/F_{n-1}$, θ_i is an approximately periodic orbit of period F_{n-1} and that $l_i \ll 1$. For the case $n = 17$ ($F_n = 2584$), Halsey et al. (1986) numerically solved (8.39), obtaining $\tau(q)$ and curve (2) of $f(\alpha)$ in Fig. 8.14. We now proceed to determine the endpoints $D(\mp\infty)$ of this curve.

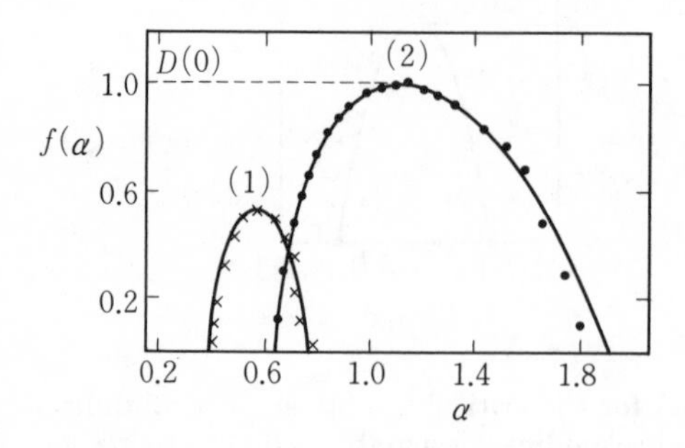

Fig. 8.14. The function $f(\alpha)$ for (1) the critical 2^∞ attractor and (2) the critical golden torus (solid curves). The data points represent experimental results for a Benard convection system [Jensen et al. (1985)]

The sparsest region, that which determines $D(-\infty)$, is situated in the neighborhood of the inflection point $\theta = 0$. The length $l_0(n) \equiv |f^{F_{n-1}}(0)|$ (mod 1) shrinks as n is increased according to the relation $l_0(n+1)/l_0(n) = \alpha_{\mathrm{GM}}^{-1}$ [Shenker (1982)]. Thus the natural scale in this neighborhood is $l_{-\infty} \propto \alpha_{\mathrm{GM}}^{-n}$. Then, the most dense region, which determines $D(\infty)$, is that to which the neighborhood of $\theta = 0$ is mapped under $f(\theta) \approx \Omega + (2\pi^2/3)\theta^3$. Hence the natural scale here decreases by a factor of $\alpha_{\mathrm{GM}}^{-3}$ when n is increased by 1, implying $l_\infty \propto \alpha_{\mathrm{GM}}^{-3n}$. Because the measure of a single interval is given by $\mu_i \propto (\rho_{\mathrm{G}})^n$, we obtain

$$D(-\infty) = \frac{\ln \rho_{\mathrm{G}}}{\ln \alpha_{\mathrm{GM}}^{-1}} \approx 1.898, \quad D(\infty) = \frac{\ln \rho_{\mathrm{G}}}{\ln \alpha_{\mathrm{GM}}^{-3}} \approx 0.632 \,.$$

These values agree quite well with those corresponding to the endpoints of the curve $f(\alpha)$ given in Fig. 8.14. The maximum of $f(\alpha)$ is $D(0) = 1$.

(3) Comparison with Experiment. In Fig. 8.14, the curves $f(\alpha)$ obtained in (1) and (2) are compared with experimental data taken from a Bénard convection system. Each point of the data set appearing in comparison with curve (2) was obtained by considering the cross-section of the critical golden torus shown in Fig. 7.5 consisting of 2500 orbit points, and then determining the function $\mu_X(l)$ by counting the number of such points $N(B)$ which appear in each small box $B_X(l)$. The q-potential $\tau(q)$ was then found using (8.28), and $f(\alpha)$ was plotted according to (8.31) [Jensen et al. (1985)]. With the exception of the right-most portion of this plot, the agreement displayed here is quite good. This right-most region corresponds to the sparsest region of the system (that in which the orbit points appear most sparsely). For this reason, in order to obtain precise results in this region it is necessary to consider a very large number of orbit points. Curve (1) in this figure compares the form of $f(\alpha)$ obtained theoretically with that obtained experimentally for a critical 2^∞ attractor. The agreement here is quite good.

8.3 Theory Regarding the Slope of $\psi(\Lambda)$

A chaotic orbit X_t eventually lies on the closure of some unstable manifold W^{u}, and the expansion rate $\lambda_1(X_t)$ of such an orbit yields information concerning the local structure of this W^{u} and other coexisting invariant sets. The spectrum $\psi(\Lambda)$ represents one example of this type of information. The information which can be most directly extracted from $\psi(\Lambda)$ is that concerning the linear segment (8.23), expressing the existence of tangency structure, and the linear segment (8.24), reflecting collision with the saddle and the resulting bifurcation which leads to a change in the attractor. These linear segments play a central role in the unification of the geometrical and statistical descriptions of chaos. We now consider the theory involved in this unification, as established by Horita et al. (1988b) and authors of subsequent works.

8.3.1 The Slope s_α Due to the Folding of W^{u} for Tangency Structure

Let us study the tangency orbit $X_{\pm j} = F^{\pm j}(X_T)$ consisting of the points at which $W^{\mathrm{u}}(X^*)$ is tangent to $W^{\mathrm{s}}(X^*)$, as shown in Fig. 7.3. Here, we take X_T to be that point along this orbit at which the relative sizes of the curvatures of $W^{\mathrm{s}}(X^*)$ and $W^{\mathrm{u}}(X^*)$ become reversed, and at which there is a transition from type II tangency structure to type I tangency structure. The neighborhood of X_T possesses a form like that shown in Fig. 8.15. Let us assume that the functional form of the curve W^{u} in the small interval $-\delta < x < \delta$ shown in this figure is given by $y = w(x) = |x|^z/z$ $(z > 1)$.

If W^{u} in the neighborhood of X_T is subjected to t iterations of F, it undergoes the folding depicted in Fig. 8.15, becoming shortened by a factor l_{s} (< 1) along the x direction and elongated by a factor l_{u} (> 1) along

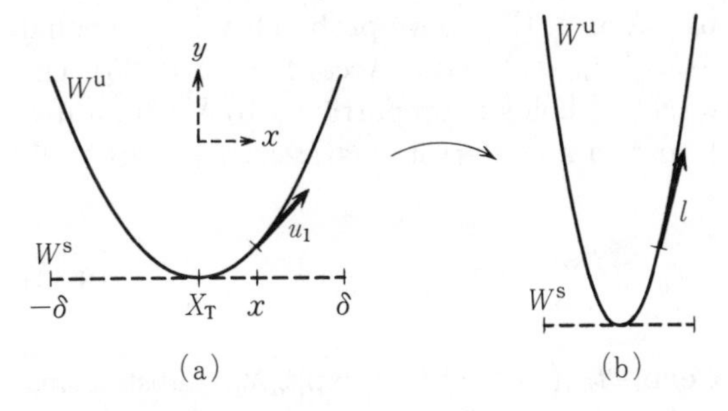

Fig. 8.15. The folding and bending of W^{u} for type I tangency structure (see Fig. 6.7) and the resulting change suffered by the unit tangent vector u_1. Under t forward mappings, the length of this vector along the x direction shrinks by a factor $l_{\mathrm{s}} = \mathrm{e}^{t\lambda_{\mathrm{s}}}$ (< 1) and stretches by a factor $l_{\mathrm{u}} = \mathrm{e}^{t\lambda_{\mathrm{u}}}$ (> 1). For $t \gg 1$, $\lambda_{\mathrm{s}} = \Lambda_\infty(X_{\mathrm{T}})$ and $\lambda_{\mathrm{u}} = \Lambda'_\infty(X_{\mathrm{T}})$

the y direction. In undergoing this transformation, the unit vector $u_1 = (u_x, u_y)$ along W^{u} located at some point x in the neighborhood of X_T becomes transformed into the tangent vector $l = (l_x, l_y)$. The length of this vector is given by

$$l^2(x,t) = [l_{\mathrm{s}}^2 + \{w'(x)\}^2 l_{\mathrm{u}}^2]/[1 + \{w'(x)\}^2] \,. \tag{8.40}$$

This expression can be obtained from the relations $l_x = l_{\mathrm{s}} u_x$, $l_y = l_{\mathrm{u}} u_y$, $u_y/u_x = w'(x)$, and $l^2 = (l_{\mathrm{s}}^2 + w'^2 l_{\mathrm{u}}^2) u_x^2$. If we solve (8.40) for w'^2, using the equality $w'^2 = |x|^{2(z-1)}$, we obtain

$$|x| = \left\{ \frac{l^2(x,t) - l_{\mathrm{s}}^2}{l_{\mathrm{u}}^2 - l^2(x,t)} \right\}^{1/2(z-1)} \approx \left\{ \frac{l(x,t)}{l_{\mathrm{u}}} \right\}^{1/(z-1)} , \tag{8.41}$$

where we have assumed that $l_{\mathrm{u}} \gg l \gg l_{\mathrm{s}}$. Then, note that $\Lambda_{\min} = \lim_{t\to\infty}(1/t)\ln l_{\mathrm{s}}(t)$.

Let us now consider a chaotic orbit of length n passing through the point x in the neighborhood of X_T. Note that the initial point of this orbit is also in the neighborhood of some point of tangency between W^{u} and W^{s}. Let us call this the 'initial tangency point'. Then, we define the number of mappings under which the initial tangency point is mapped to X_T as t_b $(0 \le t_b \le n)$ and the number of mappings corresponding to the remainder of the orbit as $t_a = n - t_b$. The coarse-grained expansion rate (6.20) for this orbit can then be written

$$\Lambda_n(X_0) = (1/n)[\ln l(x, t_a) + t_b \lambda_b(x; t_b)] \,. \tag{8.42}$$

Here, λ_b is the coarse-grained expansion rate before the initial point is mapped to x. Note that since the regions through which the orbit passes before reaching x are characterized by type II tangency structure, $\lambda_b > 0$. The distribution corresponding to $\lambda_b(x; t_b)$ gives rise to the hyperbolic portion of $\psi(\Lambda)$. Let us designate the probability that both $\Lambda_n \le \Lambda$ and $\lambda \le \lambda_b < \lambda + \mathrm{d}\lambda$ for some Λ and λ by $\mu(\Lambda_n \le \Lambda) \times P_b(\lambda; t_b)\mathrm{d}\lambda$. Then, if we define that value of $|x|$ for which $\Lambda_n = \Lambda$ and $\lambda_b = \lambda$ as δ, the above probability is also equal to that for which $|x| \le \delta$ and $\lambda \le \lambda_b < \lambda + \mathrm{d}\lambda$. According to (8.33), the probability that the inequality $|x| \le \delta$ holds is proportional to δ^{α_1} [in a dissipative system, $\alpha_1 = \alpha_1(X_T)$, and in a conservative system, $\alpha_1 = \alpha(X_T)$], and therefore the relation

$$\mu(\Lambda_n \le \Lambda) \propto \int \mathrm{d}\lambda \sum_{t_b=0}^{n} \delta^{\alpha_1} P_b(\lambda; t_b) \tag{8.43}$$

holds. Now, setting $P_b(\lambda; t_b) \propto \exp[-t_b g(\lambda)]$ and $l_{\mathrm{u}} = \exp[t_a \lambda_{\mathrm{u}}]$, substituting (8.41) into the expression $\delta = |x|$, and using $l(x, t_0) = \exp[n\Lambda - t_b \lambda]$ yields

$$\mu(\Lambda_n \le \Lambda) \propto \int \mathrm{d}\lambda \sum_{t_b=0}^{n} \mathrm{e}^{n\hat{\alpha}(\Lambda - \lambda_{\mathrm{u}}) - t_b\{g(\lambda) + \hat{\alpha}\lambda - \hat{\alpha}\lambda_{\mathrm{u}}\}} \,, \tag{8.44}$$

where $\hat{\alpha} \equiv \alpha_1/(z-1)$. As in Fig. 8.7, the function $g(\lambda)$ here represents the hyperbolic portion of $\psi(\Lambda)$, and the value of λ at which $g(\lambda) + \hat{\alpha}\lambda$ is a minimum, $\lambda = \Lambda_\alpha$, lies to the left of the Liapunov number Λ^∞. Then, using the value assumed by the integrand at $\lambda = \Lambda_\alpha$ to approximate the integral over λ in (8.44), and using the approximation $\sum_{t_b=0}^{n} \approx n\int_0^1 \mathrm{d}r$, where $t_b = rn$, for $\Lambda \le \Lambda_\alpha$ we obtain

$$\mu(\Lambda_n \le \Lambda) \propto \exp[-n\{-\hat{\alpha}(\Lambda - \Lambda_\alpha) + g(\Lambda_\alpha)\}] , \tag{8.45}$$

where we have assumed $\hat{\alpha}(\lambda_u - \Lambda_\alpha) > g(\Lambda_\alpha)$. Taking the derivative of this with respect to Λ yields the probability density (6.27), and hence the spectrum $\psi(\Lambda)$ possesses the form shown in (8.23). The linear segment of this spectrum has slope s_α, given by

$$s_\alpha = -\hat{\alpha} = -\alpha_1/(z-1) . \tag{8.46}$$

Then, because $g'(\Lambda_\alpha) = -\hat{\alpha}$, this linear segment is tangent to $g(\Lambda)$ at $\Lambda = \Lambda_\alpha$. For a dissipative system this attractor in the neighborhood of X_T assumes a smooth, string-like shape, with $\alpha_1(X_T) = 1$ and $z = 2$. Thus, as in Fig. 8.7, we have $s_\alpha = -1$. For a conservative system the probability measure is proportional to the Lebesgue measure, and for a two-dimensional map, $\alpha_1 = \alpha(X_T) = 2$, and $z = 2$, implying $s_\alpha = -2$.

8.3.2 The Slope s_β Resulting from Collision with the Saddle S

As shown in Fig. 8.16, when the point $[\Lambda_{\max}, \psi(\Lambda_{\max})]$ representing the value of the spectrum $\psi(\Lambda)$ corresponding to the Liapunov number of the saddle S, $\Lambda_{\max} = \Lambda_\infty(S)$, appears to the right of the spectrum $g(\Lambda)$, which represents the system before collision with S, there exists a linear segment [defined by (8.24)] of $\psi(\Lambda)$ tangent to $g(\Lambda)$ at Λ_β. With $\Lambda_\beta \le \Lambda \le \Lambda_{\max}$, this slope is given by

$$s_\beta = \frac{\psi(\Lambda_{\max}) - g(\Lambda_\beta)}{\Lambda_{\max} - \Lambda_\beta} = \frac{\psi(\Lambda_{\max})}{\Lambda_{\max} - \Lambda^0} , \tag{8.47}$$

where Λ^0 is the value at which this segment intersects the Λ axis. From (8.47) we have

$$\Lambda^0 = \frac{\Lambda_\beta \psi(\Lambda_{\max}) - \Lambda_{\max} g(\Lambda_\beta)}{\psi(\Lambda_{\max}) - g(\Lambda_\beta)} . \tag{8.48}$$

We now find $\psi(\Lambda_{\max})$, while also determining the manner in which this linear segment forms.

The characteristic feature of the bifurcation resulting from the collision of the attractor with S is, as shown in Figs. 7.2b and d, the accumulation of the tangency points $X_j = F^j(X_T)$ at S. The point S to which we refer in the present discussion corresponds to X^* in Fig. 7.2b. Figures 8.17 and 8.18a express the nature of this accumulation. In order to understand this situation, let us consider the Jacobian matrix $DF(S)$. Then let us define $\lambda_1(S) \equiv$

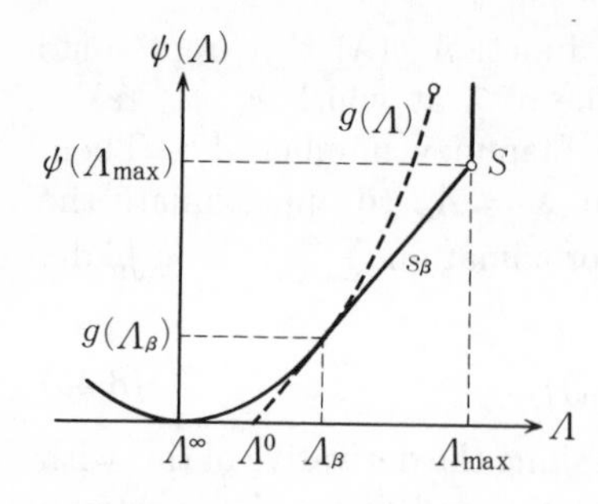

Fig. 8.16. Formation of the slope s_β linear segment resulting from collision with the saddle S possessing the maximal Liapunov number $\Lambda_{\max} = \Lambda_\infty(S)$

$\ln|\nu_1(S)|$ and $\lambda_2(S) = \ln|\nu_2(S)|$, where $\nu_1(S)$ and $\nu_2(S)$ are the eigenvalues of $DF(S)$ along the invariant manifolds $W^{\mathrm{u}}(S)$ and $W^{\mathrm{s}}(S)$, respectively. We assume $\lambda_1(S) > 0 > \lambda_2(S)$. Now, let us assume that the form of the unstable manifold W^{u} in the neighborhood of the point of tangency $(x, y) = (1, 0)$ is given by $y = w_1(x) \equiv c|x-1|^z (z > 1)$ (see Fig. 8.18a). Consider the τth image of this neighborhood under the map. This image also constitutes a piece of W^{u} that is tangent to W^{s}. Suppose that this new point of tangency is $(\delta, 0)$ $(0 < \delta \ll 1)$. We write the form of W^{u} in this neighborhood of $(\delta, 0)$ as $y = w_\delta(x)$. Then, writing the length expansion factor along the y direction as a result of these τ iterations of the map as $l_{\mathrm{u}} = \exp[\tau\lambda_1(S)]$ and that along the x direction as $\delta = l_{\mathrm{s}} = \exp[\tau\lambda_2(S)]$, following from the similarity relation $w_\delta(x)/l_{\mathrm{u}} = c|(x/\delta) - 1|^z$, we have

$$y = w_\delta(x) = c\delta^{-\{z+r(S)\}}|x-\delta|^z , \tag{8.49}$$

where $r(S) \equiv \lambda_1(S)/|\lambda_2(S)| < 1$. Next, let us choose the point $X_0 = (x_0, \epsilon)$ on the piece of W^{u} described by (8.49) as our initial point and consider the coarse-grained expansion rate over τ mappings $[\tau = \ln\epsilon^{-1}/\lambda_1(S)]$ corresponding to this point, $\Lambda = \Lambda_\tau(X_0)$. Here, we suppose that $\epsilon = w_\delta(x_0) \ll 1$. The unit vector tangent to the curve described by (8.49) at x_0 is transformed under these τ mappings to a tangent vector of length $l(x_0, \tau)$ [see (8.40)]. Then, using $(x_0 - \delta)^z = \epsilon c^{-1}\delta^{z+r(S)}$, we have

$$w'_\delta(x_0) = zc^{1/z}(\epsilon/\delta^v)^{(z-1)/z} , \tag{8.50}$$

where $v \equiv \{z + r(S)\}/(z-1)$. We have $l = \mathrm{e}^{\tau\Lambda} = \epsilon^{-\Lambda/\lambda_1(S)}$, $l_{\mathrm{u}} = \mathrm{e}^{\tau}\lambda_1(S) = \epsilon^{-1}$ and $l_{\mathrm{s}} = \mathrm{e}^{\tau\lambda_2(S)} = \epsilon^{1/r(S)}$, and thus substituing (8.50) into (8.40), we obtain

$$\epsilon^{-2\Lambda/\lambda_1(S)} = \frac{\epsilon^{2/r(S)} + z^2c^{2/z}(\epsilon/\delta^v)^{2(z-1)/z}\epsilon^{-2}}{1 + z^2c^{2/z}(\epsilon/\delta^v)^{2(z-1)/z}} . \tag{8.51}$$

When $\epsilon \gg \delta^v$, the right-hand side of this equation is approximately equal to ϵ^{-2}, yielding $\Lambda \approx \lambda_1(S)$.

Let us now investigate the probability measure $\mu(\tau)$ associated with the initial point X_0 which gives $\Lambda_\tau(X_0) \approx \lambda_1(S)$. The point $X_0 = (x_0, \epsilon)$ in question must satisfy the relation $\epsilon = \mathrm{e}^{-\tau\lambda_1(S)} \gg \delta^v$, and therefore $\mu(\tau)$ must

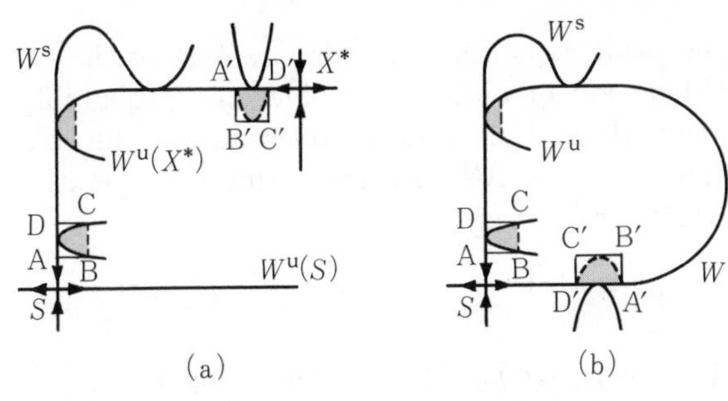

Fig. 8.17. The accumulation of tangency points $X_{\pm j} = F^{\pm j}(X_T)$ occurring at the bifurcation point. In the heteroclinic case depicted in (**a**), the rectangle ABCD in the neighborhood of S is transformed into the rectangle A′B′C′D′ in the neighborhood of X^* under a certain number of iterations of F^{-1}. In the homoclinic case shown in (**b**), ABCD is transformed into A′B′C′D′ in the neighborhood of S under F^{-1} [Horita et al. (1988b)]

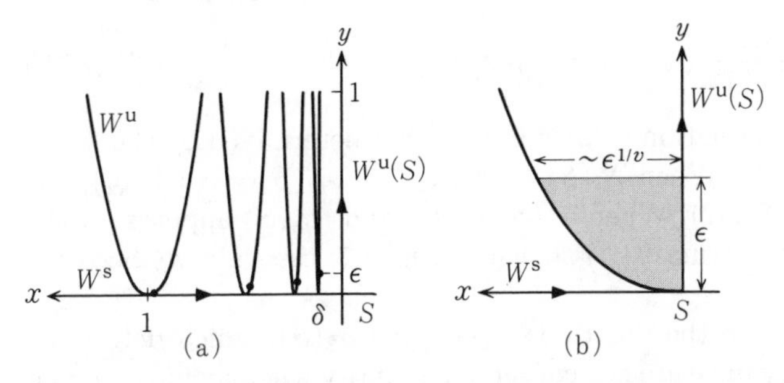

Fig. 8.18. (**a**) The convergence of the tangency points X_T to the saddle S. The piece of W^u corresponding to each tangency point is shrunk along the x direction by the factor $l_s = e^{\tau\lambda_2(S)}$ (< 1) and stretched along the y direction by the factor $l_u = e^{\tau\lambda_1(S)}$ (> 1). (**b**) The *shaded region* above the curve $(y \sim x^v)$ in the neighborhood of S represents points for which $y \gg x^v$. For the orbit with initial point $X_0 = (x_0, \epsilon)$ in this shaded region, the relation $\epsilon \gg \delta^v$ holds

be proportional to the natural probability measure of the shaded region in Fig. 8.18b. Thus, using the partial local dimensions $\alpha_1(S)$ and $\alpha_2(S)$ at S along $W^u(S)$ and $W^s(S)$, respectively, and setting $l_1 = \epsilon$ and $l_2 = \epsilon^{1/v}$ in (8.33) gives

$$\mu(\tau) \propto \epsilon^{\alpha_1(S)}\epsilon^{\alpha_2(S)/v} \propto e^{-\tau\psi_m(S)} , \tag{8.52}$$

where $\psi_m(S) \equiv h(S)\alpha_1(S)\lambda_1(S)$ and $h(S) \equiv 1+\alpha_2(S)/v\alpha_1(S)$. This measure $\mu(\tau)$ is proportional to the probability that $\Lambda_\tau(X_0)$ is equal to $\lambda_1(S) = \Lambda_\infty(S)$, and thus $\psi_m(S) = \psi(\lambda_1(S))$.

Let us write the expansion rate (6.20) for a chaotic orbit of length n $(> \tau)$ whose initial point is such an X_0 as

$$\Lambda_n(X_0) = (1/n)[\tau\lambda_1(S) + (n-\tau)\lambda_a(X_0, n-\tau)] , \tag{8.53}$$

where λ_a is the coarse-grained orbit expansion rate for any part of the orbit separated from the neighborhood of S. Now, consider the probability $P(\Lambda;n)\mathrm{d}\Lambda \times P_a(\lambda;n-\tau)\mathrm{d}\lambda$ that Λ_n and λ_a simultaneously satisfy the conditions $\Lambda \leq \Lambda_n < \Lambda + \mathrm{d}\Lambda$ and $\lambda \leq \lambda_a < \lambda + \mathrm{d}\lambda$. Integrating this quantity with respect to λ and using (8.52), we obtain

$$P(\Lambda;n) \propto \int \mathrm{d}\lambda\, P_a(\lambda;n-\tau)\mathrm{e}^{-\tau\psi_m(S)} . \tag{8.54}$$

From (8.53) we have $\tau = n(\Lambda-\lambda)/[\lambda_1(S)-\lambda]$, and thus using $P_a(\lambda;t) \propto \exp[-tg(\lambda)]$ yields

$$P(\Lambda;n) \propto \int \mathrm{d}\lambda\, \exp\left[-n\left\{\frac{\lambda_1(S)-\Lambda}{\lambda_1(S)-\lambda}g(\lambda) + \frac{\Lambda-\lambda}{\lambda_1(S)-\lambda}\psi_m(S)\right\}\right] .$$

Using the maximal value of the integrand (assumed when $\lambda = \Lambda_\beta$) to approximate this integral, in the limit as $n \to \infty$ we obtain the spectrum

$$\psi(\Lambda) = \frac{\psi_m(S) - g(\Lambda_\beta)}{\lambda_1(S) - \Lambda_\beta}(\Lambda - \Lambda_\beta) + g(\Lambda_\beta) . \tag{8.55}$$

The line defined by this relation coincides with that determined by the linear segment given by (8.24). When $\lambda_1(S) = \Lambda_{\max}$ and $\psi_m(S) = \psi(\Lambda_{\max})$, the slope here is given by (8.47). When $\lambda_1(S) < \Lambda^\infty$ and $\psi_m(S)$ appears on the left-hand side of the spectrum $\psi(\Lambda)$ shown in Fig. 8.7, the slope s_β becomes negative.

Let us next determine the functions $\alpha_1(S)$ and $\alpha_2(S)$, which determine $\psi_m(S)$. The probability measure is a conserved quantity when moving along a periodic orbit. For this reason, from (8.33), for a small box $l_1 \times l_2$ centered at S, the relation $\mu_S(l_1,l_2) = \mu_S(l_\mathrm{u}l_1, l_\mathrm{s}l_2)$ holds, and we obtain $l_\mathrm{u}^{\alpha_1(S)} l_\mathrm{s}^{\alpha_2(S)} = 1$. Then, using $l_\mathrm{u} = \mathrm{e}^{\tau\lambda_1(S)}$ and $l_\mathrm{s} = \mathrm{e}^{\tau\lambda_2(S)}$ we have

$$\alpha_2(S) = r(S)\alpha_1(S) , \tag{8.56}$$

and hence $h(S) = 1 + r(S)/v$.

To determine $\alpha_1(S)$, let us study the convergence to S of the tangency points and corresponding tangency structure depicted in Fig. 8.17. In the heteroclinic case shown in Fig. 8.17a, the tangency points in question are those at which $W^\mathrm{s}(S)$ contacts the unstable manifold $W^\mathrm{u}(X^*)$ of the other saddle point shown there. Because in the limit in which these points converge to S, the corresponding images of the rectangle ABCD shown in the figure ($\overline{\mathrm{AB}} \sim l$, $\overline{\mathrm{BC}} \sim l^{1/z}$) converge to a one-dimensional band, the probability measure $\mu(l)$ contained within this rectangle satisfies $\mu(l) \propto l^{\alpha_1(S)}$. Also, the probability measure contained within the rectangle A′B′C′D′ (a pre-image of ABCD) in the neighborhood of X^* ($\overline{\mathrm{A'B'}} \sim l'$, $\overline{\mathrm{B'C'}} \sim l'^{1/z}$, $l' \propto l$) is proportional to this value. This measure is a smooth function along $W^\mathrm{u}(X^*)$, and thus $\mu(l) \propto l'^{1/z} \times l'^{\alpha_2(X^*)}$, yielding the relation $\alpha_1(S) = (1/z) + \alpha_2(X^*)$. Furthermore, we can set $\alpha_1(X^*) = 1$, and from (8.56) we have $\alpha_2(X^*) = r(X^*)$. Therefore

$$\alpha_1(S) = (1/z) + r(X^*) . \tag{8.57}$$

In the homoclinic case depicted in Fig. 8.17b, the tangency points in question accumulating at S are those at which $W^{\mathrm{s}}(S)$ contacts $W^{\mathrm{u}}(S)$. Applying the above reasoning to the rectangles ABCD and A′B′C′D′ that appear in this figure, and noting that the probability measure contained in the latter is a smooth function at all points along W^{u} except S, we obtain $\alpha_1(S) = (1/z) + \alpha_2(S)$. Substituting this into (8.56) and solving for α_1 yields

$$\alpha_1(S) = (1/z) + r(S)/z\{1 - r(S)\} . \tag{8.58}$$

In this way we are able to express the local dimensions in terms of more fundamental quantities, the eigenvalues of the Jacobian matrix.

To summarize the above discussion, we have

$$\psi(\Lambda_{\max}) = \psi(\Lambda_\infty(S)) = h(S)\alpha_1(S)\Lambda_\infty(S) \tag{8.59}$$

$$h(S) = 1 + (z-1)r(S)/\{z + r(S)\} , \tag{8.60}$$

where $r(S) = \Lambda_\infty(S)/\{\Lambda_\infty(S) + R(S)\}$, with $R(S) \equiv -\ln|J(S)|$. The local dimension $\alpha_1(S)$ is given in the case of heteroclinic accumulation by (8.57) and in the case of homoclinic accumulation by (8.58). For a one-dimensional map ($J = 0$), $r = 0$, implying $\alpha_1(S) = 1/z$ and

$$\psi(\Lambda_{\max}) = (1/z)\Lambda_\infty(S) . \tag{8.61}$$

Table 8.1 presents various quantities predicted by the theory outlined above together with the numerically calculated value of s_β for bifurcations of the Hénon map, the quadratic map, and the circle map treated in Sect. 8.1. In the situations described there, $z = 2$. Agreement between the theoretical and numerically calculated values is reasonably good. The discrepancy exhibited between these values is due mainly to the finite nature of the value of N used in the numerical studies.

The relations (8.59) and (8.60) also hold in the case of collision with a period Q saddle $S_i = F^Q(S_i)$. However, in this case, in place of $\lambda_1(S)$ and $\lambda_2(S)$, we consider the logarithms of the absolute values of the eigenvalues of the Jacobian matrix $DF^Q(S_i)$, $\overline{\lambda}_1(S_i) = \Lambda_\infty(S_i)$ and $\overline{\lambda}_2 = -\Lambda_\infty(S_i) - R(S_i)$, with $R(S_i) \equiv -Q^{-1}\sum_{t=1}^{Q} \ln|J(S_t)|$, and treat the convergence of the tangency points and corresponding tangency structure to S_i under the map $\phi(X) \equiv F^Q(X)$.

As seen here, the slopes s_α and s_β both arise in the case of type I tangency structure and can be thought of as representing the universality exhibited by nonhyperbolic, physical systems.

8.4 The Relation Between $f(\alpha)$ and $\psi(\Lambda)$

The fractal structure function $f(\alpha)$ for a chaotic attractor differs qualitatively from that shown in Fig. 8.14 for the case of a critical attractor. In the former case, $f(\alpha)$ has a form like that shown in Fig. 8.10. That is to say, the tangency structure and bifurcations of chaos produce singular fractal structure, and the corresponding spectrum possesses linear segments of slopes q_A and q_B corresponding to s_α and s_β of Figs. 8.7 and 8.8. In order to elucidate the nature of this fractal structure of chaos, it is first necessary to discuss the relation between $f(\alpha)$ and $\psi(\Lambda)$, as determined by Morita et al. (1988).

Let us consider the natural probability measure $\mu_t(l_1, l_2)$ associated with a small cell of size $l_1 \times l_2$ centered at a point X_t belonging to a chaotic orbit X_j $(j = 0, 1, \ldots n)$. Under the map F, this cell is transformed into a cell of size $l_1 \exp[\lambda_1(X_t)] \times l_2 \exp[\lambda_2(X_t)]$. Here, $\lambda_2(X_t) = -\lambda_1(X_t) + \ln|J(X_t)|$ is the orbital expansion rate along the direction $u_2(X_t)$, transverse to $u_1(X_t)$. As implied by the conservation of probability measure, we have $\mu_t(l_1, l_2) = \mu_{t+1}\left(l_1 e^{\lambda_1(X_t)}, l_2 e^{\lambda_2(X_t)}\right)$. Writing $l_k e^{\lambda_k(X_t)}$ as l_k, this becomes $\mu_{t+1}(l_1, l_2) = \hat{M}_t \mu_t(l_1, l_2)$, where $\hat{M}_t$ is the differential operator

$$\hat{M}_t \equiv \exp\left[-\lambda_1(X_t) l_1 \frac{\partial}{\partial l_1} - \lambda_2(X_t) l_2 \frac{\partial}{\partial l_2}\right] . \tag{8.62}$$

Here we have used the identity $\exp[ax(\partial/\partial x)]x^i = (e^a x)^i$.

For the purpose of introducing the q-potential $\tau_k(q)$, let us consider

$$\{\mu_{n+1}(l_1, l_2)\}^{q-1} = \hat{M}_n \hat{M}_{n-1} \cdots \hat{M}_0 \{\mu_0(l_1, l_2)\}^{q-1} . \tag{8.63}$$

Then, defining the projection operator P by $PG = \langle G \rangle$ $(P^2 = P)$ and using $\hat{M}_t = \hat{M}_t P + \hat{M}_t Q (Q \equiv 1 - P)$, we obtain

$$\begin{aligned} \hat{M}_n \hat{M}_{n-1} \cdots \hat{M}_0 &= \sum_{t=0}^{n} \hat{M}_n Q \hat{M}_{n-1} Q \cdots \hat{M}_{n-t} P \hat{M}_{n-t-1} \cdots \hat{M}_1 \hat{M}_0 \\ &\quad + \hat{M}_n Q \hat{M}_{n-1} Q \cdots \hat{M}_0 Q . \end{aligned} \tag{8.64}$$

We can thus write the average of (8.63) as

$$\begin{aligned} \langle \{\mu_{n+1}(l_1, l_2)\}^{q-1} \rangle &= \sum_{t=0}^{n} \left\langle \hat{M}_n Q \hat{M}_{n-1} Q \cdots \hat{M}_{n-t} \right\rangle \langle \{\mu_{n-t}(l_1, l_2)\}^{q-1} \rangle \\ &\quad + \left\langle \hat{M}_n Q \hat{M}_{n-1} Q \cdots \hat{M}_0 Q \{\mu_0(l_1, l_2)\}^{q-1} \right\rangle . \end{aligned} \tag{8.65}$$

Since $\langle \{\mu_t(l_1, l_2)\}^{q-1} \rangle \propto l_1^{\tau_1(q)} l_2^{\tau_2(q)}$, the quantity $\hat{M}_t$ in the first term here can be written as the c number

$$M_t \equiv \exp[-\lambda_1(X_t)\tau_1(q) - \lambda_2(X_t)\tau_2(q)] . \tag{8.66}$$

Considering the second term in (8.65), in the limit as $n \to \infty$ the correlation between $\lambda_k(X_n)$ in $\hat{M}_n$ and the quantity $Q\{\mu_0(l_1, l_2)\}^{q-1}$ at the initial time vanishes, and therefore this second term can be ignored. This can be thought to hold at least in the case of a $\lambda_1(X_t) > 0$ hyperbolic system. Here, shifting time by $n - t$, we obtain

$$\sum_{t=0}^{\infty} \langle M_t Q M_{t-1} Q \cdots M_0 \rangle = 1 \,. \tag{8.67}$$

Next, let us consider $\Xi_n \equiv \langle M_{n-1} M_{n-2} \cdots M_0 \rangle$ with $\Xi_0 \equiv 1$. If we then replace $\hat{M}_t$ in (8.64) with M_t and operate with the resulting object on 1, we find that the second term vanishes, because $Q \cdot 1 = 0$. Furthermore, taking the average of this quantity, the left-hand side gives Ξ_{n+1}, and we obtain $\Xi_{n+1} = \sum_{t=0}^{n} \phi_t \Xi_{n-t}$, where $\phi_n \equiv \langle M_n Q M_{n-1} Q \cdots M_0 \rangle$. Defining $\tilde{\Xi}(z) \equiv \sum_{n=0}^{\infty} \mathrm{e}^{-zn} \Xi_n$, this can be written $\tilde{\Xi}(z) = 1/\left\{1 - \mathrm{e}^{-z}\tilde{\phi}(z)\right\}$. Then (8.67) gives $\tilde{\phi}(z=0) = 1$, implying that $\tilde{\Xi}(0)$ is infinite. However, for $z > 0$, $\tilde{\Xi}(z)$ is finite, and it is not the case that Ξ_n diverges exponentially with n. Hence $\lim_{n\to\infty} (1/n) \ln \Xi_n = 0$. Inserting (8.66) into M_t appearing in Ξ_n, this gives

$$\lim_{n\to\infty} (1/n) \ln \langle \exp[-n\Lambda_n(X)\tau_1(q) - n\Lambda_n'(X)\tau_2(q)] \rangle = 0 \,, \tag{8.68}$$

where $\Lambda_n'(X_0) \equiv (1/n) \sum_{t=0}^{n-1} \lambda_2(X_t)$. In many dynamical systems, $J(X_t)$ is a constant, defined by (6.4). In this case, substituting $\Lambda_n'(X) = -\Lambda_n(X) - R(R \equiv -\ln|J|)$, and using the q-potential $\Phi(q) \equiv \Phi_\infty(q)$ [see (8.2)], (8.68) can be written as

$$\Phi(1 + \tau_1(q) - \tau_2(q)) = R\tau_2(q) \,. \tag{8.69}$$

This expresses the important relation between $\tau_k(q)$ and $\Phi(q)$.

Referring to Fig. 8.7, the form of the *hyperbolic phase* corresponding to the region defined by $\Lambda \geq \Lambda_\alpha$ and $q < q_\alpha = 2$ is determined by $\lambda_1(X_t) > 0$ hyperbolic structure. This hyperbolic phase possesses no singular behavior along the $u_1(X)$ direction, tangent to the attractor, and we have

$$D_1(q) = \alpha_1(q) = f_1(\alpha_1) = 1, \quad \tau_1(q) = q - 1 \,, \tag{8.70}$$

with the stipulation that $q > q_\beta$ at the bifurcation point. For the hyperbolic phase corresponding to $q < q_{\mathrm{A}}$ (q_{A} defined below), (8.69) becomes

$$\Phi(q - \tau_2(q)) = R\tau_2(q) \,, \tag{8.71}$$

where $R = -\ln|J|$ and q_{A} is a constant determined by $q_{\mathrm{A}} - \tau_2(q_{\mathrm{A}}) = q_\alpha = 2$, or

$$q_{\mathrm{A}} = q_\alpha + \tau_2(q_{\mathrm{A}}) = 2 + R^{-1}\Phi(2) > 1 \,. \tag{8.72}$$

Then, taking the derivative of (8.71) with respect to q and using (8.3) and (8.35), for the hyperbolic phase with $q < q_{\mathrm{A}}$ we obtain

$$\alpha_2(q) = \Lambda(q - \tau_2(q))/\{\Lambda(q - \tau_2(q)) + R\} \,. \tag{8.73}$$

Using the relations $\alpha_2(1) = D_2(1)$ and $\tau_2(1) = 0$, this gives the well-known relation

$$D(1) = 1 + \{\Lambda^\infty/(\Lambda^\infty + R)\} \,. \tag{8.74}$$

Then using the equations (8.8) and (8.37) corresponding to $\psi(\Lambda)$ and $f_2(\alpha_2)$, from (8.71) and (8.73), for $\alpha_2 \geq \alpha_2(q_A - 0) = \Lambda_\alpha/(\Lambda + R)$, we obtain

$$f_2(\alpha_2) = \alpha_2 - \frac{1-\alpha_2}{R}\psi\left(\frac{R\alpha_2}{1-\alpha_2}\right) . \tag{8.75}$$

For the hyperbolic phase with $\alpha \geq \alpha_A \equiv \alpha(q_A - 0)$ this implies

$$f(\alpha) = \alpha - \frac{2-\alpha}{R}\psi\left(R\frac{\alpha-1}{2-\alpha}\right) , \tag{8.76}$$

where $\alpha_A = 1 + \{\Lambda_\alpha/(\Lambda_\alpha + R)\}$. This determines the relationship between $f(\alpha)$ and $\psi(\Lambda)$ and represents the generalization of the relation (8.74) between $D(1)$ and Λ^∞. For $\alpha \leq \alpha_A$, as shown in Fig. 8.10, a linear segment of slope q_A appears.

For the situation described by Fig. 8.8, corresponding to the bifurcation resulting from collision with the saddle S_i, we have $\Lambda \geq \Lambda_\beta (> \Lambda^\infty)$ and $q < q_\beta\,(< 0)$, and a linear segment with slope $s_\beta = 1 - q_\beta$ results. In this case, the tangency points and corresponding tangency structure accumulate at S_i, and for this reason $\alpha_1(S_i)$ is shifted from (8.70), as described by (8.57) and (8.58). Then, with

$$q_B = q_\beta + \tau_2(q_B) = q_\beta + R^{-1}\Phi(q_\beta) < 1 , \tag{8.77}$$

the validity of (8.71), (8.73), and related expressions is limited to the case in which $q > q_B$. Similarly, the validity of (8.76) is limited to the case $\alpha \leq \alpha_B \equiv \alpha(q_B + 0) = 1 + \{\Lambda_\beta/(\Lambda_\beta + R)\}$. As discussed below, the $\alpha \geq \alpha_B$ case is accompanied by a slope q_B.

According to the above discussion, for the region defined by $\alpha_A \leq \alpha \leq \alpha_B$ centered at $\alpha = \alpha(1) = D(1)$ (shown in Fig. 8.10) the relation (8.76) holds. The existence of a direct relationship between $\psi(\Lambda)$ and $f(\alpha)$, expressed by this equation, implies that the existence of $\psi(\Lambda)$, as that of $f(\alpha)$, is due to the presence of the self-similar nested structure of the chaotic attractor. The length scales appearing in the expressions for the probability densities (8.26) and (6.27) can be considered as being connected through the relation $l \sim \mathrm{e}^{-n}$. This kind of self-similarity is the result of many repeated applications of a single, simple mapping rule.

8.4.1 The Linear Segment of $f(\alpha)$ Resulting from the Folding of W^u in the Presence of Tangency Structure

The function $f(\alpha)$ can be thought of as possessing the linear segment

$$f(\alpha) = q_A(\alpha - \alpha'_A) + f(\alpha'_A) \quad (\alpha_A \geq \alpha \geq \alpha'_A) \tag{8.78}$$

corresponding to the linear segment of $\psi(\Lambda)$ defined by (8.23). Here q_A is given by (8.72). Because $f(\alpha)$ given in (8.76) is such that $f'(\alpha_A) = q_A$, the linear piece defined by (8.78) connects smoothly to the hyperbolic piece defined by (8.76). In fact, the numerical results shown in Fig. 8.19a describing $f(\alpha)$

for the Hénon attractor (Fig. 7.3) display a linear piece of slope $q_A \approx 2.3$. This agrees with the predictions provided by (8.72): $\Phi(2) \approx 0.36, R \approx 1.20$, and thus $q_A \approx 2.3$. As described in Sect. 8.3.1, this linear region appears as a result of the type of folding of W^u shown in Fig. 8.15. Also, note the relations $\alpha'_A = (2\alpha_A - 1)\alpha_A/(1+\alpha_A)$ and $f(\alpha'_A) = q_A(\alpha'_A - 2) + 3$ [Ott et al. (1989)].

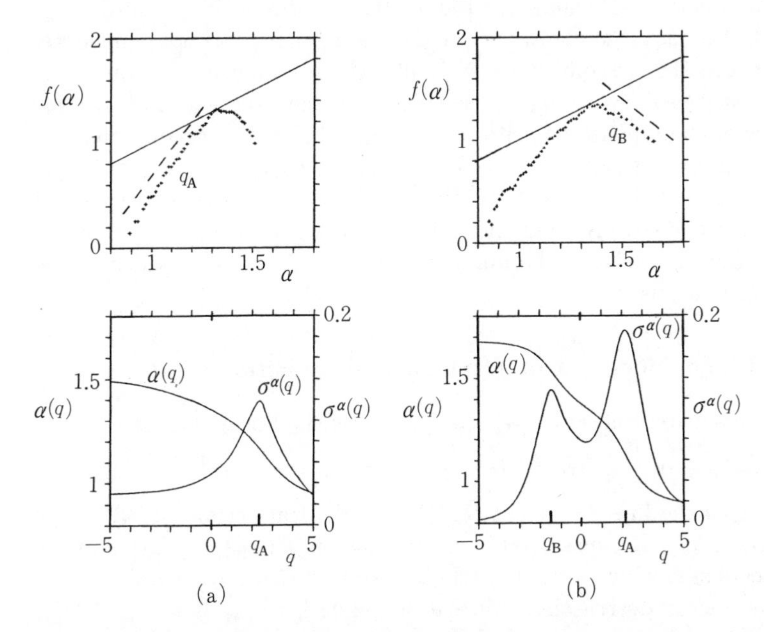

Fig. 8.19. The fractal structure functions for the Hénon map (see Figs. 7.2c and d) as obtained numerically (here $l \sim 10^{-4}$ and $N = 10^6$). The *solid lines* in the graphs of $f(\alpha)$ are the 45° lines of slope 1, while the *broken lines* are of slope q_A and q_B. The figures in (**a**) and (**b**) correspond to the cases $a = 1.4$ and $a = a_d$, respectively. The numerical values determined here are $q_A \approx 2.3$ in case (**a**) and $q_A \approx 2.1$ and $q_B \approx -1.49$ in case (**b**). The variance $\sigma^\alpha(q)$ possesses sharp peaks at $q = q_A$ and $q = q_B$ [Hata et al. (1989)]

For a one-dimensional map, in the limit as $R \to \infty$ we have $q_A = 2, \alpha_A = 1, \alpha'_A = 1/2$ and $f(\alpha'_A) = 0$. With these values, (8.78) becomes

$$f_*(\alpha) = 2\alpha - 1 \quad (1 \geq \alpha \geq 1/2) \,, \tag{8.79}$$

and we obtain $\tau(q) = q - 1$ for $q \leq 2$, and $\tau(q) = q/2$ for $q \geq 2$. In fact, for the one-dimensional quadratic map (8.14), from (8.16), in the $l \to 0$ limit the natural probability measure (8.25) satisfies $\mu_x(l) \propto l$ for $x \neq \pm 2$ and $\mu_x \propto l^{1/2}$ for $x = \pm 2$. With these expressions, and using (8.28) and (8.31), we can determine $\tau(q)$ and $f_*(\alpha)$. We have $f_*(\alpha) = -\infty$ for values of α satisfying either $\alpha > 1$ or $\alpha < 1/2$. In addition, $\alpha(q) = 1$ for $q < 2$ and $\alpha(q) = 1/2$ for $q > 2$, and thus α changes discontinuously from 1 to 1/2 at $q = 2$. This kind of q-phase transition occurs quite generally for $f(\alpha)$ given by (8.78). In application to the case $q = q_A \mp \epsilon$, the variational principle stipulating that $q\alpha - f(\alpha)$ assume a minimum value implies that $\mp\epsilon\alpha$ is minimal, and thus for $\alpha_A \geq \alpha \geq \alpha'_A$, this gives $\alpha(q) = \alpha_A$ for $q = q_A - 0$ and $\alpha(q) = \alpha'_A$ for $q = q_A + 0$. Therefore, when q is increasing, $\alpha(q)$ changes discontinuously from α_A to α'_A at $q = q_A$, signaling the q-phase transition. However, the function $\alpha(q)$ shown in Fig. 8.19 is *not* discontinuous at $q = q_A$. This is due to the fact that the value of l used here, $\sim 10^{-4}$, is not sufficiently small to capture this discontinuity. Actually, the q-variance function $\sigma^\alpha(q)$ possesses a sharp peak at $q = q_A$ whose height diverges as $l \to 0$. This insures the existence of the q-phase transition at $q = q_A$. Analytical predictions give $\Lambda^\infty \approx 0.445$, implying $D(1) \approx 1.27$. Furthermore, $\alpha_{\max} \approx \alpha(X^*) \approx 1.352$, and therefore the curve to the right side of the peak in Fig. 8.19a should in fact fall off more steeply than the function shown here. This discrepancy results from the finiteness of the number of orbits ($N = 10^6$) used in these numerical calculations.

8.4.2 The Linear Segment of $f(\alpha)$ Caused by Bifurcation

The fractal structure function $f(\alpha)$ also possesses the linear segment

$$f(\alpha) = q_B(\alpha - \alpha_B) + f(\alpha_B) \quad (\alpha_B \leq \alpha \leq \alpha'_B)\,, \tag{8.80}$$

corresponding to the linear segment (8.24) of $\psi(\Lambda)$ [Hata et al. (1989)]. Here q_B is given by (8.77). Because $f(\alpha)$ in (8.76) satisfies $f'(\alpha_B) = q_B$, this linear region connects smoothly to the hyperbolic region (8.76) at $\alpha = \alpha_B$.

At the attractor destruction bifurcation point $a = a_d$ (see Fig. 7.2d), $f(\alpha)$ possesses a linear piece of slope $q_B \approx -1.49$, as shown in Fig. 8.19b. This bifurcation is due to collision with the saddle S, and thus, as shown in Fig. 8.17a, the special feature of this bifurcation is the accumulation of tangency points and corresponding tangency structure at S under forward mapping, and accumulation at the other saddle point, X^*, under inverse mapping. Here, theoretically, we obtain the value $D(1) \approx 1.291$ from (8.74) with $\Lambda^\infty \approx 0.495$. Also, from (8.56) and the relation $\Lambda_\infty(X^*) \approx 0.655$, we obtain $\alpha(X^*) \approx 1.356$, and from Table 8.1 we have $\alpha(S) \approx 1.281$ and $\alpha_2(S) \approx 0.425$. In fact, even larger local dimensions result under the process in which the tangency structure accumulates at S, causing the linear segment (8.80) to appear. In the neighborhood of S, as shown in Fig. 8.18a, pieces of the unstable manifold $W^u(X^*)$, constituting the tangency structure, accumulate at $W^u(S)$, and thus the local dimension α^u on $W^u(S)$ (excluding the point S itself) is

$$\alpha^u = 1 + \alpha_2(S) \approx 1.425\,. \tag{8.81}$$

For $a = a_d - 0$ (i.e., just prior to the bifurcation), $\alpha_{\max} \approx \alpha(X^*)$, but at $a = a_d + 0$ this larger local dimension $\alpha^u(X)$ appears to the right of the pre-bifurcation spectrum $f_0(\alpha)$. This implies that in the graph of the post-bifurcation $f(\alpha)$, there exists a line passing through $(\alpha^u, f(\alpha^u))$ that is tangent to the graph of $f_0(\alpha)$. This tangent line creates the linear segment (8.80). The slope of the latter, q_{B}, is given by (8.77). With this, and using $q_\beta = 1 - s_\beta \approx -0.85$ (see Table 8.1) and $\Phi(q_\beta) \approx -1.06$, we obtain $q_{\mathrm{B}} \approx -1.73$. This theoretical prediction is not far from the numerical value shown in Fig. 8.19b, $q_B \approx -1.49$.

The linear segment (8.80), corresponding to the case in which $\alpha \geq \alpha_{\mathrm{A}}$, is due to the q-phase transition of $\alpha_2(q)$. In the case of a one-dimensional map ($R = \infty$), the equality $\alpha_2(q) = 0$ holds, and thus this linear segment does not exist. However, as discussed in Sect. 7.3.1, this kind of bifurcation, resulting from collision with a saddle point, is a universal phenomenon that occurs even for one-dimensional maps, and, in fact, a linear segment of $\psi(\Lambda)$ like (8.24), which corresponds to the linear segment of $f(\alpha)$ described by (8.80), exists even in this case (see Figs. 8.3 and 8.5). For this reason, we can say that $\psi(\Lambda)$ is superior to $f(\alpha)$ for the purpose of capturing the universality inherent in chaotic bifurcations.

9. Chaotic Bifurcations and Critical Phenomena

The nature of the behavior exhibited by a chaotic system is determined by the infinite number of invariant sets existing in its chaotic region. The great variety of chaotic phenomena that we observe results from the limitless variation in the types of invariant sets contained by the systems we encounter. The nature of the invariant sets that appear in any given system and the resulting behavior that it exhibits depend both on the type of system in question and the values of the various parameters characterizing it. For a nonequilibrium open system, as the values of such parameters are changed, the qualitative nature of the system's behavior is seen to assume many forms, as it experiences the emergence, development and bifurcation of chaos.

In this chapter, we first consider two types of forced oscillators, and in this context we investigate the manner in which the form and structure of the chaotic attractor in the two-dimensional phase space changes in response to both band-merging and attractor-merging crisis bifurcations and the new types of statistical structure that appear along with these changes. Then, as an example of self-organized criticality, we consider several aspects related to the cascades of 2^n-band splitting and F_m attractor merging, including the self-similar time series of associated critical attractors, the similarity exhibited by the band attractor spectrum $\psi(\Lambda)$, and the form that characterizes the disappearance of chaotic two-dimensional fractality.

9.1 Crisis and Energy Dissipation in the Forced Pendulum

The most important chaotic bifurcations are exhibited universally by both one-dimensional and two-dimensional maps, and, as seen in the previous chapter, the universality associated with these bifurcations can be understood directly through the spectrum $\psi(\Lambda)$. We now study how the geometric structure of the two types of crises seen in forced pendulums can be understood from $\psi(\Lambda)$.

When $\gamma = 1/\sqrt{15}$ and $\omega = 0.65$, the attractor of the Poincaré map (6.3) corresponding to the forced pendulum (6.1) changes discontinuously from that of three-band chaos (Fig. 9.1a) to that of fully extended chaos

(Fig. 9.1b) covering the entire interval $\theta \in (-\pi, \pi)$ at the bifurcation point $a = a_{\mathrm{w}} \equiv 0.728384\ldots$. This bifurcation, like the band-merging crisis occurring at $a = a_{\mathrm{w}}$ for the one-dimensional map described by the bifurcation diagram in Fig. 7.10, is due to collision with a three-point saddle, $S_i = F^3(S_i)$ $(i = 0, 1, 2)$. The special characteristic of this type of bifurcation is the subsuming into the attractor of a *Cantor repellor* (a repellor assuming the form of a Cantor set)[1] that exists among the three bands in the closures of the unstable manifolds $W^{\mathrm{u}}(S_i)$. [2] In addition, at this bifurcation, points of tangency $X_{i,\pm j}$ $(j = 0, 1, 2\ldots)$ between the stable manifold $W^{\mathrm{s}}(S_i)$ and the attractor converge to S_i (under F^3). Accordingly, this bifurcation is homoclinic [Tomita et al. (1989)].

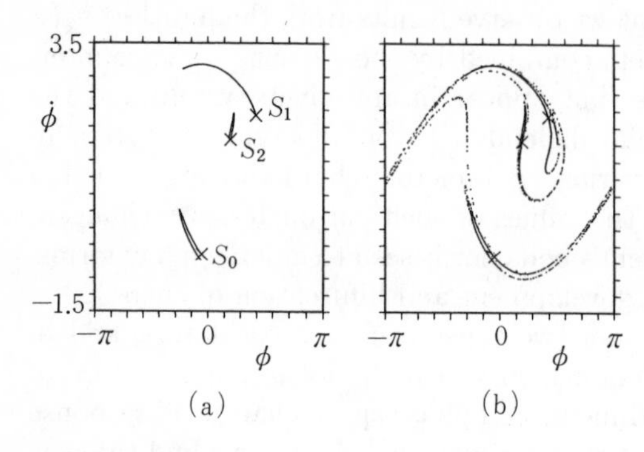

Fig. 9.1. The band-merging crisis ($\gamma = 15^{-1/2}$, $\omega = 0.65$) for the forced pendulum (6.1). **(a)** The collision of the three-band attractor with the three-point saddle S_i ($a = a_{\mathrm{w}} - 4.8 \times 10^{-5}$). **(b)** Subsuming of the Cantor repellor as a result of the collision ($a = a_{\mathrm{w}} + 4.2 \times 10^{-3}$)

9.1.1 The Slope s_δ Induced by the Cantor Repellor

The orbital expansion rate $\lambda_1(X_i)$ at each point along a chaotic orbit $X_i \equiv (\phi(t_i), \dot{\phi}(t_i))$ can be determined from the equation of motion (6.2) in the

---

[1] The concept of a 'repellor' is analogous to that of an 'attractor', and its most simple example is a saddle, analogous to a limit cycle. A Cantor repellor is a 'repelling' Cantor set [see Guckenheimer and Holmes (1986), p. 36] consisting of an uncountably infinite number of unstable orbits. A repelling set is a closed, invariant set. In general, orbits in the neighborhood of such a set move away from it, but in the case in which it is situated inside an attractor, such orbits will repeatedly move away from and then approach this set. When the topological entropy K_0 is positive, Cantor repellors exist, but when the system is within a window, they become removed from attractors and, in addition, their Lebesgue measures vanish. For this reason, in such a situation they are quite difficult to detect. The most simple type of Cantor repellor is responsible for the 'burst' behavior exhibited in intermittent chaos. Such a system is analyzed in Sect. A.3 of the Appendix.

[2] We remind the reader that the unstable manifolds referred to here are those with respect to F^3. Prior to bifurcation, there are three such manifolds. With respect to F, however, these all belong to a single invariant manifold.

following manner. Making the infinitesimal variation $\phi \to \phi + \xi$, $\dot{\phi} \to \dot{\phi} + \zeta$, we obtain the variational equations

$$\dot{\xi} = \zeta, \quad \dot{\zeta} = -\gamma\zeta - \{\cos\phi(t)\}\xi \,. \tag{9.1}$$

The function $\phi(t)$ here can be determined by numerical integration of (6.1). First, with the initial point X_0 and unit vector $(\xi(t_0), \zeta(t_0))$, (6.1) and (9.1) are numerically integrated to $t_1 = 2\pi/\omega$, thereby determining X_1 and $\lambda_1(X_0) \equiv \ln\left[|\xi(t_1)|^2 + |\zeta(t_1)|^2\right]^{1/2}$. Then, we begin from X_1 and the unit vector $(\xi(t_1)\exp[-\lambda_1(X_0)], \zeta(t_1)\exp[-\lambda_1(X_0)])$, and we numerically integrate (6.1) and (9.1) to $t_2 = 2 \times (2\pi/\omega)$, thus determining X_2 and $\lambda_1(X_1) = \ln\left[|\xi(t_2)|^2 + |\zeta(t_2)|^2\right]^{1/2}$. Repeating this operation, the unit vector $(\xi(t_i), \zeta(t_i))$ eventually approaches a direction parallel to the unstable manifold. The orbit thus determined rapidly converges so as to move tangentially to the attractor, and thus, discarding transient behavior, we obtain the orbital expansion rate:

$$\lambda_1(X_i) = \ln[|\xi(t_{i+1})|^2 + |\zeta(t_{i+1})|^2]^{1/2} \,. \tag{9.2}$$

From this, we can derive the coarse-grained orbital expansion rate (6.20), and thus other quantities, such as the distribution function (6.23) and the partition function (8.1b), can be found.

Figure 9.2 displays various quantities, which can be obtained in the manner outlined above, describing the statistical structure of systems in the three cases $a < a_{\mathrm{w}}$, $a = a_{\mathrm{w}} - 0$ and $a = a_{\mathrm{w}} + 0$. The graphs in Fig. 9.2a are similar to those shown in Fig. 8.7. The situation depicted here is that of an $a < a_{\mathrm{w}}$, pre-merging system that possesses a linear segment of slope s_α, as determined by the tangency structure. In the limit as $n \to \infty$, the peak of $\sigma_n(q)$ located at $q = q_\alpha = 1 - s_\alpha$ diverges. The graphs in Fig. 9.2b describe the system just prior to merging ($a = a_{\mathrm{w}} - 0$), with a linear segment of slope $s_\beta \approx 1.67$, as determined by collision with the saddles S_i. In this case $\sigma_n(q)$ has two peaks, at $q = q_\alpha$ and $q = q_\beta (= 1 - s_\beta)$, which both diverge as $n \to \infty$. (In the case shown here, $n = 21$.) The situation depicted here is similar to that corresponding to Fig. 8.8. Using (8.47), (8.58) and (8.59), we obtain $s_\beta \approx 1.54$, a value not too different from that found numerically. The graphs in Fig. 9.2c describe the system just after merging ($a = a_{\mathrm{w}} + 0$), displaying a linear segment of small slope, $s_\delta \approx 0.09$. The mechanism responsible for the formation of the slope s_δ ($< s_\beta$) region is described in Fig. 9.3. In Fig. 9.3a there exists a Cantor repellor spectrum $\psi_{\mathrm{R}}(\Lambda)$ around $\Lambda = \Lambda_{\mathrm{R}}$, situated to the right of the attractor spectrum $\psi_{\mathrm{A}}(\Lambda)$. As a result of collision with the S_i, the repellor is incorporated into the attractor, and, as shown in Fig. 9.3b, the line with slope s_δ forms a common tangent to $\psi_{\mathrm{A}}(\Lambda)$ and $\psi_{\mathrm{R}}(\Lambda)$. Thus we have $s_\delta \approx \psi_{\mathrm{R}}(\Lambda_{\mathrm{R}})/(\Lambda_{\mathrm{R}} - \Lambda^\infty) < s_\beta$.

For $a = a_{\mathrm{w}}(1 + \epsilon)$ (with $\epsilon > 0$, i.e., beyond the merging bifurcation), as ϵ grows, the probability for a chaotic orbit to be found in the vicinity of the Cantor repellor grows, as shown in Fig. 9.4, and the slope s_δ 'linear' piece

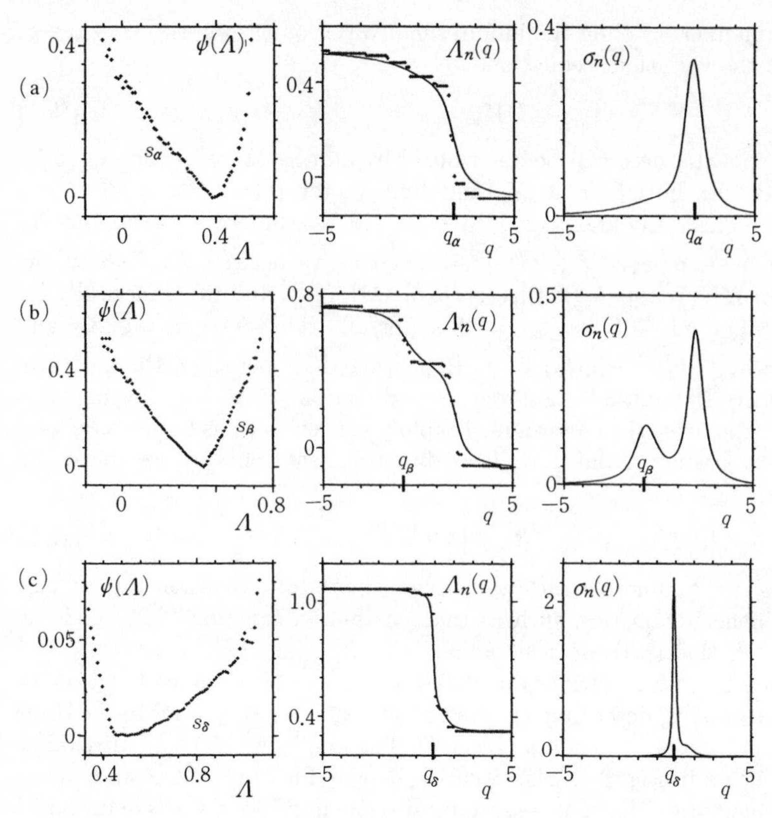

Fig. 9.2. Statistical structure functions on either side of merging. (**a**) Using $a = 0.728$, the value $s_\alpha \approx -0.91$ is found from numerical computation ($n = 18$, $N = 1.2\times 10^5$). (**b**) Graphs just before merging (Fig. 9.1a). Here, we have $s_\beta \approx 1.67$ ($n = 21$, $N = 1.6 \times 10^6$). (**c**) Graphs just after merging (Fig. 9.1b). Here, $s_\delta \approx 0.09$ ($n = 100$, $N = 3.2 \times 10^5$) [Tomita et al. (1989)]

bends downward, as shown in Fig. 9.3c. The fact that the corresponding piece in Fig. 9.2c is slightly curved can be attributed to this fact. (In that case, $\epsilon \approx 5.7 \times 10^{-4}$.) In order to understand this bending behavior, let us study the time series of the third-order map $X_{3t} = F^{3t}(X_0)$ $(t = 0, 1, 2\ldots)$. For $a < a_{\rm w}$, X_{3t} remains forever in one of the three distinct bands, but for $\epsilon > 0$ the duration spent within a given band becomes finite. The average such duration τ is characterized by the relation $\tau \sim \epsilon^{-\alpha_1(S_i)}$ [Grebogi et al. (1987)]. In the case considered presently, $\alpha_1(S_i) \approx 0.676$ [see (1) of Table 9.1]. The average time spent in the region of a Cantor repellor is $\tau_{\rm R} = 1/\psi_{\rm R}(\Lambda_{\rm R})$, and provided that $\epsilon \ll 1$, we have $\tau \gg \tau_{\rm R}$. An orbit X_{3t} will visit the Cantor repellor only in sporadic bursts. Following each such burst, it is transferred again to one of the bands, where it then remains for some interval (of average length τ). This process is referred to as *intermittent transferring*. In such a

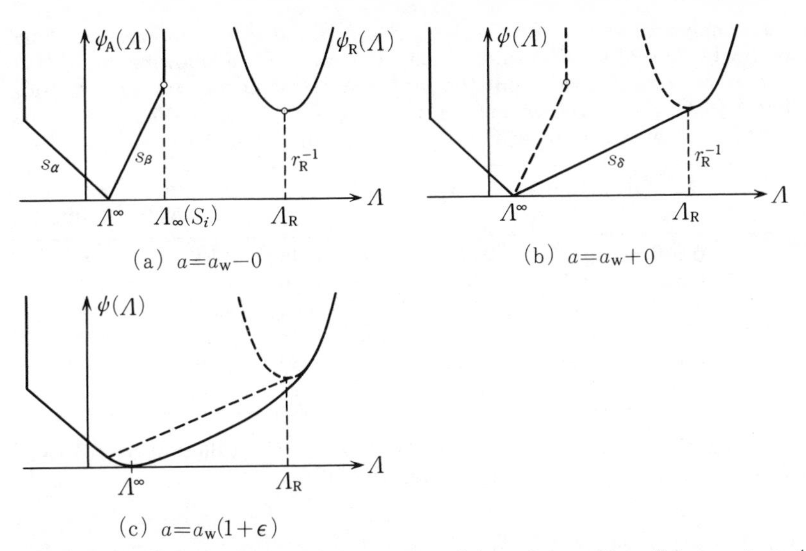

Fig. 9.3. The form of $\psi(\Lambda)$ just before **(a)** and just after **(b)** merging. **(c)** The downward curving of the slope s_δ 'linear' piece just after merging ($a = a_\mathrm{w}(1+\epsilon)$ with $0 < \epsilon \ll 1$)

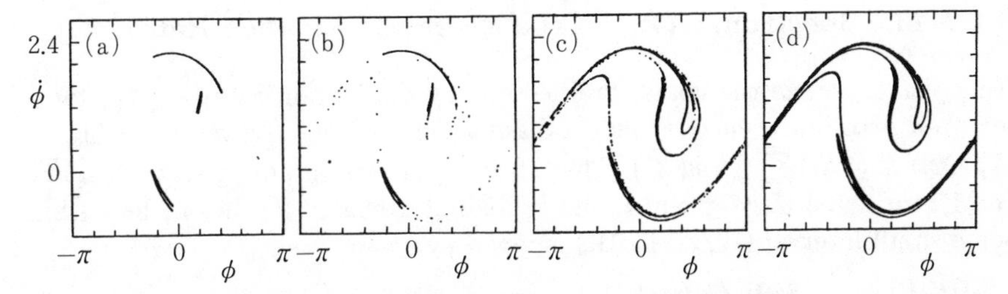

Fig. 9.4. A representation of the increase in the probability of the system remaining in the vicinity of the Cantor repellor. **(a)** $a = a_\mathrm{w} - 0.8 \times 10^{-6}$; **(b)** $a = a_\mathrm{w} + 1 \times 10^{-6}$; **(c)** $a = a_\mathrm{w} + 0.0004$; **(d)** $a = a_\mathrm{w} + 0.02$

system, $\sigma_\infty(q)$ possesses a peak at some $q = \tilde{q}$, and in the limit as $\epsilon \to 0$ this peak diverges as $\tilde{q} \to q_\delta = 1 - s_\delta$. To express the bending of the linear piece in question, if we write the difference between $\psi(\Lambda)$ and the tangent line at $\Lambda = \tilde{\Lambda} \equiv \Lambda(\tilde{q})$ as

$$\Delta\psi(\Lambda) \equiv \psi(\Lambda) - \{(1-\tilde{q})(\Lambda - \tilde{\Lambda}) + \psi(\tilde{\Lambda})\}\,, \tag{9.3}$$

for $\epsilon \ll 1$, the dynamical similarity relation

$$\Delta\psi(\Lambda) = \tau^{-\eta} B(\Lambda - \tilde{\Lambda}) \quad (\Lambda^\infty \lesssim \Lambda < \Lambda_\mathrm{R}) \tag{9.4}$$

holds. Here, η is a positive constant, and $B(y)$ is a nonnegative universal function, independent of ϵ. This is the physical picture describing the chaotic behavior existing in a system that has undergone a band-merging bifurcation.

Table 9.1. Analytical values of s_β, $\alpha_1(S_i)$, $\Lambda_\infty(S_i)$, and related quantities, along with numerical values of s_β. The values shown here are for the following cases*. (1) Band merging ($a = a_{\mathrm{w}}$) for the oscillator (6.1). (2) Attractor merging ($a = a_{\mathrm{m}}$) for the oscillator (6.1). (3) Attractor merging ($K = K_{\mathrm{c}} - 0$, $\Omega = \Omega_{\mathrm{c}}$) for the annulus map. (4) Attractor merging ($K = K_{\mathrm{c}} - 0$, $\Omega = \Omega_{\mathrm{c}}$) for the oscillator (9.6)

		$\Lambda_\infty(S_i)$	$\alpha_1(S_i)$	$\alpha(S_i)$	$h(S_i)$	Λ^{o}	s_β (Analyt.)	s_β (Num.)
(1)		0.880	0.676	0.852	1.115	0.45	1.54	1.67
(2)		0.627	0.727	0.954	1.135	0.30	1.58	1.59
(3)	$\rho = 0/1$	1.453	0.815	1.130	1.162	0.630	1.67	1.45
	$\rho = 1/2$	0.706	0.654	0.807	1.105	0.335	1.38	1.29
(4)	$\rho = 1/2$	0.719	0.760	1.020	1.146	0.30	1.49	1.50
	$\rho = 2/3$	0.507	0.683	0.866	1.118	0.23	1.40	1.30

* All the values above correspond to homoclinic cases. The values of $\alpha_1(S_i)$ were obtained from (8.58) and those for $\alpha_2(S_i)$ from (8.56).

The mechanism responsible for the similarity relation (9.4) is described for the simpler case of type I intermittent chaos in Sect. 3 of the Appendix.

9.1.2 The Spectrum $\psi(W)$ of the Energy Dissipation Rate

First, let us define the coarse-grained energy dissipation rate $W_n(X_0)$ for a length n orbit in terms of the local dissipation rate $w(X_i)$ given by (6.5b) as $W_n(X_0) \equiv (1/n)\sum_{i=0}^{n-1} w(X_i)$. Then, from the coarse-graining statistical law (6.24), for some large given n, the probability density for W_n to lie within some small interval $(W, W + \mathrm{d}W)$ is given by

$$P(W;n) \equiv \langle \delta(W_n(X) - W) \rangle = \exp[-n\psi(W)]P(\bar{W};n) . \tag{9.5}$$

The quantity $\overline{W} \equiv W_\infty(X_0)$ assumes the value $\overline{W}_{\mathrm{f}}$ for almost all X_0 corresponding to chaotic orbits. Following the presentation of Sect. 8.1, we introduce the q-potential of W_n, $\Phi_n^W(q)$, the q-average $W_n(q) \equiv \mathrm{d}\Phi_n^W(q)/\mathrm{d}q$ and the q-variance $\sigma_n^W(q) \equiv -\mathrm{d}W_n(q)/\mathrm{d}q$. We thus obtain the variational principle $\Phi_\infty^W(q) \equiv \min_W \{\psi(W) + (q-1)W\}$, and we find that the spectrum $\psi(W)$ is downward convex and that it possesses the minimum value 0 at $W = \overline{W}$. The geometric meaning of $W_n(X_0)$ can be understood from the equation (7.6) for a general two-dimensional map: $W_n(X_0)$ represents the average over an orbit of length n of the quantity $[\phi(t_{i+1}) - \phi(t_i)]^2$.

Figure 9.5a displays the form of $\psi(W)$ corresponding to the merging process shown in Fig. 9.4. Here, Fig. 9.4b represents the system just after merging. This graph has a linear segment of slope $s_W \approx -1.6$ to the left of the minimum, which represents a common line of tangency between $\psi(W)$ and the spectrum $\psi_{\mathrm{R}}(W)$ of the Cantor repellor, whose minimum is at $W \approx 0.28$. The presence of this line brings about the q-phase transition of $W_\infty(q)$ that occurs at $q_W = 1 - s_W \approx 2.6$. As ϵ becomes larger, the average dissipation rate $\overline{W}$ decreases. This is due to the fact that, as described by Figs. 9.4b–d, the Cantor repellor region becomes subsumed within the attractor, and, as a result, the average of $[\phi(t_{i+1}) - \phi(t_i)]^2$ decreases. However, the variance of corresponding fluctuations, $\sigma_\infty^W = 1/\psi''(\overline{W})$, increases. In fact, the variance for the situation described in Fig. 9.4c in this figure is approximately three times that for Fig. 9.4a. In this way, the dissipation rate expresses the randomness and the size of the change undergone by the coordinate $x_i = \phi(t_i)$ per iteration on the attractor.

In order to study the properties of the randomness and the envelope of orbital motion, let us now define the quantity $w_T(X_i) \equiv \tau^2\dot{\phi}^2(t_{i+1}) = [\phi(t_{i+1}) - \phi(t_i)]^2$. This quantity can also be used in the case of a one-dimensional map ($J = 0$). Figure 9.5b corresponds to the region around the band-merging crisis point $a = a_\text{w} = 1.79032\ldots$ for the one-dimensional map (7.22). This figure displays the spectrum $\psi_T(W)$ of $w_T(X_i)$. Figure (ii) here, representing the system just after merging, possesses a linear segment with negative slope, $s_W \approx -0.04$. The graph displayed here is qualitatively the same as that shown in Fig. 9.5a(ii), but the variance for the case represented by Fig. 9.5b(iii) is 40 times that in case (i). We can thus conclude that this qualitative form is universal for band-merging crises.

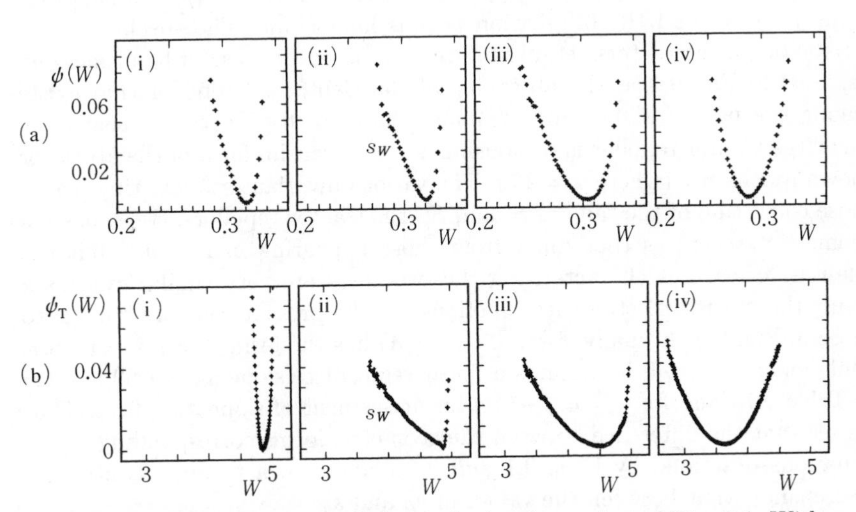

Fig. 9.5. (**a**) The spectrum $\psi(W)$ for the band merging of Fig. 9.4. With $n = 100$, $N = 5 \times 10^5$, the numerical computation gives $s_W \approx -1.6$. (**b**) The spectrum $\psi_\text{T}(W)$ on either side of the band merging point $a = a_\text{w}$: (i) $a = a_\text{w} - 3 \times 10^{-5}$; (ii) $a = a_\text{w} + 8 \times 10^{-5}$ $s_W \approx -0.04$; (iii) $a = a_\text{w} + 0.001$; (iv) $a = a_\text{w} + 0.03$ [Mori et al. Mori et al. (1991)]

9.1.3 The Formation of the Attractor Form in Figure 6.1

Figure 9.6 depicts the process of attractor merging. This demonstrates the manner in which the distinct form of the attractor shown in Fig. 6.1 is produced and indicates the types of invariant sets that it contains. In the case of the forced pendulum (6.1), for $a < a_m = 2.6465274\ldots (\gamma = 0.22, \omega = 1)$ the rotation number $\rho = \langle \dot{\phi} \rangle / \omega$ is locked at $\rho = \pm 1$, and, correspondingly, the two attractors $\tilde{A}_{+1}$ and $\tilde{A}_{-1}$ coexist. The two bands located in the upper and lower parts of Fig. 9.6a constitute $\tilde{A}_{+1}$ and $\tilde{A}_{-1}$, respectively. The attractor $\tilde{A}_{+1}$ is on the closure of the unstable manifold $W^{\mathrm{u}}(S_i)$ corresponding to the period 6 saddle $\{S_0, S_1 \ldots S_5\}$. The evolution equation (6.1) is invariant under the simultaneous transformation $\phi \to -\phi$, $t \to t + (\pi/\omega)$, under which $\rho \to -\rho$ and $\tilde{A}_{\pm 1} \to \tilde{A}_{\mp 1}$. It follows that $\tilde{A}_{+1}$ and $\tilde{A}_{-1}$ possess the same structure with regard to topology and measure. Let us suppose that under this transformation the S_i are mapped to some S_i'. The boundaries of the basins of $\tilde{A}_{+1}$ and $\tilde{A}_{-1}$ are then given by the stable manifolds of the S_i and S_i', $W^{\mathrm{s}}(S_i)$ and $W^{\mathrm{s}}(S_i')$. As shown in Fig. 9.6, at $a = a_m - 0$, $\tilde{A}_{+1}$ and $\tilde{A}_{-1}$ collide with the S_i and S_i', and for $a > a_m$ the attractors merge into a phase-unlocked fully extended attractor, extending over all values of ϕ. At $a = a_m$, points of tangency $X_{i,\pm j}$ between $\tilde{A}_{+1}$ and $W^{\mathrm{s}}(S_i)$ accumulate at S_i as $j \to \infty$, and the bifurcation is thus homoclinic. The special characteristic of this attractor-merging bifurcation, as in the case of band merging depicted in Fig. 9.1, is the absorbing of the Cantor set repellor that exists among the bands of $\tilde{A}_{+1}$ and $\tilde{A}_{-1}$ into the attractor. In the present case, this large Cantor repellor is responsible for the graceful form of the attractor shown in Fig. 6.1 (where $a = 2.7$). The various invariant sets existing here – those consisting of the saddles S_i and S_i', the Cantor repellors, etc. – possess geometric structures that differ from those appearing in Fig. 9.1, but the qualitative features characterizing the two cases are very similar, and as a result the statistical structure functions for the present case are similar to those in Fig. 9.2. Actually, for $a < a_m$, $\psi(\Lambda)$ has the form shown in Fig. 8.7, while for $a = a_m - 0$ it contains a linear segment of slope $s_\beta \approx 1.59$ [see (2) in Table 9.1], and for $a = a_m + 0$, a linear segment of slope $s_\delta \approx 0.11$. Thus we see that the difference between the geometric forms corresponding to the situations described by Figs. 9.1 and 9.6 reduces to a merely quantitative discrepancy, that between the values of s_β and s_δ.

9.2 Fully-Extended Chaos That Exists After Attractor Merging

In Sect. 7.4.2 we studied the situation in which band chaos on a torus characterized by a constant rotation number locked to some rational number p/q is destroyed and phase-unlocked chaos extending over the entire torus appears, and in Sect. 8.1.3 we discussed the statistical structure of this bifurcation. In

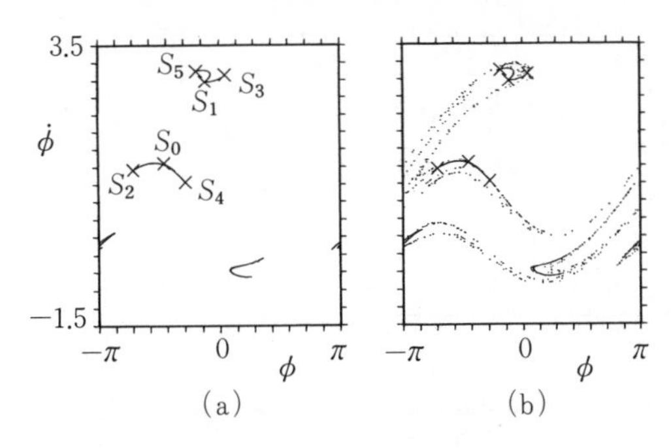

Fig. 9.6. The attractor-merging crisis in the pendulum (6.1) ($\gamma = 0.22$, $\omega = 1$). In (**a**), $a = a_m - 2\times 10^{-4}$, and in (**b**), $a = a_m + 1\times 10^{-3}$. At $a = 2.7$ the attractor is that shown in Fig. 6.1

this section we treat the geometric and statistical structure that characterize the fully extended chaos appearing after this bifurcation.

9.2.1 Attractor Merging in the Annulus Map

Setting $J = 0.1$, for (Ω, K) in the neighborhood of $(\Omega_c, K_c) \equiv (0.20026\ldots, 3.38335\ldots)$, the annulus map (7.10) possesses a phase diagram similar to that for the circle map (see Fig. 7.21). Also, the bifurcation diagram corresponding to the route leading from the coexistence region of the attractors $A_{0/1}$ and $A_{1/2}$ (the darkly shaded region of Fig. 7.21), with rotation numbers $\rho_{0/1}$ and $\rho_{1/2}$, to the U region lying just above it is similar to the bifurcation diagram in Fig. 7.22, corresponding to the neighborhood of K_c.

The attractor for the system that we consider is shown for the situations just before and just after bifurcation in Fig. 9.7. In the former case, shown in Fig. 7.9a, the single-band attractor $\tilde{A}_{0/1}$ is on the left side, and the two-band attractor $\tilde{A}_{1/2}$ is on the right side. These attractors correspond, respectively, to the single square in Fig. 8.4a and the two squares in Fig. 8.4b, describing the circle map. Increasing K, $\tilde{A}_{0/1}$ collides with the stable manifold of the one-point saddle (dashed line) at $K = K_c$, and $\tilde{A}_{1/2}$ collides with the stable manifold of the two-point saddle (dot–dashed line). When K exceeds this value, the attractors cross over the corresponding manifolds. As a consequence of these collisions, the points of tangency $X_{\pm j}$ ($j = 0, 1, 2\ldots$) on either attractor accumulate at the corresponding saddles, designated by the symbols "×" in Fig. 9.7a, and we see that the bifurcation is homoclinic. In particular, as in the cases depicted in Figs. 7.4 and 8.18a, the accumulation of points X_{+j} creates a two-dimensional fractal structure in the attractor appearing in the neighborhood of each of these saddles. After merging, as shown in Fig. 9.7b, the system is characterized by the inclusion of the Cantor repellor existing among the three bands into the attractor, which now exhibits phase-unlocked chaos and extends over the entire interval $\theta \in (0, 1)$. The torus on which the attractor exists is, in both Figs. 9.7a and b, a wrinkled torus possessing a Cantor set multiplex string-like structure along the r-axis. For the circle map ($J = 0$), a one-dimensional map, this fractal structure becomes the folding process undergone by the interval.

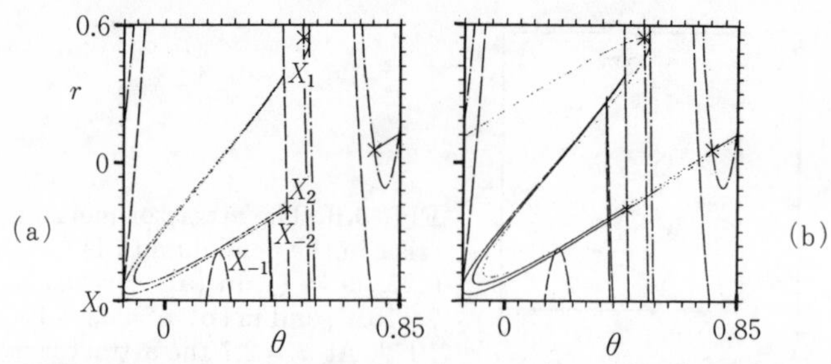

Fig. 9.7. Attractor-merging crisis in the annulus map (7.10). **(a)** Here $(\Omega, K) = (\Omega_c, K_c - 0)$. This figure displays the attractor $\tilde{A}_{0/1}$ which has collided with the stable manifold (*dashed lines*) of the one-point saddle represented by the small "×" and the attractor $\tilde{A}_{1/2}$ which has collided with the stable manifold (*dot–dashed lines*) of the two-point saddle represented by the two large "×". **(b)** Here $(\Omega, K) = (\Omega_c - 7.5 \times 10^{-3}, K_c + 6.7 \times 10^{-3})$. This shows the inclusion of the Cantor repellor into the attractor resulting from the collision [Tomita et al. (1988)]

The statistical structures of $\tilde{A}_{0/1}$ and $\tilde{A}_{1/2}$ just prior to bifurcation (see Fig. 9.7a) are of the same type as that in Fig. 8.8 and are qualitatively equivalent to those shown in Figs. 8.5a and b, respectively. Quantities relevant to these attractors, including s_β, are listed in (3) of Table 9.1.

Figure 9.8 displays the statistical structure of this system just after merging (corresponding to Fig. 9.7b). The linear segment of slope s_δ appearing on the right side of $\psi_n(\Lambda)$ is formed in the following manner. The spectrum $\psi_1(\Lambda)$ of $\tilde{A}_{0/1}$ appears to the right of the spectrum $\psi_2(\Lambda)$ of $\tilde{A}_{1/2}$. (A similar situation is depicted in Fig. 8.5.) Then, the appearance of the Cantor repellor spectrum $\psi_R(\Lambda)$ to the right of $\psi_1(\Lambda)$, as shown in Fig. 9.9, causes the formation of a line commonly tangent to $\psi_1(\Lambda)$ and $\psi_R(\Lambda)$. The slope of this line is given by $s_\delta \approx \{\psi_R(\Lambda_R) - \psi_1(\Lambda_1)\}/(\Lambda_R - \Lambda_1)$.

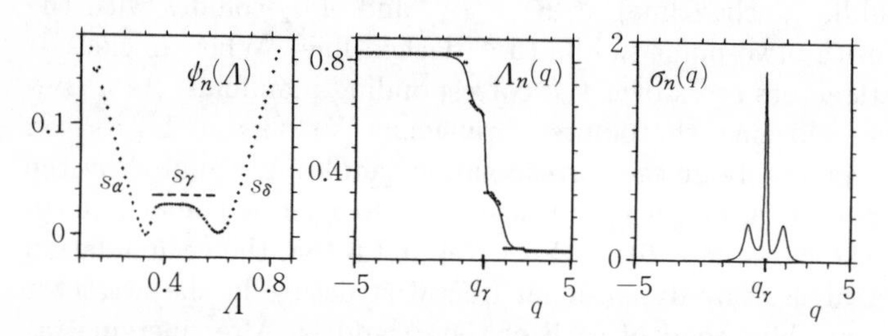

Fig. 9.8. The statistical structure functions appearing just after attractor merging (see Fig. 9.7b) ($n = 100$, $N = 4 \times 10^8$). The data shown here yield the values $s_\delta \approx 0.82$ and $q_\gamma = 1 - s_\gamma \approx 1.00$

The fact that $\psi_n(\Lambda)$ shown in Fig. 9.8 has two minima (as opposed to the single minimum shown in Fig. 9.9) results from the finite length of the orbit ($n = 100$) used in the numerical calculations. This point can be un-

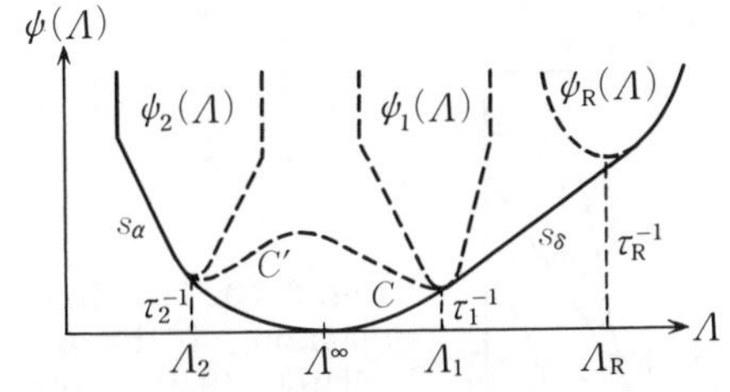

Fig. 9.9. The form assumed by the composition of the spectra $\psi_1(\Lambda)$, $\psi_2(\Lambda)$ and $\psi_R(\Lambda)$ corresponding, respectively, to the attractors $\tilde{A}_{0/1}$, $\tilde{A}_{1/2}$ and the Cantor repellor in a system just beyond the merging point: $K = K_c(1+\epsilon)$, $0 < \epsilon \ll 1$. The values Λ_1, Λ_2 and Λ_R correspond to the minima of the three individual spectra. The curve connecting $\psi_1(\Lambda)$ and $\psi_2(\Lambda)$ is the downward convex solid curve C when $n > \tau_1, \tau_2$, and the upward convex broken curve C' when $n < \tau_1, \tau_2$

derstood as follows. Referring to Fig. 9.9, defining $\tau_k \equiv 1/\psi_k(\Lambda_k)$ $(k = 1, 2)$, the probability $p_k(n)$ that an orbit of length n will remain within a given $\psi_k(\Lambda)$ is $p_k(n) \sim \int d\Lambda e^{-n\psi_k(\Lambda)} \sim e^{-n\psi_k(\Lambda_k)}$, and hence we can write $p_k(n) = (1/\tau_k)\exp[-n/\tau_k]$. Here, $\langle n \rangle = \tau_k$, and τ_k expresses the average length of a visit to the corresponding region. As described in Sect. 9.1.1, because we can write $\tau_1 \sim \epsilon^{-0.815}$ and $\tau_2 \sim \epsilon^{-0.654}$, using the values appearing in (3) of Table 9.1, if $\epsilon \ll 1$ we have $\tau_1, \tau_2 \gg \tau_R$. Thus we see that chaotic orbits exhibit intermittent hopping between $\tilde{A}_{0/1}$ and $\tilde{A}_{1/2}$. In this case, if $n > \tau_1, \tau_2$, the condition for coarse graining is satisfied, and $\psi(\Lambda)$ is, as depicted by the solid line C in Fig. 9.9, downward convex. However, when $n < \tau_1, \tau_2$, the individual minima of $\psi_1(\Lambda)$ and $\psi_2(\Lambda)$ appear, and $\psi(\Lambda)$, as depicted by the broken line C', possesses two minima. For $\epsilon \to 0$ (from the positive side), $\tau_1, \tau_2 \to \infty$, and the line C is defined by $\psi(\Lambda) = 0$ $(\Lambda_1 \geq \Lambda \geq \Lambda_2)$. As a result, a slope $s_\gamma = 0$ linear segment appears. The presence of this segment is reflected by the sharp peak at $q_\gamma = 1 - s_\gamma = 1.0$ in the q-variance $\sigma_n(q)$ shown in Fig. 9.8. The appearance of this slope s_γ linear piece characterizes the chaos exhibited just after attractor merging.

9.2.2 Attractor Merging in the Forced Pendulum

If we add a constant external force of magnitude Ω to the forced pendulum (6.1), we obtain

$$\ddot{\theta} + \gamma\dot{\theta} + K\sin\theta = \Omega\{1 + \cos(\omega t)\}\,, \tag{9.6}$$

for which the symmetry of (6.1) is broken. Here we have $0 \leq \theta < 2\pi$, and we set $\gamma = 0.22$, $\omega = 1$. Let us consider the attractor of the map (6.3) for the case (9.6), whose orbit $X_i = (\theta(t_i), \dot{\theta}(t_i))$ is to be obtained through numerical integration. The corresponding phase diagram in the Ω–K plane is shown in Fig. 9.10. In the first figure shown here, the Arnold tongues for rotation numbers $\rho = \langle\dot{\theta}\rangle/\omega = 0/1, 1/2, 2/3$, and $1/1$ are shown. The second figure

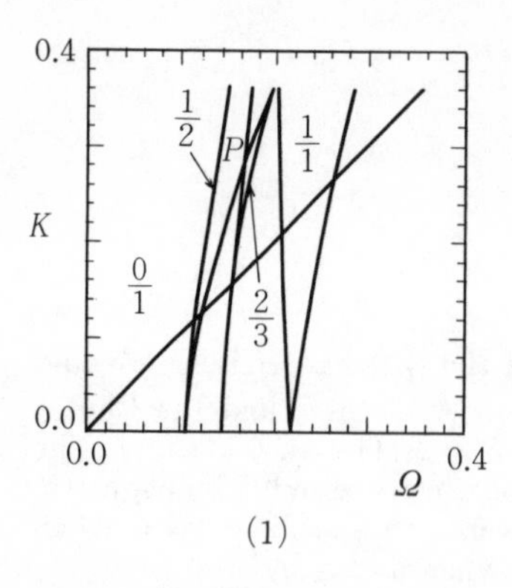

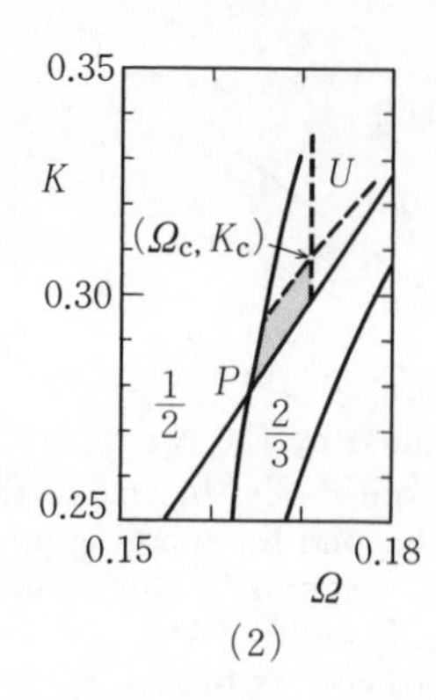

Fig. 9.10. The phase diagram of the oscillator (9.6) (note the similarity with the phase diagrams in Figs. 7.17 and 7.21 for the circle map)

provides a close-up view of the neighborhood of the point P. The shaded region here represents the set of points that are common to the $\rho = 1/2$ attractor $A_{1/2}$ and the $\rho = 2/3$ attractor $A_{2/3}$. In the region labeled U just above this, the system exhibits fully extended chaos. The attractor-merging point leading to this phase-unlocked chaos is given by $\Omega_c = 0.170589\ldots$, $K_c = 0.307148\ldots$.

Figure 9.11a shows the collision of the two-band attractor $\tilde{A}_{1/2}$ with the two-point saddle $S_0^{(1/2)}, S_1^{(1/2)}$ and the three-band attractor $\tilde{A}_{2/3}$ with the three-point saddle $S_0^{(2/3)}, S_1^{(2/3)}, S_2^{(2/3)}$. Figure 9.11b describes the system just after this merging, showing the attractor for the case $K = 0.3074$, $\Omega = 0.170664$ in the U region. Here, the Cantor repellor existing among the five bands is contained within the attractor. Close-up views of the neighborhoods of the saddles $S_0^{(1/2)}$ and $S_0^{(2/3)}$ appear in Fig. 9.12. As shown in Fig. 9.12a, the tangency points $X_{\pm j}$ between the attractors and corresponding stable manifolds accumulate at these saddles, and thus the bifurcation is homoclinic.

The statistical structures of $\tilde{A}_{1/2}$ and $\tilde{A}_{2/3}$ for the situation just prior to bifurcation shown in Fig. 9.11a are of the same type as that shown in Fig. 8.8, and are qualitatively equivalent to Figs. 8.5a and b. Relevant numerical values, including that for s_β, are listed in (4) of Table 9.1. The statistical structure for the system just after bifurcation is of the same form as that shown in Fig. 9.8, possessing linear segments of slopes s_δ and s_γ. In this case, $s_\delta \approx 0.50 < s_\beta$ and $s_\gamma = 1 - q_\gamma \approx 0.00$. The spectrum $\psi(\Lambda)$ in the present case has the same form as that in Fig. 9.9, where, in the present context, $\psi_1(\Lambda)$ and $\psi_2(\Lambda)$ are the spectra of $\tilde{A}_{1/2}$ and $\tilde{A}_{2/3}$, respectively. We thus have, using the value of $\alpha_1(S_i)$ appearing in (4) of Table 9.1, $\tau_1 \approx 0.15 \times \epsilon^{-0.760}$ and $\tau_2 \approx 2.28 \times \epsilon^{-0.683}$. For $\epsilon > 0$, the segment characterized by the slope s_δ prior to bifurcation becomes curved downward. If we express this curved form by $\Delta\psi(\Lambda)(\Lambda_1 \lesssim \Lambda < \Lambda_R)$, as in (9.3), the similarity relation (9.4) holds. Here, $\tau = \tau_1$. Figure 9.13 presents a demonstration of this similarity relation. In addition, from the data represented by this figure we obtain the value

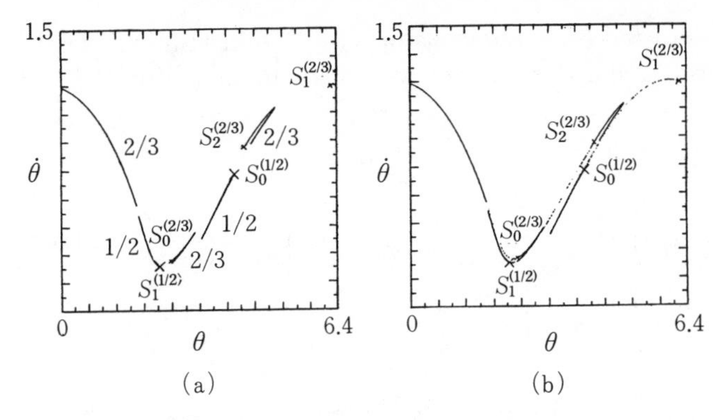

Fig. 9.11. The attractor-merging crisis for the pendulum (9.6). **(a)** Here $(\Omega, K) = (\Omega_c, K_c)$. This figure displays the attractor $\tilde{A}_{1/2}$, which has collided with the two-point saddle $S_i^{(1/2)}$, represented by the two large symbols "×", and the attractor $\tilde{A}_{2/3}$, which has collided with the three-point saddle $S_i^{(2/3)}$, represented by the three small symbols "×". **(b)** Here $(\Omega, K) = (\Omega_c + 7.4 \times 10^{-5}, K_c + 2.5 \times 10^{-4})$. This shows the inclusion of the Cantor repellor into the attractor, which results from the collision [Murayama et al. (1990)]

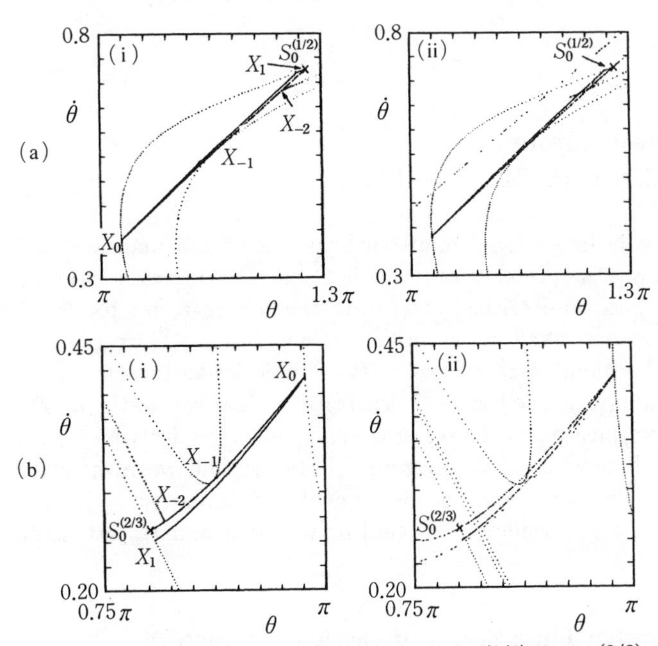

Fig. 9.12. Detail of the neighborhoods of $S_0^{(1/2)}$ and $S_0^{(2/3)}$. **(a)** The (i) tangency points $X_{\pm j}$ and (ii) intersection of the stable manifold of $S_0^{(1/2)}$ (*dotted curve*) and $\tilde{A}_{(1/2)}$. **(b)** The (i) tangency points $X_{\pm j}$ and (ii) intersection of the stable manifold of $S_0^{(2/3)}$ (*dotted curve*) and $\tilde{A}_{(2/3)}$

$\eta \approx 1.0$. In the case depicted here, the value of n (= 250) used is sufficiently large to satisfy $n > \tau_1, \tau_2$, and there is thus in a single minimum of $\psi(\Lambda)$.

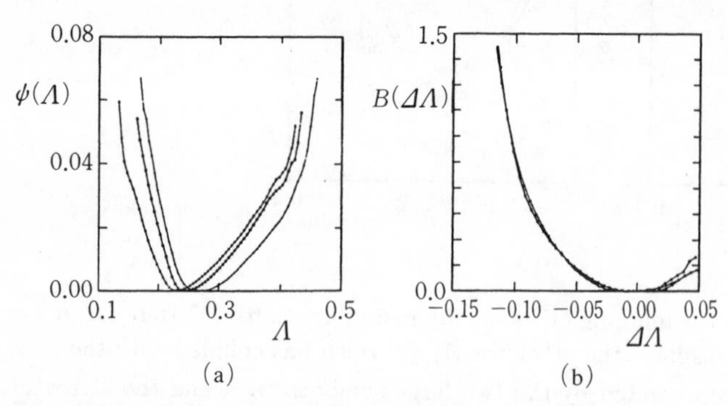

Fig. 9.13. Verification of the similarity relation (9.4) for the bending of the slope s_δ 'linear' piece that occurs for $0 < \epsilon \ll 1$ ($n = 250$, $N = 6 \times 10^7$). The three curves, corresponding [from left to right in (**a**)] to $\tau = 100$, 69 and 54, can be distinguished, but if we set $B(\Delta\Lambda) = \tau^\eta \Delta\psi(\Lambda)$ ($\Delta\Lambda \equiv \Lambda - \tilde{\Lambda}$), their principal parts very nearly coincide, as seen in (**b**) ($\eta \approx 1.0$) [Murayama et al. (1990)]

9.3 Critical Phenomena and Dynamical Similarity of Chaos

Certain aspects of two-dimensional maps such as the Hénon map and the annulus map, including the presence of periodic attractors that exist prior to the appearance of chaos and critical attractors that characterize its emergence point, can be described with one-dimensional maps, but because the corresponding chaotic attractors possess two-dimensional fractality, one-dimensional maps cannot be used in their description. However, in the neighborhood of this emergence point there exists universal similarity which does not depend on the dimension of the system. In this section we investigate this kind of chaotic similarity and the dynamical self-similarity of critical attractors in terms of the cascade of 2^m-band bifurcations and F_m attractor merging.

9.3.1 The Self-Similar Time Series of Critical Attractors

(1) The 2^m Time Series of the Critical 2^∞ Attractor. For orbits of the quadratic map (7.22) on the critical 2^∞ attractor occurring at $a = a_\infty$, both the time series of the variance (6.28) of $S_n \equiv n\Lambda_n(x_0)$ and its envelope diverge logarithmically as $n \to \infty$, as seen in Fig. 9.14. This is due to the fact, that at the emergence point of chaos, mixing vanishes, and memory of the initial state remains forever. In addition, the time series of Fig. 9.14 exhibits the progressive development of blocks (of width $\Delta \log_2 n = 1$) of self-similar inverted nested structure. The mth ($m = 1, 2 \ldots$) block, situated between $n = M \equiv 2^m$ and $n = 2^{m+1}$, is composed of $M+1$ points. The function

$$w_m(i/M) \equiv \left\langle \{S_{M+i}(X)\}^2 \right\rangle / \{\ln \alpha_{\mathrm{PD}}\}^2 , \quad (i = 0, 1, \cdots, M)$$

that describes the structure of each block, assumes the form shown in Fig. 9.15. Here, the mth block consists of $2^m - 1$ triangular pieces, each of height $h \equiv (\ln \alpha_{\mathrm{PD}})^2 \approx 0.8417$, and the time series of the variance $\langle S_n^2 \rangle$ displays this self-similar inverted nested structure in each interval $\log_2 n = m$–$m+1$.

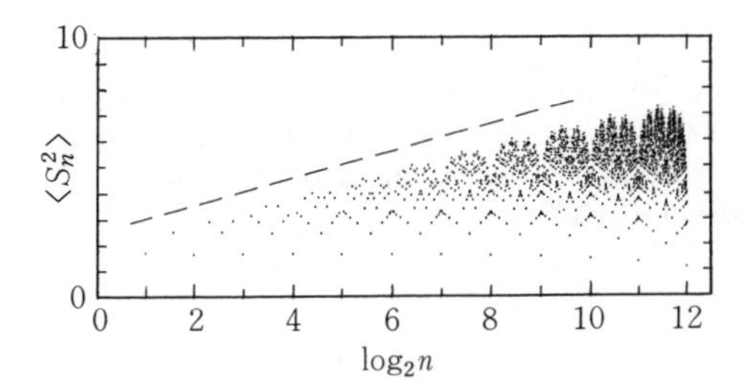

Fig. 9.14. The variance $\langle S_n^2 \rangle$ of the coarse-grained orbital expansion rate for the critical 2^∞ attractor as a function of $\log_2 n$ ($N = 2^{16}$) [Hata et al. (1989)]

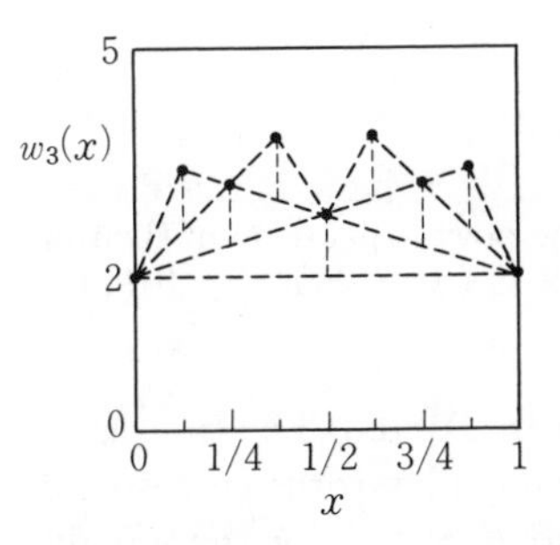

Fig. 9.15. The structure function w_3 corresponding to the range of values satisfying $\log_2 n = 3$–4 in Fig. 9.14, plotted as a function of $x = i/2^3$. This function assumes the form of seven triangles connecting nine distinct points. From the peak of each triangle there emanates a vertical line segment of length 1 that extends to the midpoint of that triangle's base

The time series for $\beta_n(x_0) \equiv n\Lambda_n(x_0)/\ln n$ ($n \gg 1$) is given in Fig. 9.16. This series, consisting of blocks of self-similar inverted nested structure, neither converges nor diverges in the $n \to \infty$ limit. The width of the mth such block is given by $\Delta n \sim 2^m$, and this block consists of smaller blocks of widths 2^i ($i = 0, 1, \ldots, m-1$). (Note that $\sum_{i=0}^{m-1} 2^i = 2^m - 1$.) In this way, at the 2^∞ attractor the time series of β_n explicitly displays a 2^m-cascade corresponding to the 2^m-bifurcation cascade that gives rise to this attractor. The power spectrum of this time series,

$$S_\beta(\omega) \equiv \frac{1}{T} \left| \sum_{n=n_1}^{n_1+T} \beta_n(x_0) \exp[-2\pi \mathrm{i} \omega n] \right|^2 , \tag{9.7}$$

reflects its self-similarity. As seen in Fig. 9.17, the functional form of $S_\beta(\omega)$ consists of blocks whose widths (along ω) converge geometrically to 0 as $\omega \to 0$, with the width of each successive block being smaller by a factor of 1/2 in this limit.

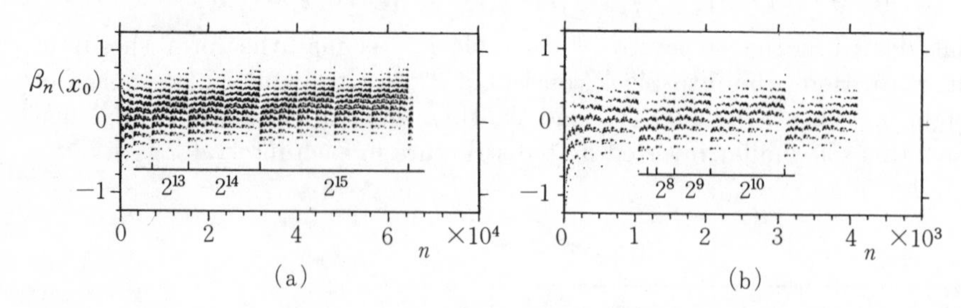

Fig. 9.16. β_n as a function of $n\,(= 2 \sim 2^{16})$ for the critical 2^∞ attractor. Here, $x_0 = f^{1000}(a) = 1.393616\ldots$ is the 1000th image of $a = f(0)$. **(b)** is a close-up of the early time portion of **(a)**. This exhibits the development of the self-similar inverted nested structure with blocks of widths 2^m

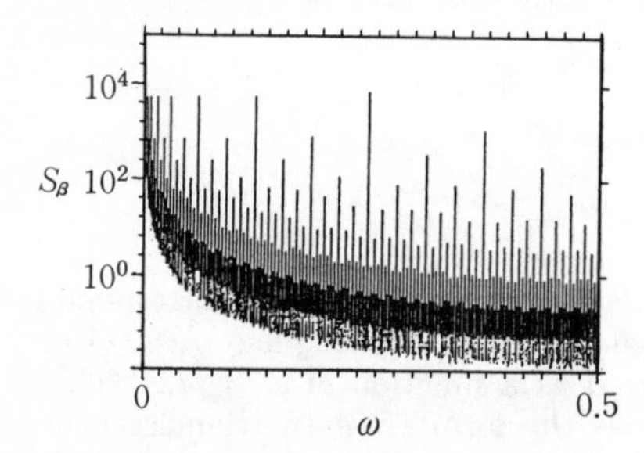

Fig. 9.17. The power spectrum of the time series in Fig. 9.16 ($n_1 = 256$, $T = 8192$)

(2) The F_m Time Series of the Critical Golden Torus. The time series of $\beta_n(X_0) \equiv n\Lambda_n(X_0)/\ln n$ for orbits on the critical golden torus corresponding to the parameter values $J = 0.5$, $\Omega = \Omega_\infty$, and $K = K_\infty$ for the annulus map (7.10) is shown in Fig. 9.18. This series consists of blocks of self-similar inverted nested structure whose widths (in n) are given by the Fibonacci numbers F_m [see (7.47)]. The block of width $\Delta n = F_m$ consists of smaller blocks of widths F_i ($i = 0, 1, \ldots, m-2$). As n increases, blocks of widths given by successively increasing Fibonacci numbers appear. In this way, for a critical torus, with rotation number given by the inverse of the golden ratio (7.47), the time series of β_n explicitly describes the F_m cascade. The self-similarity of this time series is reflected by the power spectrum (9.7). As shown in Fig. 9.19, this power spectrum consists of self-similar blocks whose widths converge geometrically to 0 as $\omega \to 0$, with the ratio of the widths of successive blocks given by $\rho \approx 0.618$. The time series discussed here in both cases (1) and (2) are clear examples of self-organized criticality.

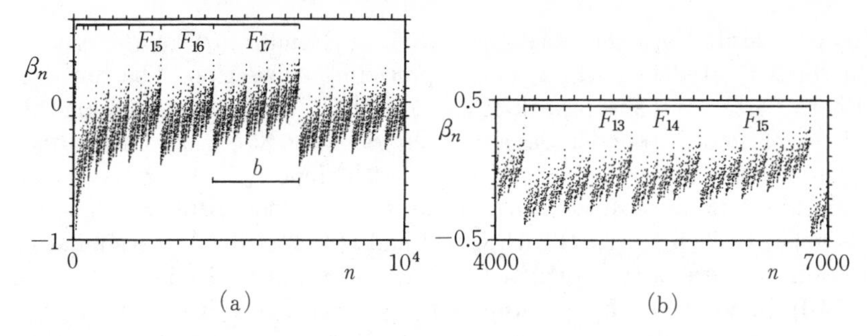

Fig. 9.18. β_n for the critical golden torus as a function of $n\,(= 2 \sim 10^4)$. Here, $(\theta_0, r_0) = (0.0878329, 0.0134251)$. **(b)** is a close-up view of the interval "b" of **(a)** [Horita et al. (1989)]

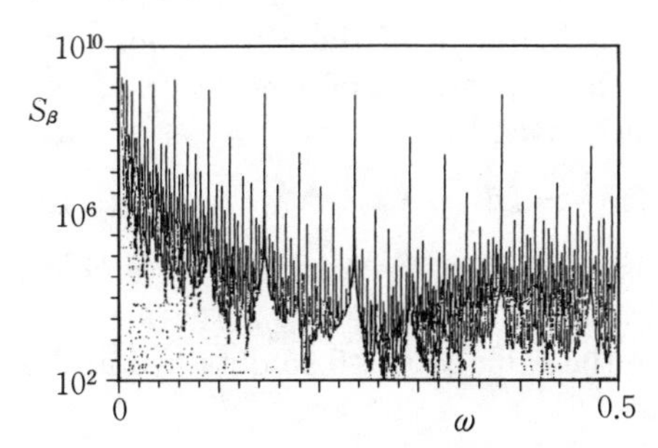

Fig. 9.19. The power spectrum of the time series in Fig. 9.18 ($n_1 = 1600$, $T = 2^{14}$)

9.3.2 The Algebraic Structure Functions of the Critical Attractor

In the cases considered in the previous subsection, the self-similar time structure of $\beta_n(X_0) \equiv n\Lambda_n(X_0)/\ln n$ does not depend on the initial point X_0. In that case, the value of $\beta_n(X_0)$ for a given n depends randomly on X_0. Let us now consider the corresponding probability density (the probability of a particular value of β being realized), $P(\beta; n) \equiv \langle \delta(\beta_n(X) - \beta) \rangle$. The appearance of the denominator $\ln n$ in the definition of β_n indicates that $P(\beta; n)$ can be obtained from (6.27) by replacing n that appears there with $\ln n$. That is, we can write $P(\beta; n)$ in a form similar to (6.30),

$$P(\beta; n) = n^{-\psi_\beta(\beta)} P(\bar{\beta}; n) \quad (n \gg 1)\,, \tag{9.8}$$

where $\overline{\beta} \equiv \langle \beta_n(X) \rangle = 0$ and $\psi_\beta(\beta) \geq \psi_\beta(0) = 0$. Also, in the present context, we replace n by $\ln n$ in the statistical thermodynamic formulas appearing in Sect. 8.1. For example, the q-potential of β_n then becomes

$$\Phi_\beta(q) \equiv -\lim_{n\to\infty} (1/\ln n) \ln \left[\int \mathrm{d}\beta\, n^{(1-q)\beta} P(\beta; n) \right]\,, \tag{9.9}$$

where $\beta(q) \equiv \Phi_\beta'(q)$ and $\sigma_\beta(q) \equiv -\beta'(q)$. From the variational principle $\Phi_\beta(q) = \min_\beta \{\psi_\beta(\beta) + (q-1)\beta\}$, we know that $\psi_\beta(\beta)$ is downward convex and possesses a minimum at $\beta = 0$.

Figure 9.20 displays the functions $\psi_\beta(\beta)$, $\beta(q)$ and $\sigma_\beta(q)$ for the critical golden torus treated in part (2) of the preceding subsection. The function $\sigma_\beta(q)$ here possesses a sharp peak at $q = q_c \approx 1.95$. In the $n \to \infty$ limit this peak diverges, and $\beta(q)$ undergoes a q-phase transition at $q = q_c$. This implies the existence of a linear segment of $\psi_\beta(\beta)$ with slope $s_c = 1 - q_c \approx -0.95$. The function $\beta_n(\theta_0) = n\Lambda_n(\theta_0)/\ln n$ of the critical golden torus, which corresponds to the circle map (7.44) with $K = 1$ and $\Omega = \Omega_\infty$, displays a similarity of the same nature as that depicted in Figs. 9.18 and 9.19 for the annulus map (7.10). In addition, the spectrum of this $\beta_n(\theta_0)$ is also characterized by a linear segment with slope s_c and the same kind of q-phase transition as that described by Fig. 9.20. These properties do not depend on the dimension of the system and can be considered as universal characteristics of critical golden tori. For a one-dimensional map, the existence of a linear segment of $\psi_\beta(\beta)$ is due to the influence of the neighborhood of the critical point x_T [note that $f'(x_T) = 0$], which induces the appearance of a negative β, and thus this slope s_c is, in analogy to (8.46), given by

$$s_c = -\alpha_{\max}/(z-1), \quad \alpha_{\max} = \alpha_1(x_T) = D(-\infty)\,. \tag{9.10}$$

For the circle map with $K = 1$, x_T is the inflection point $\theta = 0$, and we have $z = 3$. Therefore using $D(-\infty)$ from part (2) of Sect. 8.2.3 yields $s_c \approx -0.949$. This is in good agreement with the value $s_c \approx -0.95$ obtained in the numerical simulations discussed above.

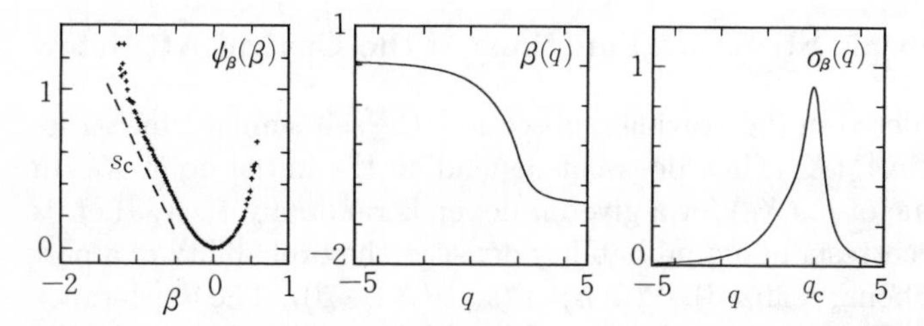

Fig. 9.20. The algebraic structure functions for the critical golden torus ($n = 8000$, $N = 3 \times 10^6$). These data yield $s_c = 1 - q_c \approx -0.95$ [Horita et al. (1989)]

The algebraic structure functions for β_n corresponding to the critical 2^∞ attractor treated in part (1) of the previous subsection also have forms similar to those given in Fig. 9.20. Here, $\sigma_\beta(q)$ possesses a sharp peak at $q = q_c \approx 1.70$, and $\psi_\beta(\beta)$ displays a linear segment of slope $s_c = 1 - q_c \approx -0.70$. For the quadratic map, $z = 2$, and thus using $D(-\infty)$ from part (1) of Sect. 8.2.3 yields $s_c \approx -0.755$. This agrees well with results of numerical computations [see Hata et al. (1989) for details].

9.3.3 The Internal Similarity of Bands for the Spectrum $\psi(\Lambda)$

The critical attractor is, as depicted in Fig. 9.21, observed in the triangular shaped region (critical region) situated directly above the emergence point of chaos, $\epsilon_k = 0$. In this figure, the horizontal axis corresponds to ϵ_k of (7.40) or ϵ_m of (7.50), Q_m represents the number of bands ($Q_m = 2^m \gg 1$ or $Q_m = F_m \gg 1$), and n is the time scale under consideration. This figure also applies to the Hénon map and the annulus map. If we extend the graphical presentation of the variance (6.28) given in Fig. 9.14 to the case $\epsilon_k > 1$, for $n \gg 1$,

$$n^2 \left\langle \{\Lambda_n(X) - \Lambda^\infty\}^2 \right\rangle \propto \begin{cases} \ln n & (n < Q_m) \\ n^\zeta & (n \gg Q_m) \end{cases}, \tag{9.11}$$

where $n^2\langle\cdots\rangle$ represents the envelope of $n^2\langle\cdots\rangle$, and $0 \le \zeta \le 1$. As seen in Fig. 9.21, for values satisfying $\epsilon_k > 0, n^{-1} \ll Q_m^{-1}$, the system is within the chaotic region. Here, an orbit of length n displays chaotic motion in each band superimposed on a periodic hopping among all bands. Each band is visited every Q_m time steps, and thus if $n/Q_m \gg 1$, intra-band chaos is seen to appear. However, with n fixed, taking the $\epsilon_k \to 0$ ($Q_m \to \infty$) limit, the width of each band goes to 0, and the hopping among bands comes to define the critical attractor discussed in the previous two subsections. In the present case, as indicated by the broken line labeled "(1)" in Fig. 9.21, with ϵ_k fixed, decreasing the value of n^{-1} from the critical region into the chaotic region, the n dependence of the envelope (9.11) of the variance changes from the $\ln n$ dependence of the critical region to the n^ζ dependence of the chaotic region.

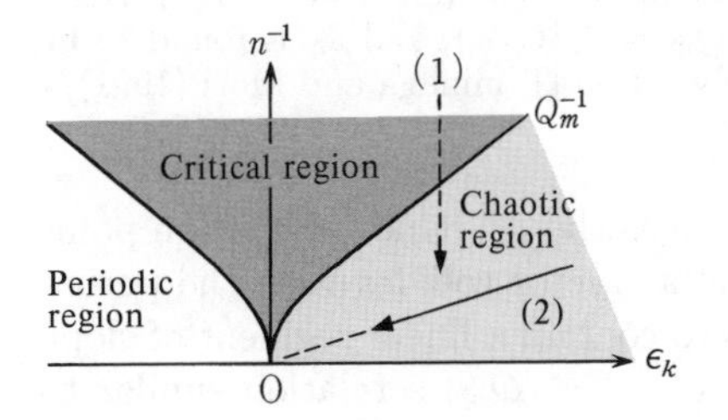

Fig. 9.21. The critical region ($n < Q_m$) and chaotic region ($n \gg Q_m$, $\epsilon_k > 0$)

(1) The 2^m-Band Bifurcation Cascade. The Hénon map (7.1) with $b = 0.5$ and $a > a_\infty \equiv 0.94977288\ldots$ displays a 2^m-band bifurcation cascade a_m ($m = 1, 2 \ldots \infty$). Let us now consider the internal band structure of the Q_m bands that appear in the range $a_m \ge a > a_{m+1}$ ($Q_m \equiv 2^m \gg 1$). As in the case corresponding to (7.40) and (7.41), we have $\epsilon_k \equiv (a - a_\infty)/a_\infty \propto \delta^{-m}$ and $Q_m \propto \epsilon_k^{-\kappa}$. Also, the Liapunov number can be written

$$\Lambda^\infty = c_k^\infty / Q_m \propto \epsilon_k^\kappa , \tag{9.12}$$

where $c_0^\infty \approx 0.6836$ when $k = 0$ and $c_{3/4}^\infty \approx 0.382$ when $k = 3/4$. Now, as indicated by the line labeled "(2)" in Fig. 9.21, let us approach $\epsilon_k = 0$ along a path within the chaotic region. In this situation, the spectrum $\psi(\Lambda)$ can be thought of as satisfying the *dynamical similarity relation*

$$\psi(\Lambda) = (1/Q_m)V_k(Q_m(\Lambda - \Lambda^\infty)) . \tag{9.13}$$

Here, $V_k(y)$ is a universal function, independent of ϵ_k, defined for each value of the parameter k that satisfies the relation $0 \le k < 1$. This relation expresses the internal band similarity discussed at the end of Sect. 7.3.3.

Figure 9.22 exhibits $V_k(y)$ for $k = 3/4$ obtained by plotting $Q_m\psi(\Lambda)$ as a function of $y \equiv Q_m(\Lambda - \Lambda^\infty)$ for the Hénon map with $b = 0.5$. This graph was generated using the values 8, 16, and 32 for Q_m, and, as can be seen, these three plots are nearly coincident, verifying the similarity relation (9.13). The function $V_k(y)$ here possesses a linear segment of slope $s_\alpha \approx -1.0$ existing in the $y < 0$ region, caused by the folding of W^{u} due to the tangency structure. This function thus belongs to the same type as $\psi(\Lambda)$ shown in Fig. 8.7. When the minimum is rounded, as in the case depicted in the figure, the exponent ζ of the variance assumes a value close to 1.

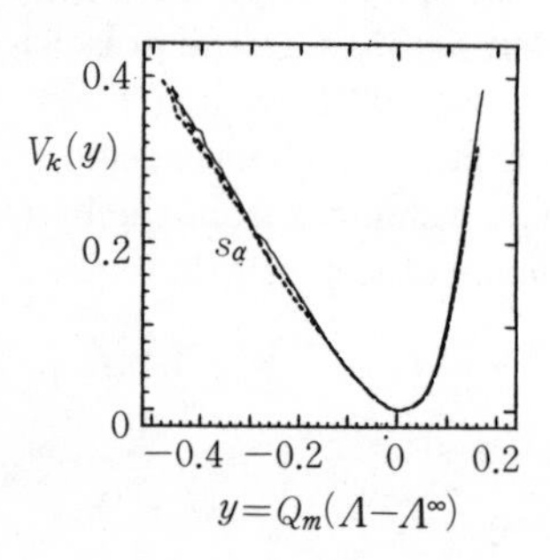

Fig. 9.22. The function $V_k(y)$ corresponding to the $k = 3/4$ sequence of points in 2^m-band bifurcation of the Hénon map ($b = 0.5$ and $N = 5 \times 10^7$). Here, $n = 30Q_m$ ($Q_m = 8, 16, 32$), and s_α is found to be approximately -1.00 [Tominaga and Mori (1991)]

When $k = 0$, ϵ_k converges to the value 0 as the 2^m-band bifurcation point $a = a_m$ is approached. At this point, the attractor collides with the period 2^{m-1} saddle points X_i^*, and $\psi(\Lambda)$ comes to contain a linear segment of slope s_β [see (8.47)]. Then we can write $\Lambda_\infty(X_i^*) = c^*/Q_m$, a relation similar to (9.12). Hence the function r in (8.60) satisfies $r(X_i^*) \approx c^*/RQ_m$. Also, $z = 2$ here, and thus $h(X_i^*) \approx 1$, $\alpha_1(X_i^*) \approx 1/2$, and the two-dimensional fractality disappears. Then, setting $\Lambda^0 \approx \Lambda^\infty$, (8.47) and (8.59) lead to

$$s_\beta \approx s_\beta^* \equiv c^*/2(c^* - c_0^\infty) \equiv 1.3880 , \tag{9.14}$$

where $c^* \approx 1.0685$. The $y < 0$ region is characterized by a slope $s_\alpha = -1$ linear segment, and therefore $V_0(y)$ becomes

$$V_0(y) = \begin{cases} -y & (y < 0) \\ s_\beta^* y & (0 < y < \Delta c) , \end{cases} \tag{9.15}$$

where $\Delta c \equiv c^* - c_0^\infty \approx 0.3849$. Figure 9.23 displays the form of $V_0(y)$ obtained numerically for the quadratic map with $a = a_4$ ($Q_m = 16$) and $\epsilon_0 \approx 0.9 \times 10^{-3}$. Here $s_\beta \approx 1.37$, and these results are in good agreement with (9.14) and (9.15). A sharply pointed minimum of $\psi(\Lambda)$ such as that shown here implies a value of ζ near 0. Thus, in this case the variance has no n dependence [see Mori et al. (1989a)].

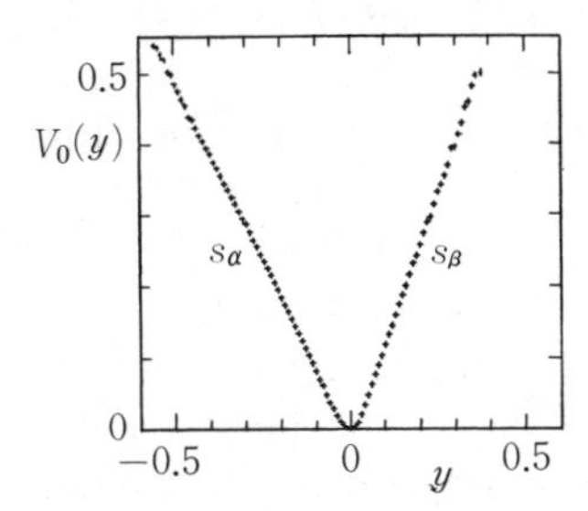

Fig. 9.23. The function $V_0(y)$ corresponding to the $k = 0$ sequence of points in 2^m-band bifurcation of the quadratic map ($N = 10^6$). Here, $n = 15Q_m$ ($Q_m = 16$, $a = a_4 = 1.402492\ldots$) and the data yield $s_\alpha \approx -1.00$ and $s_\beta \approx 1.37$

Taking the derivative of (9.13) with respect to Λ and using the relation $\psi'(\Lambda) = 1 - q$, which can be obtained from the variational principle (8.7), we obtain $1 - q = V_k'(Q_m\Lambda - c_k^\infty)$. Then, writing the inverse relation of $z = V_k'(y)$ as $y = L_k(z)$, this becomes

$$\Lambda(q) = (1/Q_m)\{L_k(1-q) + c_k^\infty\} . \tag{9.16}$$

The quantity $L_k(z)$ here is a nondecreasing universal function of z satisfying $L_k(0) = 0$. Equation (9.16) is a generalization of (9.12). According to the similarity relations (9.13) and (9.16), in terms of the scaled quantities $\hat{n} \equiv n/Q_m$ and $\hat{\Lambda} = Q_m\Lambda$, the fluctuations in the coarse-grained expansion rate $\Lambda_n(X_0)$ become independent of ϵ_k for each k.

As seen above, the scaling functions $V_k(y)$ and $L_k(z)$ are universal functions, independent of the dimension of the map for each value of k, and they can thus be determined by the one-dimensional map obtained by setting $J = 0$ in (7.8). The similarity relations (9.13) and (9.16) describing the internal structure of bands thus express the universality of chaos.

(2) The F_m Attractor-Merging Cascade. At a value $T_m^- \equiv (\Omega_m, K_m - 0)$ just before the attractor-merging point T_m in the attractor-merging cascade of the circle map, the F_m-band attractor a_m^- and the F_{m+1}-band attractor a_m^+ coexist (see Fig. 7.23). The similarity relations (7.50) and (7.54) hold for the cascade of T_m^-, and in addition we have

$$\psi(\Lambda) = (1/Q_m^\pm)V^\pm(Q_m^\pm(\Lambda - \Lambda^\infty)) , \tag{9.17}$$

corresponding to (9.13) and (9.15). Here $V^\pm(y) = -y$ for $y < 0$, and $V^\pm(y) = s_\pm^* y$ for $0 < y < \Delta c_\pm$. In the present case, $s_\pm^* \equiv c_\pm^*/2(c_\pm^* - c_\pm^\infty)$, $s_+^* \approx 0.961$, $s_-^* \approx 0.926$, $\Delta c_\pm \equiv c_\pm^* - c_\pm^\infty$, $\Delta c_+ \approx 0.751$, and $\Delta c_- \approx 0.813$. This $\psi(\Lambda)$ has the same form as that given by (8.22). These similarity relations also hold for the T_m^- cascade in the case of attractor merging of the annulus map.

9.3.4 The Form Characterizing the Disappearance of Two-Dimensional Fractality

Using the quantity l_d (see discussion in Sect. 8.2 regarding the significance of the lengths l_d and l_w) representing the minimum of the distance between neighboring bands as a reference length for the spatial scale l used in the definition of the local dimension (8.25), we obtain Fig. 9.24, in analogy to Fig. 9.21, and we have the similarity relation $l_\mathrm{d} \propto \epsilon_k^\nu$ [see (7.42)]. Now, as indicated by the broken line labeled "(1)" in Fig. 9.24, if we fix ϵ_k and decrease l from the $l > l_\mathrm{d}$ critical region into the chaotic region to a position for which $l \ll l_\mathrm{d}$, the exponent $\tau(q)$ in the l dependence of the partition function (8.27) changes from ~ 1.32 to ~ 2.16 for $q = 4$, as shown in Fig. 9.25a. The undulations exhibited in the critical region reflect the discrete nature of the distribution of the bands. Figure 9.25b displays the functions $f(\alpha)$ obtained here for the critical region (broken curve) and the chaotic region (solid curve) for the quadratic map with $a = a_4 - 0$. Because both the number of bands Q_m $(= 16)$ and number of orbits N here are not sufficiently large, the curves extend farther to the right of the maxima than the analytically predicted curves [see Fig. 8.13 and (8.79)]. Despite this fact, the figure gives a fairly good representation of these functions.

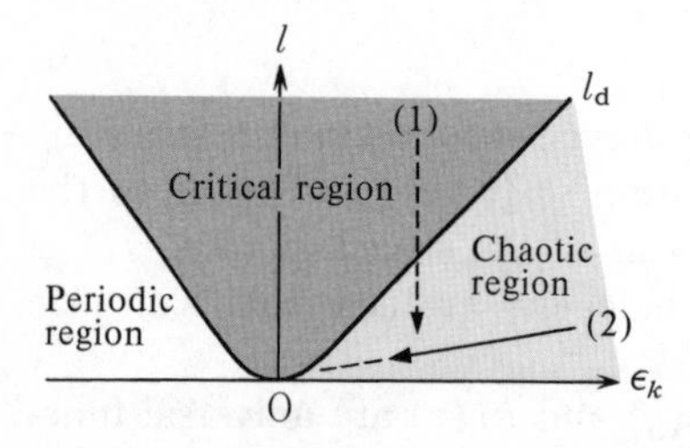

Fig. 9.24. The critical region ($l > l_\mathrm{d}$) and the chaotic region ($l \ll l_\mathrm{d}$, $\epsilon_k > 0$)

In order to study the two-dimensional fractality in the chaotic region, following Tominaga and Mori (1991), let us consider the Hénon map with $b = 0.5$ and consider the situation in which $\epsilon_k = 0$ is approached along the line labeled "(2)" in Fig. 9.24. Figure 9.26a displays $f(\alpha)$ at the 2^m-band bifurcation points $a = a_1, a_2$, and a_3, as obtained numerically, and Fig. 9.26b displays $f(\alpha)$ for $a = a_2, a_3, a_4, a_5$, and a_6 obtained analytically. Here, $D(0) > f(\alpha_\mathrm{max}) > 1$ and $\alpha_\mathrm{max} > 1$. These curves possess linear segments to the left and right of their maxima corresponding to (8.78) and (8.80). These results indicate that as $m \to \infty$ $(a_m \to a_\infty)$, $f(\alpha)$ approaches the one-dimensional spectrum $f_*(\alpha)$ of (8.79).

The form of the convergence of $f(\alpha)$ to $f_*(\alpha)$ can be derived explicitly from the similarity relations (9.13) and (9.16). Substituting (9.16) into (8.73) and using (9.13) and the relation $R(\alpha - 1)/(2 - \alpha) = \Lambda(q - \tau_2(q))$ in (8.76), for the hyperbolic phase corresponding to $q < q_\mathrm{A}$ and $\alpha \geq \alpha_\mathrm{A}$, we have

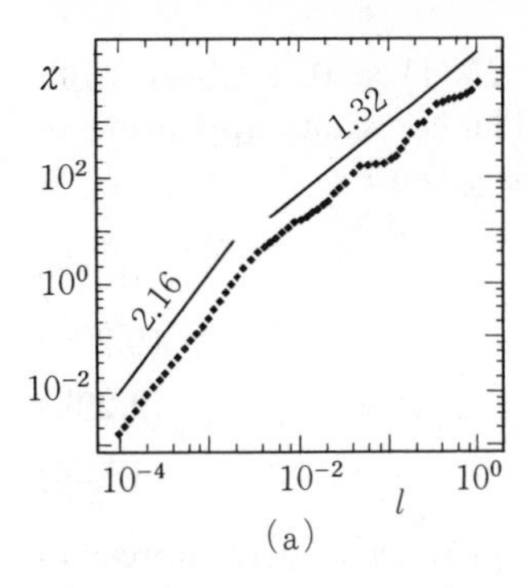

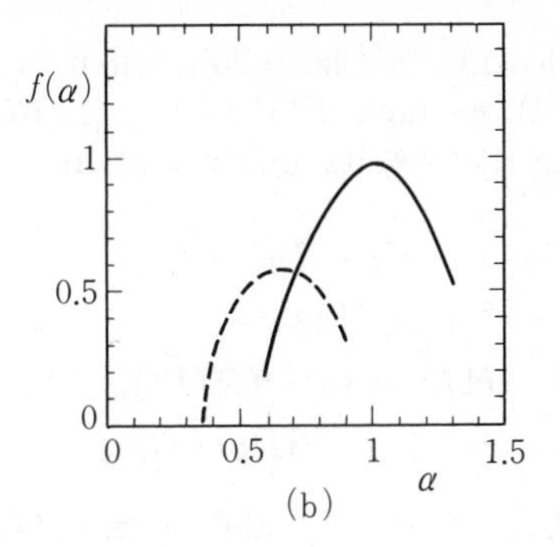

Fig. 9.25. The quadratic map ($a = 1.4024921764 \approx a_4 - 0$). **(a)** $\chi(q = 4; l)$ shown as a function of l. The data yield the results $\tau(q = 4) \approx 1.32$ for $l > l_\mathrm{d}$, and $\tau \approx 2.16$ for $l \ll l_\mathrm{d}$. **(b)** The function $f(\alpha)$ in the critical region (*broken curve*) and the chaotic region (*solid curve*) ($-3 \leq q \leq 12$)

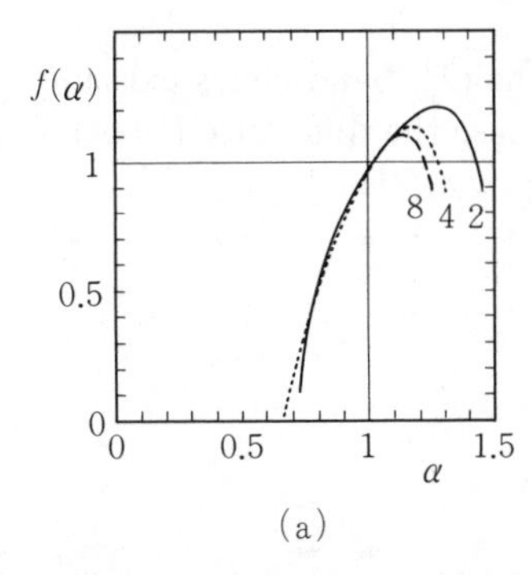

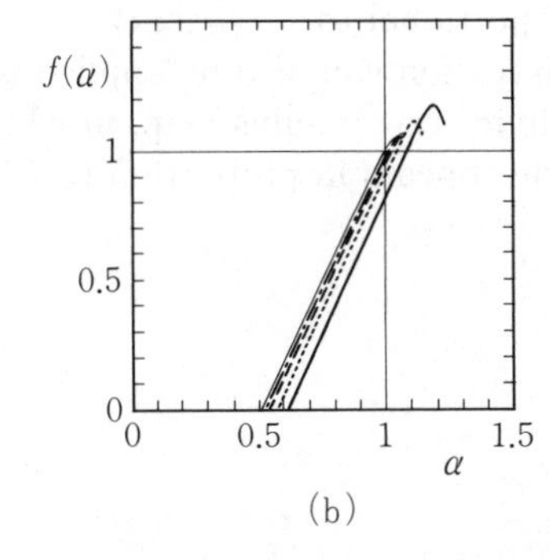

Fig. 9.26. The form of $f(\alpha)$ at several of the 2^m-band bifurcation points $a = a_m$ in the chaotic region for the Hénon map with $b = 0.5$. **(a)** Numerical results (from right to left): $a = a_1$ (*solid curve*, $Q_m = 2$); a_2 (*dotted curve*, $Q_m = 4$); a_3 (*dashed curve*, $Q_m = 8$). **(b)** Analytical results (from right to left): $a = a_2$ (*heavy solid curve*, maximum value $D(0) \approx 1.18$); a_3 (*dotted curve*, maximum value $D(0) \approx 1.11$); a_4 (*dashed curve*, maximum value $D(0) \approx 1.06$); a_5 (*dashed–dotted curve*, maximum value $D(0) = 1.03$); a_6 (*light solid curve*, maximum value $D(0) \approx 1.02$) [Tominaga and Mori (1991)]

$$\alpha(q) \approx 1 + \{L_k(1-q) + c_k^\infty\}/RQ_m \tag{9.18}$$

$$f(\alpha(q)) \approx 1 + H_k(1-q)/RQ_m\,, \tag{9.19}$$

where $H_k(z) \equiv L_k(z) + c_k^\infty - V_k(L_k(z)) > 0$. Note here that we have ignored terms of $\tau_2(q)$ contained in L_k and H_k as being of 'higher order', since $\tau_2(q) = \int_1^q \alpha_2(q)\mathrm{d}q \propto 1/RQ_m$. If we set $q = 0$ in (9.19), we obtain

$$D(0) = f(\alpha(0)) \approx 1 + H_k(1)/RQ_m\,. \tag{9.20}$$

For $f(\alpha)$ appearing in Fig. 9.26b, when $m \geq 4$, $H_0(1) \approx 0.70$. Also, from (9.18) or (8.74), we have $D(1) \approx 1 + c_k^\infty / RQ_m$. The constants appearing in (8.78) defining the $\alpha \leq \alpha_{\mathrm{A}}$ linear segment are thus given by

$$q_{\mathrm{A}} \approx 2 + (C_k + D_k)/RQ_m \tag{9.21}$$

$$\alpha_{\mathrm{A}} \approx 1 + C_k/RQ_m \tag{9.22a}$$

$$\alpha'_{\mathrm{A}} \approx (1/2) + (5/4)C_k/RQ_m \tag{9.22b}$$

$$f(\alpha'_{\mathrm{A}}) \approx \{C_k - (3/2)D_k\}/RQ_m \,, \tag{9.23}$$

where $C_k \equiv L_k(-1+0) + c_k^\infty$ and $D_k \equiv V_k(L_k(-1+0))$ are positive universal constants for each given value of k. For $f(\alpha)$ in Fig. 9.26b, we have $C_0 \approx 0.47$ and $D_0 \approx 0.23$. The existence of two-dimensionality is equivalent to the realization of the inequality $R \equiv -\ln|J| < \infty$. Therefore, when $a \to a_\infty$, the difference between $f(\alpha)$ and $f_*(\alpha)$ (i.e. two-dimensional fractality) disappears as a quantity proportional to $1/Q_m \propto \epsilon_k^\kappa$.

The above discussion can also be applied to the $Q_m^\pm$-band chaos existing in the T_m^- cascade for the annulus map, in which case the difference between $f(\alpha)$ and $f_*(\alpha)$ disappears in proportion to $1/Q_m^\pm \propto \epsilon_m^{\kappa_{\mathrm{GM}}}$.

10. Mixing and Diffusion in Chaos of Conservative Systems

The characteristic feature of the chaotic sea in a conservative dynamical system is the presence of a multitude of tori of different sizes, forming an 'islands around islands' self-similar hierarchical structure. Chaotic orbits are often trapped for long times within such structure, and as a result the long-time correlation $C_t^\lambda \propto t^{-(\beta-1)}$ $(2 > \beta > 1)$ appears. In this situation, for $\Lambda > 0$ the probability distribution of mixing $P(\Lambda; n)$ obeys the anomalous scaling relation $P(\Lambda; n) = n^\delta p(n^\delta(\Lambda - \Lambda^\infty))$, where $\delta \equiv (\beta - 1)/\beta < 1/2$, and $p(x)$ is a universal function determined by β. As determined by the folding of the unstable manifold induced by the tangency structure, for $0 > \Lambda > \Lambda_{\min}$ we have $\psi(\Lambda) = -2\Lambda$. In the case in which islands of accelerator mode tori exist, the diffusion coefficient for chaotic orbits (or fluid particles) diverges, and the corresponding probability distribution of the coarse-grained velocity of such orbits (or particles) obeys an anomalous scaling relation similar to that given above. In this chapter we investigate the universality displayed by the statistical structure of this kind for chaos in conservative systems and the self-similar time series of the last KAM torus.

10.1 The Dynamical Self-Similarity of the Last KAM Torus

In this chapter we treat the standard map (7.17) defined as the cross-section of a torus whose time series is designated by $(2\pi\theta_i, J_i)$. We will apply the concepts and methods that emerge from this study to the elucidation of the mixing and diffusion exhibited by fluids as a result of the oscillating laminar flow discussed in Sect. 6.2. The standard map provides useful models of a number of conservative dynamical systems. It possesses the special feature that, as shown in Fig. 7.6, for $K < K_C$, all chaotic seas are limited in extent within bounded regions (cells of finite extent along the J-axis), but for $K > K_C$, all KAM tori are destroyed, and widespread chaos extending over all values of J $(-\infty$ to $+\infty)$ results. In this section we first study the nature of the last KAM torus appearing at the threshold $K = K_C$.

10.1.1 The Self-Similar F_m Time Series

The last KAM torus is a golden torus of rotation number ρ_G, the inverse of the golden ratio. The value ρ_G is distinguished as being that value whose continued fraction expansion (7.47) displays maximally slow convergence. It is the irrational number which, in some sense, is most difficult to approximate by use of a fraction. For this reason, the golden torus is the torus which is most 'difficult' to destroy. Figure 10.1 displays the time series of $\beta_n(X_0) \equiv n\Lambda_n(X_0)/\ln n$ for an orbit on such a torus. The form of this time series differs from that shown in Fig. 9.18 for the critical golden torus of the annulus map, but these two time series are both composed of length F_m blocks of self-similar inverted nested structure. (Recall that $F_{m-1} + F_m = F_{m+1}$.) The power spectrum of the time series β_n in the present case has a form similar to that shown in Fig. 9.19, with the width (along ω) of successive blocks decreasing in the ratio ρ_G, converging geometrically to 0 as $\omega \to 0$. This self-organized criticality of the time series and power spectrum can be thought of as constituting the universality of the critical golden torus.

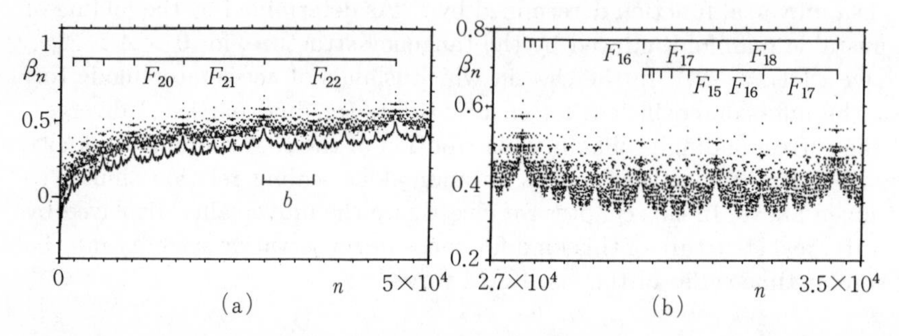

Fig. 10.1. The time series of β_n for the last KAM torus ($n = 2$–5×10^4). These data were generated with the initial point $X_0 = (\theta_0, J_0) = (-0.369427, -0.435860)$. (**b**) consists of an expanded view of the interval b in (**a**)

10.1.2 The Symmetric Spectrum $\psi_\beta(\beta)$

The algebraic structure functions obtained from the probability distribution of $\beta_n(X_0)$, (9.8), are shown in Fig. 10.2. These functions are qualitatively different from those shown in Fig. 9.20; in the present case $\sigma_\beta(q)$ displays no peak, and $\psi_\beta(\beta)$ no linear segment. In addition, $\psi_\beta(\beta)$ is symmetric under reflection: $\psi_\beta(\beta) = \psi_\beta(-\beta)$. This symmetry results from that of conservative dynamical systems described by (7.15), corresponding to the time-reversal symmetry $X = (\theta, J) \to \tilde{X} = (\theta, -J)$. In other words, the equation of motion $\tilde{X}_{i+1} = \tilde{F}(X_i)$, obtained through time reversal, is identical to that given by the inverse map, $\tilde{X}_{i+1} = F^{-1}(\tilde{X}_i)$, and therefore $D\tilde{F}(\tilde{X}_i) = DF^{-1}(X_{i+1}) = \{DF(X_i)\}^{-1}$. Considering (6.19), $u_1(\tilde{X}_i)$ [as

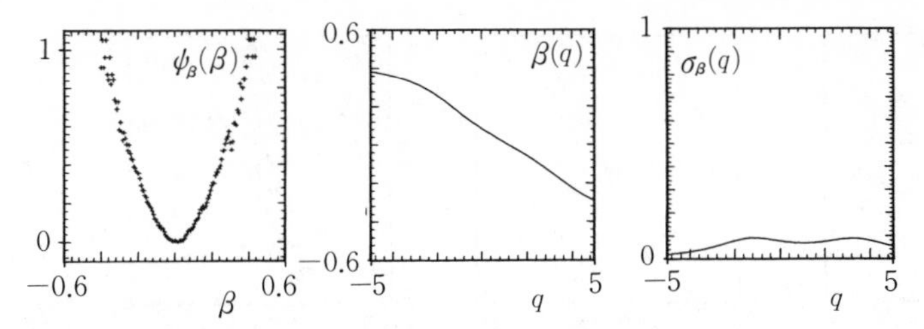

Fig. 10.2. The algebraic structure functions of β_n for the last KAM torus ($n = 2090$, $N = 10^5$) [Horita et al. (1989)]

$u_1(X_t)$ appearing in (6.19)] is tangent to the KAM torus at $\tilde{X}_i$, and thus replacing $DF(X_i)$ in (6.19) with $D\tilde{F}(\tilde{X}_i)$, for motion obtained under time reversal we have $\tilde{\beta}_n(\tilde{X}_0) = -\beta_n(X_0)$. Hence time-reversal symmetry implies $\psi_\beta(\beta) = \psi_\beta(-\beta)$, as displayed graphically in Fig. 10.2. In contrast to this situation, the annulus map (7.10) ($J \neq 1$) represents a dissipative dynamical system, for which time-reversal symmetry does not exist. This fact is reflected by the asymmetry of the $\psi_\beta(\beta)$ shown in Fig. 9.20.

10.2 The Mixing of Widespread Chaos

The residue of the Jacobian matrix for the standard map (7.17) is $R = (K/4)\cos(2\pi\theta_i)$ [see (A.5) in the Appendix]. Thus for the fixed point $X^* \equiv (\theta^*, J^*) = (0,0)$, we have $R = K/4$. Therefore in the range $0 < K < 4$, this point is neutral and elliptic, while for $K > 4$ it becomes unstable with respect to period doubling. On the other hand, the fixed point $S \equiv (1/2, 0)$ is characterized by $R = -K/4$, and thus it is an unstable saddle for all $K > 0$. In what follows we focus on the unstable manifold of the saddle S in considering the widespread chaotic sea containing this manifold in the $K > K_\mathrm{C}$ region.

For the sake of concreteness we consider two cases here, (a) $K = K_\mathrm{a} \equiv 6.9115$ and (b) $K = K_\mathrm{b} \equiv 3.86$. The most easily discernible islands of tori existing here are displayed in Fig. 10.3. In these two cases, the Liapunov numbers for chaotic orbits are (a) $\Lambda^\infty \approx 1.26$ and (b) $\Lambda^\infty \approx 0.82$.

Figure 10.4 shows the 'islands around islands' self-similar hierarchical structure mentioned above. In a typical situation, a chaotic orbit X_i ($i = 0, 1, 2 \ldots$) will become trapped within this type of structure in a region u of the chaotic sea containing some such islands, and then, after some time, it will escape, wandering about the chaotic sea to be later trapped in some new such region. This 'trapping and escaping' behavior continues indefinitely. Designating the probability of a given orbit to be trapped for a time longer than some given t in an arbitrary such region of islands u as $W(t)$, the inverse power relation

$$W(t) \propto t^{-(\beta-1)} \quad (2 > \beta > 1) \tag{10.1}$$

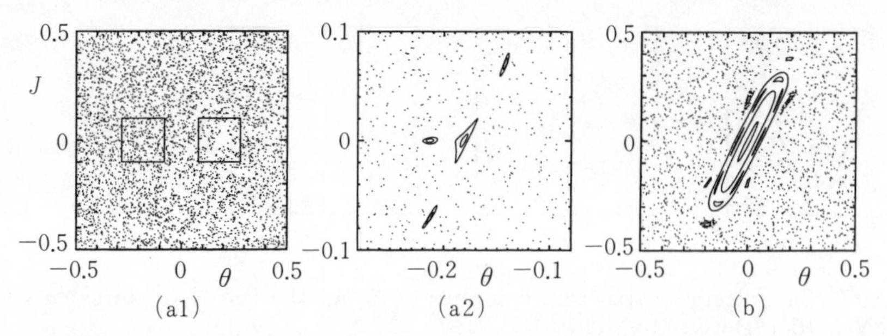

Fig. 10.3. The most easily discerned islands of tori for the standard map with **(a)** $K = K_{\mathrm{a}} \equiv 6.9115$ and **(b)** $K = K_{\mathrm{b}} \equiv 3.86$. (a2) is a close-up view of the left square in (a1). Each of these four islands consists of accelerator mode tori

is satisfied in the limit as $t \to \infty$. The time correlation function C_t^λ in (6.28) has the same time dependence in this limit:

$$C_t^\lambda \approx (\Lambda^\infty)^2 W(t) \propto t^{-(\beta-1)} . \tag{10.2}$$

Because $\lambda_1(X_i) \approx 0$ during the time in which X_i is trapped within some region of islands u, a visit to such a region in general continues for a long time. When X_i is positioned in a region u^{c} within the chaotic sea, separated by a sufficient distance from any such islands, C_t^λ exhibits exponential attenuation. This contribution to the $t \to \infty$ form of C_t^λ can thus be ignored [being much smaller than that expressed by (10.2)]. If we substitute the form (10.2) into (6.28) and integrate, the variance $S_n(X_0) \equiv n\Lambda_n(X_0)$ becomes

$$\langle \{S_n(X) - \langle S_n \rangle\}^2 \rangle \propto n^\zeta \quad (1 < \zeta = 3 - \beta < 2) , \tag{10.3}$$

where $\langle S_n \rangle = n\Lambda^\infty$. In general, the value of the exponent β of (10.1) differs for each region of islands. In this situation, the smallest such value controls the long-time behavior, and thus it is this value that appears in the time correlation (10.2) and variance (10.3)[1]. Figure 10.5 displays the results of a numerical calculation of the variance (10.3) for cases (a) $K = K_{\mathrm{a}}$ and (b) $K = K_{\mathrm{b}}$, whose phase spaces are given in Fig. 10.3. This figure verifies the validity of the asymptotic form n^ζ given by (10.3). It also yields the values (a) $\zeta \approx 1.40$, $\beta \approx 1.60$ and (b) $\zeta \approx 1.45$, $\beta \approx 1.55$. The inverse power laws (10.1) and (10.2) are thus also verified by these numerical results.

As seen here, the motion of a chaotic orbit in regions u^{c} deep within the chaotic sea appear as intermittent bursts between its prolonged lingering in island regions u. We now turn to a study of the anomalous statistical structure of these orbits, following Horita and Mori (1992).

---

[1] However, we must note here that, roughly speaking, islands whose surface areas are so small that their presence can be ignored are excluded from this consideration.

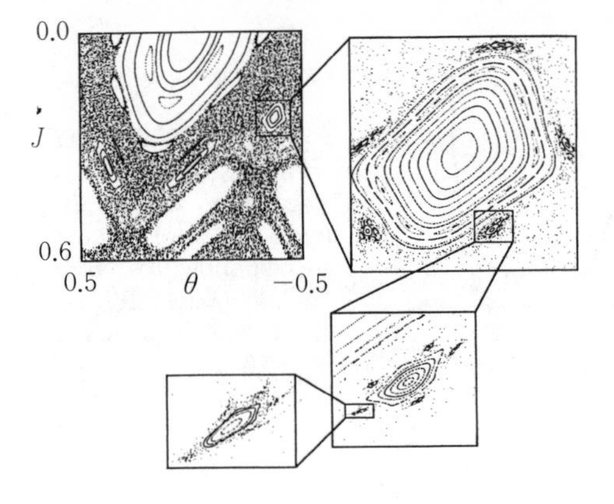

Fig. 10.4. The hierarchical 'islands around islands' structure ($K = 1.20141333$). At any level, if we choose one of the five small islands and take a closer look, we will see that it is surrounded by five smaller islands. This is the self-similar structure exhibited by such a system [Meiss (1992)]

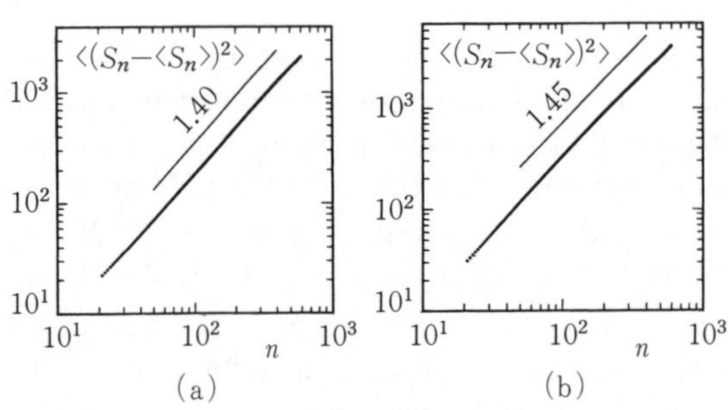

Fig. 10.5. A log–log plot of the variance (10.3) ($N = 2 \times 10^7$, $n = 21 \sim 660$). The lines of slope $\zeta = 1.40$ and 1.45 are shown for comparison [Horita and Mori (1992)]

10.2.1 The Form of $\psi(\Lambda)$ and the Breaking of Time-Reversal Symmetry

Figure 10.6a represents the results of numerical calculations of the spectrum (6.26) for the case shown in Fig. 10.3b. There are three plots shown here, corresponding to (from top to bottom) $n = 10$ (circles), $n = 20$ (triangles), and $n = 40$ (squares). In the $n \to \infty$ limit, if N (the number of orbits) remains finite, the upper limit of $\psi_n(\Lambda)$ can be observed, but if we first take $N \to \infty$ and then $n \to \infty$, the spectrum $\psi_n(\Lambda)$ assumes the form shown in Fig. 10.6b. This form is characterized by a slope $s_\alpha = -2$ segment in the interval $\Lambda_{\min} < \Lambda < 0$ and satisfies $\psi(\Lambda) = 0$ for $0 \leq \Lambda \leq \Lambda^\infty$.

The chaotic orbits exist in the closure of $W^{\mathrm{u}}(S)$, and the slope $s_\alpha = -2$ linear piece results from the folding of W^{u} induced by type I tangency structure (Fig. 6.7b). In fact, for a conservative system we have $z = 2$ and $\alpha_1 = 2$ in (8.46). This determines the slope $s_\alpha = -2$. The appearance of the linear piece in the interval $0 \leq \Lambda \leq \Lambda^\infty$, where $\psi(\Lambda) = 0$, is caused by the trapping of orbits in the hierarchical structure of the islands regions, described by (10.1). By definition (6.26), $\psi(\Lambda^\infty) = 0$, but even for $\Lambda = 0$, substituting $P(0; n)/P(\Lambda^\infty; n) \propto W(n)$ into (6.26), and taking the $n \to \infty$ limit, we have $\psi_n(0) \propto \ln n/n \to 0$. That is, $\psi(0) = \psi(\Lambda^\infty) = 0$. Thus since $\psi(\Lambda)$ is a non-negative, downward convex function, we obtain $\psi(\Lambda) = 0$ for $0 \leq \Lambda \leq \Lambda^\infty$. For $\Lambda > \Lambda^\infty$, as we will see in the following subsection,

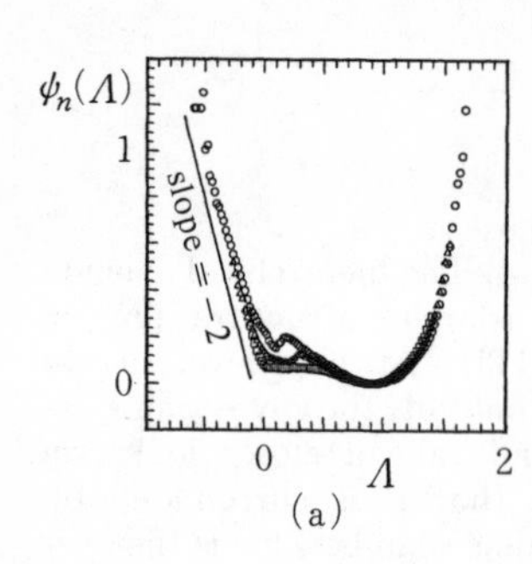

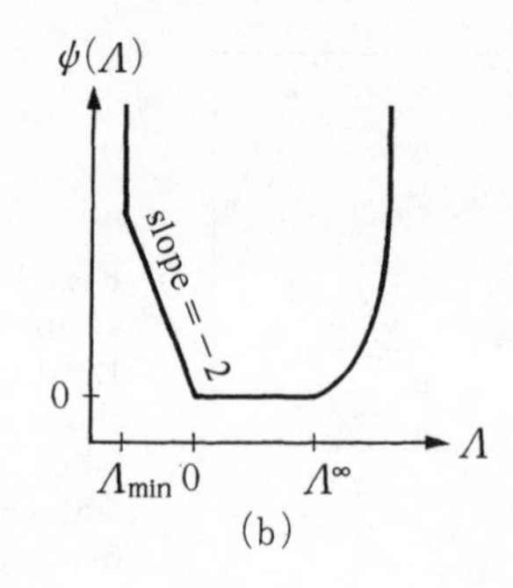

Fig. 10.6. The spectrum $\psi_n(\Lambda)$ for $K = K_\mathrm{b}$. Here, the *circles*, *triangles*, and *squares* correspond to $n = 10, 20$, and 40, respectively. In all cases, $N = 10^7$

$$\psi(\Lambda) \propto (\Lambda - \Lambda^\infty)^{1/\delta} \quad (n^\delta(\Lambda - \Lambda^\infty) \gg 1)\,, \tag{10.4}$$

where $1/\delta \equiv \beta/(\beta - 1) > 2$.

As mentioned above, $\psi(\Lambda)$ is asymmetric with respect to Λ. We now show that this results from the breaking of time-reversal symmetry suffered by conservative dynamical systems in the present context. This situation contrasts with that for the symmetric spectrum $\psi_\beta(\beta)$ of the last KAM torus.

For motion obtained under time reversal, described by $\tilde{X}_{i+1} = \tilde{F}(\tilde{X}_i) = F^{-1}(\tilde{X}_i)$, as determined by the fact that $D\tilde{F}(S) = \{DF(S)\}^{-1}$, the eigenvalues of the saddle $S = (1/2, 0)$ are changed from ν_1^Q and $\nu_2^Q = 1/\nu_1^Q$ to $\tilde{\nu}_1^Q = 1/\nu_1^Q$ and $\tilde{\nu}_2^Q = \nu_1^Q$. For this reason, the unstable manifold and stable manifolds reverse roles, and in reference to (6.19) we have $\tilde{\lambda}_1(\tilde{X}_i) = -\lambda_1(X_i)$. This gives $\tilde{\Lambda}_n(\tilde{X}_0) = -\Lambda_n(X_0)$, and thus time-reversal symmetry implies $\psi(\Lambda) = \psi(-\Lambda)$. However, the functions shown in Fig. 10.6 do not satisfy this relation. In this way, the asymmetry of $\psi(\Lambda)$ reflects the breaking of time-reversal symmetry of the system. As we explain below, this symmetry breaking is rooted in the instability of the chaotic orbits. Before proceeding, however, let us note that, as in the case described by Fig. 8.7, for dissipative dynamical systems, $\psi(\Lambda)$ is asymmetric with respect to reflection about $\Lambda = 0$.

Under the forward map $X_{i+1} = F(X_i)$, the unstable manifold W^u is stretched, and orbits in the neighborhood of this manifold are attracted to it (see Fig. 6.6), while under the backward map $X_{i-1} = F^{-1}(X_i)$, the stable manifold W^s is stretched, and orbits in its neighborhood are attracted to it. Thus for almost all neighboring orbits X_i and X_i', the vector representing their difference, $y_i = X_i' - X_i$, becomes (under repeated mappings in one direction) tangent to the invariant manifold which becomes stretched under mappings in the direction under consideration. Also, under such repeated mappings we obtain, on average, $|y_{i+1}| > |y_i|$ for the case of forward mappings and $|y_{i-1}| > |y_i|$ for the case of backward mappings. Thus under either type of mapping, the orbital expansion rate, $\lambda_1(X_i) = \ln[|y_{i+1}|/|y_i|]$ in the former case and $\lambda_1(X_i) = \ln[|y_{i-1}|/|y_i|]$ in the latter case, is positive on average, implying a positive Liapunov number: $\Lambda^\infty = \Lambda_\infty(X_0) > 0$. The motion obtained under time reversal is identically that obtained under the backward map, and thus the Liapunov number for motion seen under time reversal is also positive, and the corresponding spectrum $\psi(\Lambda)$ is also asymmetric with respect to reflection about $\Lambda = 0$ [see Mori et al. (1989)].

Two neighboring orbits located on the unstable manifold W^u, correspond under time reversal to two orbits on the stable manifold $\tilde{W}^s$ and, since their difference $\tilde{y}_i$ is tangent to $\tilde{W}^s$, we find $\tilde{\Lambda}_\infty(\tilde{X}_0) = -\Lambda^\infty$ for some $\tilde{X}_0$ on $\tilde{W}^s$ (i.e. X_0 on W^u). However, if the direction of the initial difference vector $\tilde{y}_0$ differs even very slightly from the direction of tangency to $\tilde{W}^s$, $\tilde{y}_i$ will eventually become tangent to $\tilde{W}^u$ (under the inverse map that we are considering). This implies $\tilde{\Lambda}_\infty(\tilde{X}_0) = \Lambda^\infty$ if $\tilde{X}_0$ is not on $\tilde{W}^s$. Then, since the Lebesgue measure of vectors $\tilde{y}_0$ tangent to $\tilde{W}^s$ is zero, time-reversed motion for such systems does not appear in Nature. In computer simulations also, since there is always a degree of computational error present, it is not possible to realize time-reversed motion from simulation of such systems over long intervals. In this manner, the instability of chaotic orbits causes the breaking of time-reversal symmetry and leads to irreversibility in conservative dynamical systems.

10.2.2 The Appearance of Anomalous Scaling Laws for Mixing

For $0 \le \Lambda \le \Lambda^\infty$, $\psi(\Lambda) = 0$, and the probability distribution $P(\Lambda; n)$ is determined by the logarithmic term in the statistical law (6.24) regarding time coarse-graining. This can be obtained from Feller's theorem with regard to recurrent independent events as follows. For $\Lambda > 0$, the scaling law

$$P(\Lambda; n) = n^\delta p(n^\delta \hat{\Lambda}) \quad (\hat{\Lambda} \equiv \Lambda - \Lambda^\infty) \tag{10.5}$$

is obeyed. Here, $0 < \delta \equiv (\beta - 1)/\beta < 1/2$, and $p(x)$ is a universal function determined by β. In particular, for $0 < \Lambda < \Lambda^\infty$ $(x < 0)$, in the $n \to \infty$ limit the relation

$$p(x) \propto |x|^{-\beta-1} \quad (-x = -n^\delta \hat{\Lambda} \gg 1) \tag{10.6}$$

holds. This implies $P(\Lambda; n) \propto n^{-\beta\delta}/|\hat{\Lambda}|^{\beta+1}$, and thus, comparing with (6.30), we obtain $\phi(\Lambda) = \beta - 1$ and $C(\Lambda) \propto |\hat{\Lambda}|^{-\beta-1}$. For $\Lambda > \Lambda^\infty$, as $n \to \infty$, we obtain

$$p(x) \propto \exp[-a x^{1/\delta}] \quad (x = n^\delta \hat{\Lambda} \gg 1) , \tag{10.7}$$

where a is a positive constant. This gives (10.4). Then, for finite n, the most probable value of Λ, $\overline{\Lambda}_n$, differs from Λ^∞, with $\overline{\Lambda}_n - \Lambda^\infty \propto n^{-\delta}$.

In order to construct a model to which Feller's theorem can be applied, let us suppose that the expansion rate $\lambda_1(X_i)$ of an arbitrary chaotic orbit X_i satisfies $\lambda_1(X_i) = 0$ in the region U, that part of the chaotic sea consisting of the set of all island regions u discussed above, and $\lambda_1(X_i) = \lambda(> 0)$ in the region U^c of the chaotic sea outside U. Then, if upon being trapped in U, the probability that this orbit will be trapped for a time t is given by $f(t)$, the average length of time spent trapped in U on a single visit is $\tau \equiv \sum_{t=1}^{\infty} t f(t)$.

Similarly, if the probability that this orbit will spend a time s in U^c, given that it has entered this region, is $f^c(s)$, the average length of time spent in U^c on a single visit is given by $\tau_c \equiv \sum_{s=1}^{\infty} sf^c(s)$. Then, as a result of (10.1) and the relation $W(t) = \sum_{u=t}^{\infty}(u-t+1)f(u)/(\tau+\tau_c)$, we have $f(t) \propto t^{-\beta-1}$. For U^c we have $f^c(s) \propto \exp[-s/\tau_c]$, where $\tau_c \ll \tau$. Thus corresponding to events characterized by $\lambda_1(X_i) = \lambda$ (i.e., events in which the orbit enters U^c) there is the distribution of lifetimes $f^c(s)$ and the distribution of recurrence times $f(t)$, constituting statistically independent quantities. Hence, Feller's theorem can be used to determine the form of $P(\Lambda; n)$ $(\Lambda > 0)$.

According to Feller's theorem, the probability $F(t)$ that two successive visits to U^c are separated by a time shorter than t is given by $1 - F(t) \propto \sum_{u=t}^{\infty} f(u) \propto t^{-\beta}$ $(2 > \beta > 1)$, and if such events occur independently, the number N_n of such events occurring in a time n is given in the limit as $n \to \infty$ by

$$\Pr\{N_n \geq (n/\tau) - (b_n y/\tau^{\alpha})\} \to G_\beta(y) \ . \tag{10.8}$$

Here $b_n \propto n^{1/\beta}$, $\alpha \equiv (\beta+1)/\beta$, and $G_\beta(y)$ is the distribution function of the Lévy stability law[2] with characteristic function

$$\gamma_\beta(z) = \exp[-|z|^\beta \Gamma(1-\beta)\exp\{-\mathrm{i}(\mathrm{sgn}z)\pi\beta/2\}] \ , \tag{10.9}$$

where $\Gamma(x)$ is the gamma function, and for $\beta = 2$, $G_2(y)$ is a normal distribution.

We can write $N_n = n\Lambda_n/\lambda$, and thus, with c a positive constant, we have

$$\Pr\{\Lambda_n \geq \Lambda\} \to G_\beta(y) \quad [\Lambda \equiv (\lambda/\tau) - (y/cn^\delta)] \ .$$

Taking the derivative of this expression with respect to y yields, as $n \to \infty$,

$$P(\Lambda; n)/P_{\max}(n) \to g_\beta(-cn^\delta \hat{\Lambda}) \ , \tag{10.10}$$

where $g_\beta(y) = \mathrm{d}G_\beta(y)/\mathrm{d}y$, $y = -cn^\delta \hat{\Lambda}$, and $\Lambda^\infty \approx \lambda/\tau$. The quantity $P_{\max}(n)$ is the maximum of $P(\Lambda; n)$ and is proportional to n^δ. Hence (10.5) can be obtained from (10.10). For $\beta = 1.60$ (as obtained numerically for $K = K_a$; see Fig. 10.5), the maximal value of $g_\beta(y)$ is assumed at $y = y_m \equiv -1.696\ldots$, where the relation $\overline{\Lambda}_n - \Lambda^\infty = (-y_m/c)n^{-\delta}$ $(c \approx 0.16)$ holds. From (10.9) and (10.10), (10.6) and (10.7) can be obtained.

Figure 10.7a gives a plot of numerically obtained data for $P(\Lambda; n)/P_{\max}(n)$ as a function of $x = n^\delta\ (\Lambda - \overline{\Lambda}_n)$ for the case $K = K_a$ in Fig. 10.7. (Here $n = 1000$.) The solid curve represents the analytically determined function $g_\beta(y_m - cx)$. In this figure the value $\delta = 0.375$ $(\beta = 1.60)$ is used. Because N is not sufficiently large, the data points for large negative values of x are somewhat scattered, but overall the numerical data fit the analytically determined curve quite well. Figure 10.7b presents a plot of $P(\Lambda; n)/P_{\max}(n)$ as a function of $|x| = n^\delta|\Lambda - \overline{\Lambda}_n| \gg 1$ for the range $0 < \Lambda < \overline{\Lambda}_n$. These

[2] For details, see, for example, Feller, W. (1949): Trans. Am. Math. Soc. **67**, 98–119; Wang, X.-J. (1989): Phys. Rev. A **40**, 6647–6661.

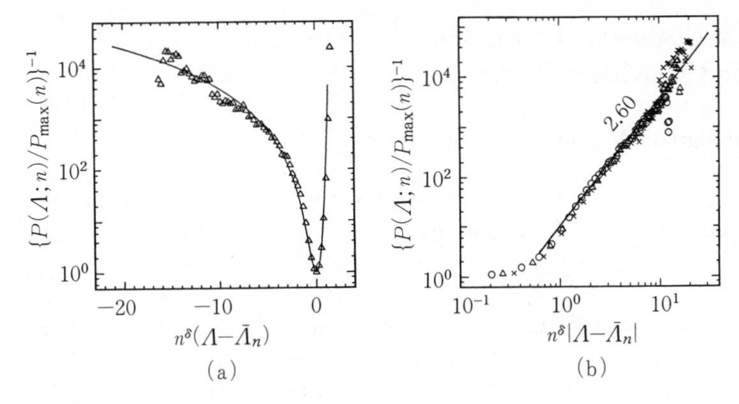

Fig. 10.7. (**a**) A comparison of the analytically obtained form of $\log_{10}\{P(\Lambda;n)/P_{\max}(n)\}^{-1}$ as a function of $x \equiv n^\delta(\Lambda - \overline{\Lambda}_n)$ ($\delta = 0.375$) with numerically generated data ($n = 1000$, $N = 2 \times 10^7$). (**b**) A comparison between numerically and analytically obtained results for $\log_{10}\{P(\Lambda;n)/P_{\max}(n)\}^{-1}$ as a function of $\log_{10}|x|$ with $0 < \Lambda < \overline{\Lambda}_n$. Here the *circles*, *triangles*, and *crosses* correspond to $n = 500, 1000$, and 2000, respectively. The slope 2.60 corresponds to the value $\beta + 1$. In both figures, $K = K_\mathrm{a}$

numerical data verify the asymptotic form (10.6), with which, from the figure, we obtain $\beta + 1 \approx 2.60$, consistent with the value $\beta = 1.60$.

For $\Lambda > \Lambda^\infty$, when $n^\delta \hat{\Lambda} \gg 1$, the asymptotic form (10.7) holds, and instead of the variance (10.3) we have

$$\left\langle \{S_n - n\Lambda^\infty\}^2 \theta(S_n - n\Lambda^\infty) \right\rangle \propto n^{\zeta_+} , \tag{10.11}$$

where $\zeta_+ = 2(1-\delta) = 2/\beta$ and $\theta(x)$ is the step function ($\theta(x) = 1$ for $x > 0$ and $\theta(x) = 0$ for $x < 0$). This is the conditional variance taking into account only fluctuations for which $\Lambda > \Lambda^\infty$, and because $2/\beta < 3 - \beta$ this contribution is overwhelmed in (10.3) by that due to fluctuations in the range $0 < \Lambda < \Lambda^\infty$. Figure 10.8 presents the numerically calculated n dependence of (10.11) for the case $K = K_\mathrm{a}$. These results yield the exponent value $\zeta_+ \approx 1.25$ ($\beta \approx 1.60$), verifying previous determinations.

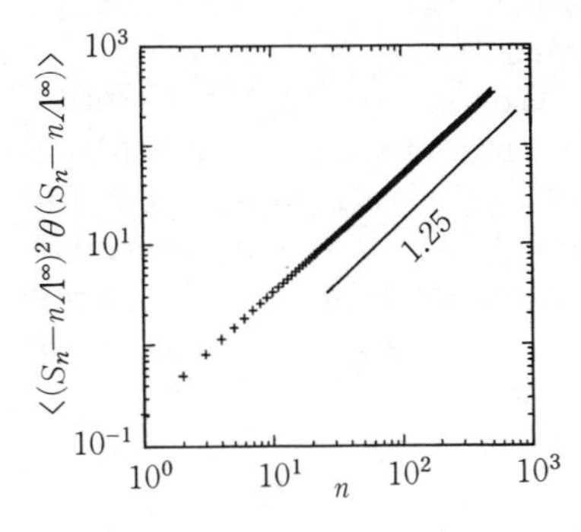

Fig. 10.8. A log–log plot of the conditional variance (10.11). The data are consistent with $\zeta_+ = 1.25$

10.3 Anomalous Diffusion Due to Islands of Accelerator Mode Tori

10.3.1 Accelerator Mode Periodic Orbits

The characteristic feature of the sea of widespread chaos is the coexistence of neutral orbits $X_i^* = (\theta_i^*, J_i^*)$ satisfying

$$J_{i+Q}^* = J_i^* \pm l, \quad \theta_{i+Q}^* = \theta_i^* \quad (i = 1, 2, \cdots, Q) , \tag{10.12}$$

where Q and l are positive integers. Under Q iterations of the mapping, such an orbit, originally at some point $X_0 = (\theta_0, J_0)$ within a given cell, is transformed to a point at the same value of θ with the value of J shifted by $\pm l$. In other words, these orbits move in the J direction with velocity $v_a = \pm l/Q$, and they are referred to as period Q, step l accelerator mode periodic orbits. This name comes from a model of cosmic ray acceleration [Lichtenberg and Lieberman (1982)] that describes the manner in which high-energy cosmic particles are created through the interaction of electrically charged particles with time-varying magnetic fields. From (10.12) and the second equation in (7.17), we obtain

$$\frac{K}{2\pi} \sum_{i=1}^{Q} \sin(2\pi\theta_i^*) = \mp l . \tag{10.13}$$

Using the residue $R = \left[2 - \mathrm{tr}\left\{DF^Q(X_i^*)\right\}\right]/4$, the range of values of K in which these neutral periodic orbits X_i^* exist can be obtained from the condition of neutral stability, $0 < R < 1$, discussed in Sect. A.1 of the Appendix.

We now discuss the accelerator mode tori of the standard map, following Ichikawa et al. (1987). When $Q = 1$, from (A.5) we have $R = (K/4)\cos(2\pi\theta_i^*)$, and from (10.13), $\sin(2\pi\theta_i^*) = \mp 2\pi l/K$, and hence $4R = \left\{K^2 - (2\pi l)^2\right\}^{1/2}$. In this case, the condition of neutral stability $0 < R < 1$ reduces to

$$2\pi l < K < \{(2\pi l)^2 + 16\}^{1/2} . \tag{10.14}$$

For $l = 1$, this yields the condition $6.2832 < K < 7.4483$. Increasing K beyond this range of values causes the accelerator mode periodic points to become unstable with respect to period doubling.

In the range $6.8927 < K < 6.9743$, there also exist $Q = 3$, $l = 3$ neutral accelerator mode periodic orbits. In fact, the center of the right-most island through which the line $J = 0$ passes in Fig. 10.3a2 (with $K = K_a = 6.9115$) is an accelerator mode periodic point with $Q = 1$ and $l = 1$, and the centers of the three islands around this island are accelerator mode periodic points with $Q = 3$ and $l = 3$. Such modes possess velocities $v_a = \pm 1$ along the J direction.

10.3.2 Long-Time Velocity Correlation

The islands of tori situated around accelerator mode periodic points possess 'islands around islands' hierarchical structure similar to that in Fig. 10.4. This hierarchical structure, as well as the chaotic orbits X_i trapped within it, move along the J direction with velocity $v_a = \pm l/Q$. Thus we have $J_n - J_0 = \pm n v_a$. The probability $W(n)$ that a chaotic orbit X_i will be trapped for a time longer than n in such a region of islands of accelerator mode tori takes the form of (10.1): $W(n) \propto n^{-(\tilde{\beta}-1)}$ $(2 > \tilde{\beta} > 1)$. With this, the variance of $J_n - J_0$ along the + and − directions of J can be written

$$\langle\{J_n(X) - J_0(X)\}^2\rangle \approx (n v_a)^2 W(n) \propto n^\eta \,, \tag{10.15}$$

where $\eta = 3 - \tilde{\beta} > 1$, and thus in the limit as $n \to \infty$, the diffusion coefficient $D \equiv \langle (J_n - J_0)^2 \rangle / 2n \propto n^{\eta - 1}$ diverges. This corresponds to the anomalous diffusion of fluids described by (6.10) with $\eta > 1$.

Figure 10.9 displays the numerically obtained form of this variance for cases (a) $K = K_a$ and (b) $K = K_b$. These data yield the values (a) $\eta \approx 4/3$ $(\tilde{\beta} \approx 5/3), D = \infty$ and (b) $\eta = 1.00, D \approx 0.056$. The value of $\tilde{\beta}$ for a given system (as in Fig. 10.9a) is due to the islands of accelerator mode tori present in that system, while in general the value of β characterizing a system is equal to the smallest in the set of all such values corresponding to individual islands of tori. Thus in general, $\tilde{\beta} \geq \beta$. In Fig.10.9b, with $K = K_b$, all 'visible' islands are $l = 0$, normal islands (those consisting of normal tori), and thus we can assume that accelerator mode tori of nonnegligible size do not exist here. When a chaotic orbit X_i becomes trapped in a region of normal islands, diffusion ceases, and $\langle (J_n - J_0)^2 \rangle \approx 0$ is realized. For this reason, the only contribution to the diffusion comes from the time during which the orbit is inside the chaotic sea, implying $\eta = 1$ and the existence of a diffusion coefficient D. However, as $K \to K_c$, the growth of the islands causes the area comprised by turnstiles (allowing for cell-to-cell motion) to shrink, leading to a decrease in the value of D. In particular, in the range $K_c < K < 2.5$, D can be written as $D \propto (K - K_c)^\xi$ (with $\xi \approx 3.01$) [MacKay et al. (1984)]. This is due to the fact that the probability of a chaotic orbit passing through a *cantorus*, remaining after the destruction of the last KAM torus, decreases in proportion to $(K - K_c)^\xi$.

In order to clarify the anomalous nature of diffusion resulting from the presence of accelerator modes, we introduce the quantity $u_t \equiv J_{t+1} - J_t$ and the *coarse-grained velocity*,

$$v_n(X_0) \equiv \frac{J_n - J_0}{n} = \frac{1}{n} \sum_{t=0}^{n-1} u_t \,. \tag{10.16}$$

Using the velocity correlation $C_t \equiv \langle u_t u_0 \rangle$, the corresponding variance can be written

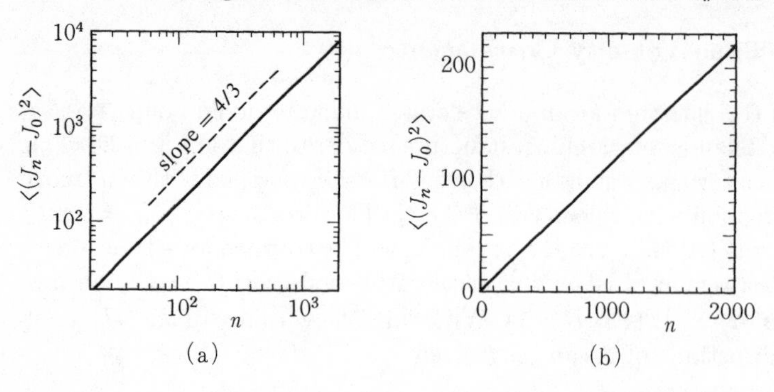

Fig. 10.9. The n dependence of the variance (10.15) determined numerically. For the case $K = K_{\rm a}$ the data yield $\eta = 4/3$, while for $K = K_{\rm b}$ they give $\eta = 1.00$, as shown in (**a**) and (**b**), respectively

$$n^2 \langle \{v_n(X)\}^2 \rangle = nC_0 + 2\sum_{t=1}^{n-1}(n-t)C_t \,. \tag{10.17}$$

(See Sect. A.2 of the Appendix). When K is large enough that the velocity correlation can be ignored (i.e., we can write $C_t = C_0 \delta_{t,0}$), we can write $C_0 = (K/2\pi)^2 \langle \sin^2(2\pi\theta) \rangle = K^2/8\pi^2$, and the diffusion coefficient becomes $D = C_0/2 = K^2/16\pi^2$. Then, if the time correlation of the velocity decays exponentially, i.e., $C_t = C_0 \exp[-\gamma t]$ ($\gamma \ll 1$), we have $D = C_0/\gamma$. However, when accelerator mode tori exist, a long-time correlation of the form $C_t \approx v_{\rm a}^2 W(t) \propto t^{-(\tilde{\beta}-1)}$ results, and the variance (10.17) comes to assume the anomalous form given by (10.15). We now describe the corresponding statistical structure, as determined by Ishizaki et al. (1991).

10.3.3 The Anomalous Nature of the Statistical Structure of the Coarse-Grained Velocity

Let us determine the $n \to \infty$ asymptotic form of the probability distribution $P(v;n) \equiv \langle \delta(v_n(X) - v) \rangle$ for the coarse-grained velocity (10.16). From the inversion invariance of the standard map, we know that $P(v;n)$ must be an even function of v, possessing a maximum at the average value $\overline{v} = 0$. As in the $K = K_{\rm b}$ case, when the velocity correlation remains positive as it decays exponentially, $\eta = 1$ results, a diffusion coefficient D exists, and the normal distribution

$$P(v;n) = (n/4\pi D)^{1/2} \exp[-nv^2/4D] \tag{10.18}$$

is realized.

When accelerator mode tori exist, we have $\eta > 1$ and $D = \infty$. As a result, the quantity

$$\psi(v) \equiv -\lim_{n\to\infty} \frac{1}{n} \ln \frac{P(v;n)}{P(0;n)} \tag{10.19}$$

is 0 for $|v| \leq v_{\rm s} \equiv l/Q$ and ∞ for $|v| > v_{\rm s}$. This result can be derived from the statistical thermodynamic formalism of Sect. 8.1. In the $n \to \infty$ limit, the partition function of the coarse-grained velocity, in analogy to (10.15), becomes

$$Z_n(q) \equiv \left\langle {\rm e}^{(1-q)nv_n(X)} \right\rangle \propto [{\rm e}^{(1-q)nv_{\rm s}} + {\rm e}^{-(1-q)nv_{\rm s}}]W(n) \,, \tag{10.20}$$

and for $q < 1$, $\Phi(q) = (q-1)v_{\rm s}$, while for $q > 1$, $\Phi(q) = -(q-1)v_{\rm s}$. Hence the q-average becomes $v(q) = v_{\rm s}$ for $q < 1$ and $v(q) = -v_{\rm s}$ for $q > 1$. Therefore, from the variational principle $\psi(v) = \max_q \{\Phi(q) - (q-1)v\}$, $\psi(v)$ can be obtained in the form given above. Here, $P(v;n)$ is given by the logarithmic term in the coarse-graining statistical law (6.24) and obeys the anomalous scaling relation

$$P(v;n) = n^{\delta}\tilde{p}(n^{\delta}v) \quad (0 < \delta \equiv (\tilde{\beta}-1)/\tilde{\beta} < 1/2) \,. \tag{10.21}$$

Here $|v| < v_{\rm s}$, and $\tilde{p}(x)$ is a universal, even function of x, independent of n.

Figure 10.10a presents numerical results for the scaling function $\tilde{p}(x)/\tilde{p}(0)$ as a function of $x = n^{\delta}v$ for $K = K_{\rm a}$ obtained with $\delta = 2/5$ ($\tilde{\beta} = 5/3$). This figure consists of plots generated by data using five different values of n. The five functions so obtained are nearly coincident, verifying the scaling relation (10.21). In contrast to this situation, as in the case $K = K_{\rm b}$, despite the existence of large islands of tori, if there are no accelerator mode tori, a diffusion constant D exists, and the normal distribution (10.18) is realized [see Ishizaki et al. (1991)].

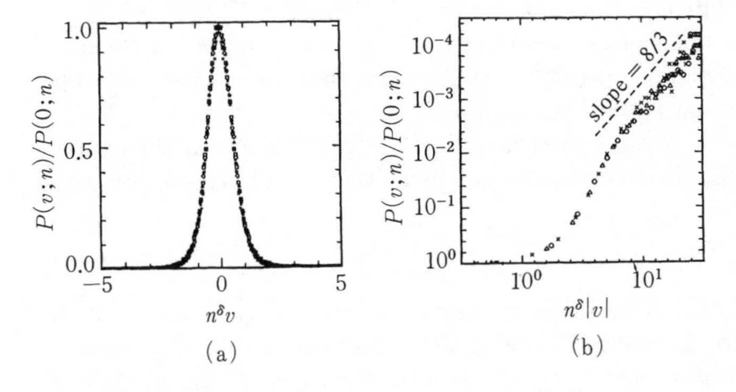

Fig. 10.10. (**a**) The numerically generated form of $P(v;n)/P(0;n)$ as a function of $x = n^{\delta}v(\delta = 2/5)$. Here the *crosses*, *squares*, *diamonds*, *triangles* and *circles* correspond to $n = 50, 100, 200, 400$, and 800, respectively. In all cases, $N = 10^5$. (**b**) The numerically generated form of $\log_{10}\{P(v;n)/P(0;n)\}$ as a function of $\log_{10}|x|$. The *crosses*, *triangles*, and *circles* correspond to $n = 400, 800$, and 1600, respectively. Here, the data are consistent with $\tilde{\beta} + 1 = 8/3$. (For both graphs, $K = K_a$)

The explicit form of $\tilde{p}(x)$ in the case in which accelerator modes exist can be obtained from Feller's theorem (10.8). Since $\tilde{p}(x)$ is an even function, in analogy to the case described by (10.6), in the limit as $n \to \infty$ we have

$$\tilde{p}(x) \propto |x|^{-\tilde{\beta}-1} \quad (|x| = n^{\delta}|v| \gg 1) . \tag{10.22}$$

Figure 10.10b represents the asymptotic form of $\tilde{p}(x)$ for $K = K_{\mathrm{a}}$ obtained numerically. The broken line here has slope $\tilde{\beta} + 1 = 8/3$. The fact that this slope matches that of the data for $|x| \gtrsim 10$ leads us to conclude that $\tilde{p}(x)$ displays its asymptotic form for such values. The somewhat large variance displayed by the data points results from the finiteness of the value N used in the numerical studies. This kind of anomalous scaling law results from the self-similar hierarchical structure of the islands of tori, which is responsible for the inverse power law time correlation $C_n \propto n^{-(\tilde{\beta}-1)}$.

10.4 Diffusion and Mixing of Fluids as a Result of Oscillation of Laminar Flow

As discussed in Sect. 6.2, in a Bénard convection system consisting of a large number of convective rolls aligned along the x-axis, widespread chaos emerges from the neighborhoods of boundaries separating cells corresponding to consecutive rolls. This is a result of the jumping of fluid particles between oscillating streamlines. Such chaotic behavior represents one type of Lagrangian turbulence [Aref (1984)]. The dye particles within a given cell, while being mixed with other fluid particles in this cell as a result of stretching and folding, will undergo diffusion from cell to cell upon entering regions near its boundaries. This diffusion takes place through turnstiles $L_{j-1,j}$ and $L_{j,j-1}$ located in such neighborhoods (see Fig. 6.5). The corresponding diffusion coefficient is three orders of magnitude larger than that for normal molecular diffusion [Solomon and Gollub (1988)].

In a certain range of the amplitude B of roll oscillations, in the context of the two-dimensional conservative map (6.9), there exist accelerator mode periodic orbits expressed by

$$x^*_{t+Q} = x^*_t \pm l, \quad z^*_{t+Q} = z^*_t , \tag{10.23}$$

and as a result, diffusion becomes anomalously large. Here, because of the translational periodicity (period 2) along the x direction, l is an even number. The mixing and diffusion seen here are qualitatively similar to those exhibited by the standard map. Let us now study these quantities by considering a numerical investigation of the equation of motion (6.7) yielding the stream function (6.8) [see Ouchi and Mori (1992)].

10.4.1 Islands of Accelerator Mode Tori Existing Within Turnstiles

Figure 10.11 is a plot of data describing the diffusion of 'dye' in terms of the quantity $D = \lim_{n\to\infty}\langle(x_n - x_0)^2\rangle/2n$ as a function of the amplitude B of roll oscillations. (Here, B ranges from 0 to 0.2.) These data were generated by following the motion of 100 particles. At $t = 0$ these particles were distributed evenly over the range $-0.5 < z < 0.5$ at $x = 0$, and their orbits X_t were recorded for 5000 time steps. The angle brackets in the expression for D represent the operation of first averaging over time along each orbit, and then taking the average of the 100 values so obtained. We will use this averaging procedure again below. According to this figure, D increases roughly as $\sqrt{B}$, while superimposed on this overall behavior there exist several small peaks, as indicated by the arrows.

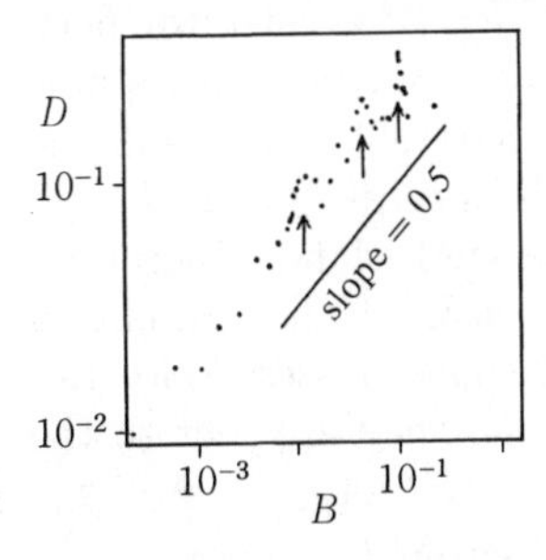

Fig. 10.11. A plot of $\log_{10} D$ versus $\log_{10} B$ ($N = 5 \times 10^5$). The *arrows* indicate the peaks at $B = 0.012, 0.045$, and 0.097

Designating the widespread chaotic sea existing in the cell R_j by C_j, let us define the time τ_B to be the *mixing time* needed for the lobes $F^t L_{-1,0}$ and $F^t L_{1,0}$ ($t = 1, 2 \ldots$) to uniformly fill C_0 (after coarse graining). Also, let us define the time t_B to be the shortest time for a fluid particle in C_{-1} to be transported to C_1 due to the lobe $F^t L_{-1,0}$. The quantity t_B is the so-called *first visiting time*. Then, referring to the surface areas of C_j and $L_{j,j\pm1}$ as $\mu(C_j)$ and $\mu(L_{j,j\pm1})$, we can write the diffusion coefficient as $D \approx (\tau_B/t_B)\mu(L_{0,1})/\mu(C_0)$. Here $\mu(C_j) = \mu(C_0)$ and $\mu(L_{j,j\pm1}) = \mu(L_{0,1})$. With an increase in B, $\mu(L_{0,1})$ and $\mu(C_0)$ increase in proportion to B and $\sqrt{B}$, respectively. Then, ignoring the B dependence of τ_B/t_B, we have $D \propto \sqrt{B}$.[3] This explains the $\sqrt{B}$ dependence shown in Fig. 10.11. In fact, outside the narrow peaks in Fig. 10.11, mixing associated with $F^t L_{-1,0}$ carries particles in C_{-1} to the turnstile $L_{0,1}$. These particles are then carried to C_1. Because this results from the mixing of $L_{-1,0}$ under F^t, we have $t_B \approx \tau_B$. However, inside these peaks, as in the case of Fig. 6.5b, particles in C_{-1} are carried into $L_{0,1}$ under only a few mappings of $L_{-1,0}$, and thus here $t_B \ll \tau_B$. As a result, diffusion is greatly enhanced. In particular, in the case that accelerator mode tori exist within turnstiles $L_{j,j\pm1}$, this enhancement continues indefinitely, implying that $D = \infty$. In this case, near the centers of these peaks, anomalous diffusion is exhibited, and if N were sufficiently large (e.g., as large as that in a real fluid experiment, perhaps $\sim 10^{20}$), the divergent behavior of D could be observed. As we will now see, in the intervals of B corresponding to the

[3] For details, see Ouchi et al. (1991) Prog. Theor. Phys. **85**, 687.

peaks in Fig. 10.11 there do in fact exist accelerator mode islands do in fact exist inside turnstiles.

Figure 10.12 displays the islands surrounding the $Q = 2$, $l = 2$ accelerator mode periodic points (x_t^*, z_t^*) $(t = 1, 2)$ for $B = B_2 \equiv 0.0404411$. The structure here consists of two large islands, each surrounded by a set of smaller islands. Figure 10.13a is a bifurcation diagram describing the period doubling of these periodic points. (Represented here are the bifurcations from period 2^1 to 2^2 and 2^2 to 2^3.) Figure 10.13b represents the system for the case $B = B_4 \equiv 0.0420660$. In the latter, the four islands existing in an arbitrary single cell, the centers of which are the four $Q = 2^2, l = 2^2$ accelerator periodic points, are shown. For this system there exist two $Q = 2$ neutrally stable accelerator mode periodic points for $B \in (0.039748, 0.041923)$. Let us define A_j^+ and A_j^- to be the regions occupied by the islands of these $Q = 2^n, l = 2^n$ accelerator mode tori and the associated 'islands around islands' hierarchical structure within the cell R_j. Here A_j^+ (A_j^-) consists of all islands corresponding to positive (negative) velocities. Such regions possess velocities $v_a = \pm l/Q = \pm 1$, and under one iteration of the map they are sent to the neighboring cells. In other words, we can write $A_j^\pm = F^{\pm 1} A_{j-1}^\pm = F^{\pm j} A_0^\pm$. Thus, by the definition of a turnstile, A_j^+ is located within $L_{j,j+1}$ and A_j^- is located within $L_{j,j-1}$, and because $FL_{j\mp 1,j} \subset R_j$ and $F^{-1}L_{j\pm 1,j\pm 2} \subset R_j$,

$$A_j^\pm \subset \left(FL_{j\mp 1,j} \cap F^{-1}L_{j\pm 1,j\pm 2}\right) . \tag{10.24}$$

Figure 10.14 shows $Q = 2, v_a = 1$ islands and the surrounding regions A_{-1}^+, A_0^+, and A_1^+ obtained numerically. These results demonstrate that $A_{-1}^+ \subset L_{-1,0}$, $A_0^+ \subset \left(FL_{-1,0} \cap F^{-1}L_{1,2}\right)$, and $A_1^+ \subset L_{1,2}$.

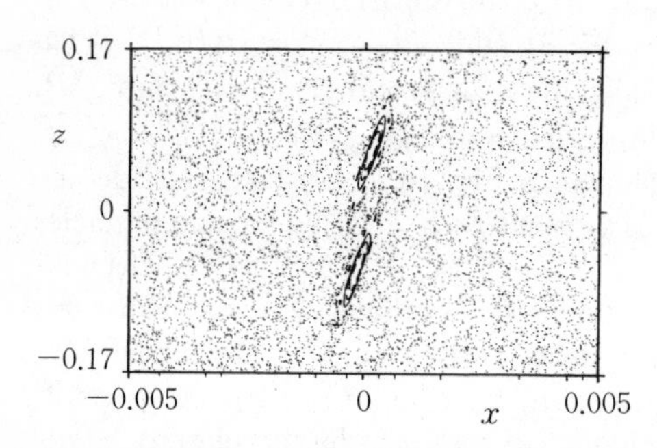

Fig. 10.12. Islands of accelerator mode tori existing at $B = B_2 \equiv 0.0404411$ $(Q = 2,\ l = 2)$ [Ouchi and Mori (1992)]

As described here, the islands of accelerator mode tori and corresponding regions $A_j^\pm$ use the turnstiles $L_{j,j\pm 1}$ as stepping stones as they move from cell to cell with velocity $v_a = \pm 1$, heading toward $\pm\infty$. Of course, a chaotic orbit X_t trapped within some A_j^+ (A_j^-) will move with velocity $v_a = 1(-1)$ during the time it is so trapped. However, in the $j \to \infty$ limit, $F^{\pm j} R_0 \cap R_j \to A_j^\pm$, and thus we can write the area of the set $F^{\pm j} R_0 \cap R_j$ as

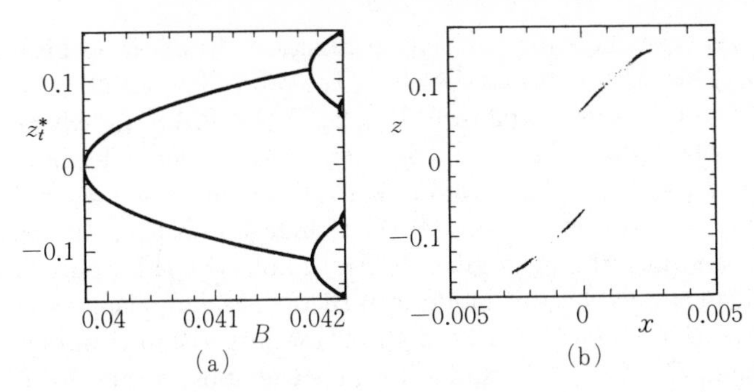

Fig. 10.13. (**a**) The period doubling of accelerator mode periodic points. (**b**) Islands of accelerator mode tori for $B = B_4 \equiv 0.042066$ ($Q = 4$, $l = 4$)

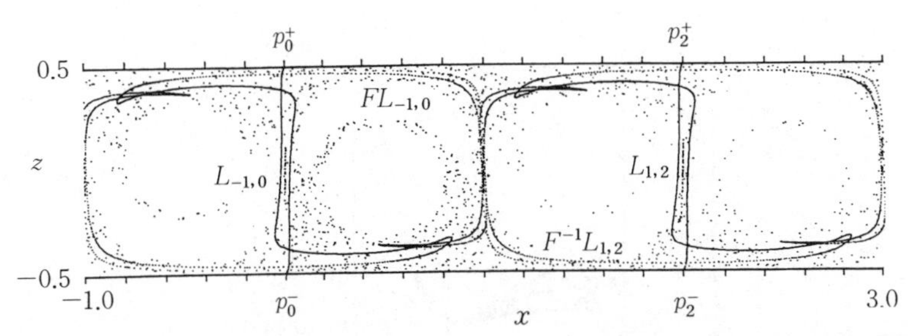

Fig. 10.14. Numerically generated visualizations of the four invariant manifolds $W^{\mathrm{u}}(p_0^-)$, $W^{\mathrm{s}}(p_0^+)$, $W^{\mathrm{u}}(p_2^-)$ and $W^{\mathrm{s}}(p_2^-)$, and the regions A_{-1}^+ contained within $L_{-1,0}$ and A_1^+ contained within $L_{1,2}$. This figure corresponds to $B = B_2$

$$\mu(F^{\pm j} R_0 \cap R_j) = \mu(A_0^{\pm}) + Cj^{-(\tilde{\beta}-1)} \quad (2 > \tilde{\beta} > 1) . \tag{10.25}$$

Due to conservation of area, $\mu(A_j^{\pm}) = \mu(A_0^{\pm})$, and C is a positive constant. The second term on the right-hand side of this equation is simply a new expression for the fact that the probability that a chaotic orbit remains in $A_j^{\pm}$ for a time greater than j is $W(j) \propto j^{-(\tilde{\beta}-1)}$.

Introducing the coarse-grained velocity $v_n(X_0) \equiv (x_n - x_0)/n$, we obtain expressions describing the present system which are similar to (10.15), (10.17), (10.21), and (10.22), and from these the anomalous diffusion resulting from the presence of accelerator modes can be understood in statistical terms. However, because such $A_j^{\pm}$ are generally small, in order to observe this anomalous diffusion experimentally, N must be sufficiently large. Verification of this behavior through numerical studies is quite difficult, and for this reason a real fluid experiment is awaited.

10.4.2 Anomalous Mixing Due to Long-Time Correlation

Anomalies in mixing of fluids result not only from islands of accelerator mode tori but also, as suggested by Fig. 6.4, from the presence of large islands of normal tori. In this case the time correlation function C_t^λ of the orbital expansion rate $\lambda_1(X_t)$ for chaotic orbits X_t obeys the power law (10.2), and the variance of the coarse-grained expansion rate $S_n(X_0) \equiv n\Lambda_n(X_0)$ obeys the anomalous power law (10.3). Figure 10.15a displays the n dependence of this variance for $B = B_3 \equiv 0.0413782$. From this graph we find $\zeta \approx 1.37$ and $\beta \approx 1.63$. In this case, $\Lambda^\infty \approx 0.92$. Figure 10.15b demonstrates that, in the $n \to \infty$ limit, $\psi_n(\Lambda)$ assumes the form given in Fig. 10.6b. For this value of B, in addition to islands of accelerator mode tori, there exist large islands of normal tori similar to those shown in Fig. 6.4b. These large islands induce a long-time correlation, $C_t^\lambda \propto t^{-(\beta-1)}$, and determine the above values for ζ and β. In contrast to the anomalous diffusion caused by accelerator modes, because these islands of tori are large, anomalous mixing in this case is easily observed.

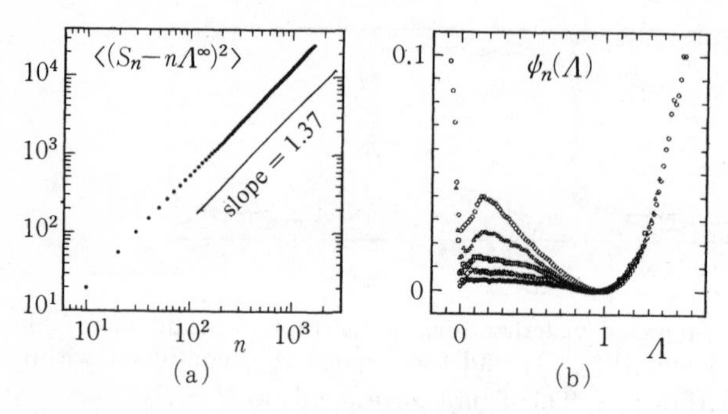

Fig. 10.15. (**a**) The variance (10.3) ($n = 10 \sim 1600$, $N = 5 \times 10^5$) for $B = B_3 \equiv 0.0413782$. The data are consistent with $\zeta = 1.37$. (**b**) The spectrum $\psi_n(\Lambda)$ for $B = B_3$. The various curves generated here correspond to (from top to bottom) $n = 100, 200, 400, 800$, and 1600

For $B = B_3$, in the limit as $n^\delta|\Lambda - \Lambda^\infty| \to \infty$, the probability distribution $P(\Lambda; n)$ becomes

$$P(\Lambda;n) \propto \begin{cases} n\exp[2n\Lambda] & (0 > \Lambda > \Lambda_{\min}) \quad (10.26) \\ n^\delta\{n^\delta(\Lambda^\infty - \Lambda)\}^{-\beta-1} & (0 < \Lambda < \Lambda^\infty) \quad (10.27) \\ n^\delta\exp[-an(\Lambda - \Lambda^\infty)^{1/\delta}] & (\Lambda > \Lambda^\infty)\,, \quad (10.28) \end{cases}$$

where in this case $\beta \approx 1.63$, $\delta = (\beta - 1)/\beta \approx 0.387$, and a is a positive constant. Actually, the $\Lambda < 0$ regions of Figs. 10.15b, 10.16a and 10.16b provide verification of (10.26), (10.27), and (10.28), respectively. The slopes of the functional forms in Figs. 10.16a and b represent $\beta + 1$ and $1/\delta$. The lines here correspond to the values $\beta + 1 = 2.63$ and $1/\delta = 2.58$. As seen, these slopes are consistent with the data.

As described above, the diffusion and mixing displayed in Benard convection subject to oscillatory motion are qualitatively the same as those occurring among chaotic orbits in the widespread chaos of the standard map. It

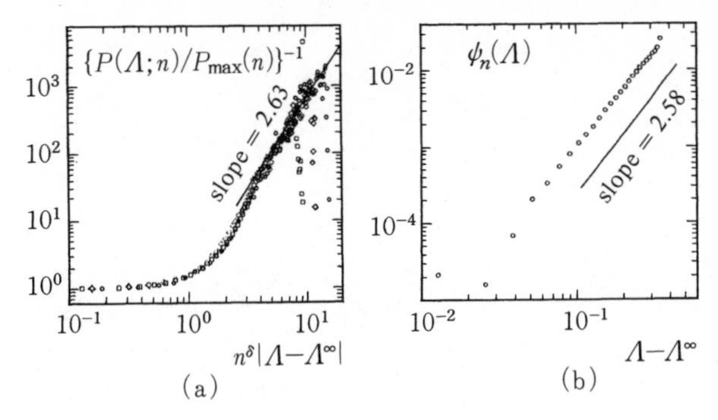

Fig. 10.16. (a) $\{P(\Lambda;n)/P_{max}(n)\}^{-1}$ as a function of $n^\delta|\Lambda - \Lambda^\infty|$ ($\delta = 0.387$) in the interval $0 < \Lambda < \Lambda^\infty$. The data shown here represent results for $n = 400, 800$ and 1600. They are consistent with $\beta + 1 = 2.63$. (b) The spectrum $\psi_n(\Lambda)$ for $\Lambda > \Lambda^\infty$ ($n = 800$). The data here are consistent with $\delta^{-1} = 2.58$. Both figures correspond to $B = B_3$ [Ouchi and Mori (1992)]

is expected that similar phenomena will also be exhibited by Taylor vortex flow in a system of rotating coaxial cylinders. The nature of the flow seen in such systems is determined by the local structure of the unstable manifold, described schematically by Fig. 6.7, and by the two types of islands of tori peculiar to conservative dynamical systems. In these systems, accelerator mode tori appear quite often in the context of maps periodic in their action variables [see Ishizaki et al. (1993)]. The existence of such tori and the resulting properties characterizing conservative systems differ markedly from anything seen in dissipative dynamical systems. They can be considered to represent the universality of widespread chaos exhibited by periodic maps of conservative dynamical systems, Lagrangian turbulence, and related systems.

Supplement II: On the Structure of Chaos

SII.1 On–Off Intermittency

The type of chaos known as 'on–off intermittency' was discovered by Fujisaka and Yamada (1985) in the situation in which there exists a bifurcation from synchronous to asynchronous motion in a system of two coupled chaotic oscillators. With the elucidation of the geometric structure of this intermittent chaos due to Platt et al. (1993) and Ott and Sommerer (1994), the importance of this type of system has come to be recognized.

Let us assume that the state variables (X, x) obey a dissipative equation of motion of the form

$$\dot{X}(t) = F(X, x) \, , \quad \dot{x}(t) = G(X, x) \, . \tag{SII.1}$$

The variables X and x here represent the sum and difference, respectively, of the state variables X_1 and X_2 of the two chaotic oscillators. If we take G to be antisymmetric in x [i.e., $G(X, -x) = -G(X, x)$],

$$\dot{X}^0(t) = F(X^0(t), 0) \, , \quad x^0(t) = 0 \tag{SII.2}$$

is a particular solution of (SII.1) representing the synchronous motion of the two oscillators. Let us suppose that the region in state space occupied by this solution in the $t \to \infty$ limit constitutes a smooth invariant manifold M. Then, let us assume that there exists some parameter ϵ such that for $\epsilon < 0$ the oscillators display synchronous motion and that the corresponding attractor consists of some invariant set S on M. For $\epsilon > 0$, this attractor extends along a surface V orthogonal to M, and a time series like that exhibited in Fig. SII.1 results. As seen here, the $x = 0$ *laminar state* ('off' state) and the $x \neq 0$ *burst state* ('on' state) appear alternatingly. We note the following:

(1) The distribution of durations τ during which the system exists in the laminar state assumes the form $f(\tau) \propto \tau^{-3/2}$ in the $\epsilon \to 0+$ limit. The average value of τ is given by $\overline{\tau} \approx \epsilon^{-1}$.

(2) If we designate the duration of a given burst by t, the average value of the logarithm of the height of such a burst is $\sqrt{t}$ [Hata (1997)], and the corresponding time series $x(t)$ is statistically self-similar.

The many unstable periodic orbits on S create an infinite number of both stable and unstable manifolds on V, producing a complex, 'riddled' structure

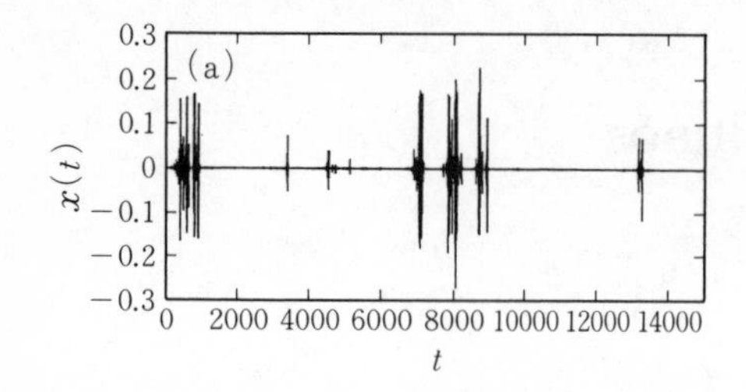

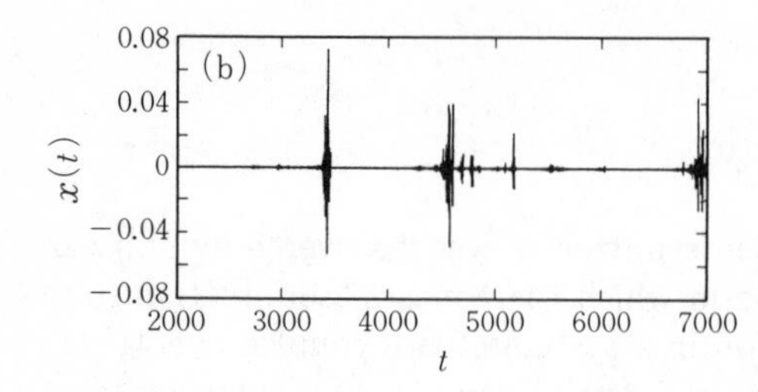

Fig. SII.1. The 'on–off' behavior of $x(t)$. (**b**) is a close-up of the region containing the small bursts in (**a**)

on this surface. When $\epsilon < 0$, V behaves in general as a stable manifold of the attractor S; for $\epsilon > 0$ it in general behaves as an unstable manifold of S.

SII.2 Anomalous Diffusion Induced by an Externally Applied Force

In order to study the nonlinear response of diffusion that occurs near tori in conservative systems, let us consider a map obtained from the standard map (7.17) by the addition of a constant external forcing term Γ:

$$\begin{cases} \theta_{i+1} = \theta_i + J_{i+1} \pmod 1 , \\ J_{i+1} = J_i - (K/2\pi)\sin(2\pi\theta_i) + \Gamma . \end{cases} \tag{SII.3}$$

That is, we consider the 'velocity' $u_t = J_{t+1} - J_t$ associated with a chaotic orbit of the *Josephson map*. Figure SII.2 displays a typical time series of the quantity $\tilde{u}_t = (1/3)\sum_{s=0}^{2} u_{t+s}$ for the case $\Gamma = 0.38$, $K = 3.8$. Here, there exist tori in the neighborhoods of fixed points and period 3 periodic points, and for a chaotic orbit trapped in the region U of the chaotic sea that contains such islands, $\tilde{u}_t \approx 0$. However, for a chaotic orbit situated deep within the chaotic sea, away from U, $\tilde{u}_t$ fluctuates randomly about the average value $\overline{v} = 0.37$ determined by the external forcing term Γ.

From the time series shown in Fig. SII.2, we see how the random motion (fluctuations) around this average $\overline{v}$ and the $\tilde{u}_t \approx 0$ laminar motion exhibited in the island region U appear alternatingly and intermittently [Ishizaki and Mori (1997)]. Here we have:

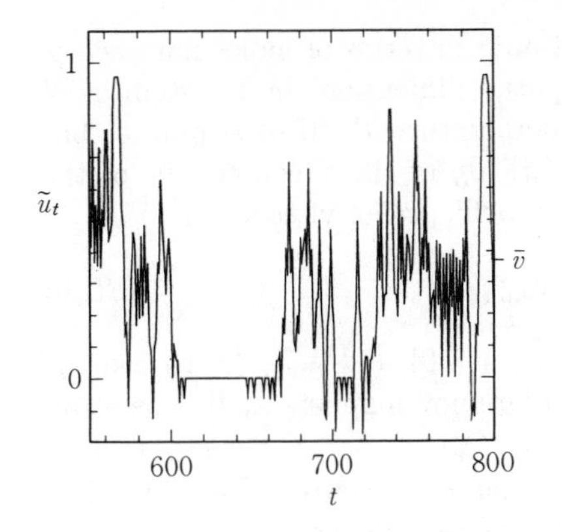

Fig. SII.2. The behavior of the 'velocity' $\tilde{u}_t$. At points where $\tilde{u}_t$ assumed values less 14% of the average velocity $\overline{v} = 0.37$, the system was assumed to be trapped in an island region, and the value of $\tilde{u}_t$ was set to zero

(1) The distribution $f(\tau)$ corresponding to the duration τ of the $\tilde{u}_t \approx 0$ laminar state assumes the inverse power-law form $f(\tau) \propto \tau^{-\tilde{\beta}-1}$ ($\tilde{\beta} = 1.65$).

(2) Contrastingly, the distribution $f^{\mathrm{c}}(t)$ corresponding to the duration t of the state of random motion about $\overline{v}$ has the exponential form $f^{\mathrm{c}}(t) \propto \exp[-t/\tau^{\mathrm{c}}]$.

Thus a time series for such a system represents a Lévy process, and its statistical nature is described by Feller's theorem, discussed in Chap. 10. In fact, the probability distribution of the coarse-grained velocity (10.16) assumes the form

$$P(v;n) \propto \begin{cases} n^{\delta}\left\{n^{\delta}(\overline{v}-v)\right\}^{-\tilde{\beta}-1} & (0 < v < \overline{v}) \\ n^{\delta}\exp\left[-an(v-\overline{v})^{1/\delta}\right] & (v > \overline{v}) \end{cases} \qquad \text{(SII.4)}$$

in the $n \to \infty$ limit. Here, $\delta = (\tilde{\beta}-1)/\tilde{\beta} = 0.394$, and a is a positive constant. When $\Gamma = 0$ and $K = 3.8$, we have $\overline{v} = 0$, $\delta = 0.5$, and $D = 0.057$, and in this case of no external force, the normal distribution results.

The type of anomalous diffusion ($\eta = 3-\tilde{\beta} > 1, D = \infty$) considered above also appears in the context of Bénard convection discussed in Sects. 6.2 and 10.4 and can be thought of there as resulting from the addition of a constant external forcing term Γ to the right-hand side of $\dot{x}$ in (6.7).

SII.3 Transport Coefficients and the Liapunov Spectrum

A system in a nonequilibrium steady state must be supplied with kinetic energy on a macroscopic scale from some external source. Through transport processes whose effects are reflected by quantities such as the coefficient of viscosity and the thermal conductivity, this energy is eventually dissipated in the form of microscopic molecular thermal kinetic energy. Thus, nonequilibrium steady states represent dissipative dynamical systems.

In order to describe the viscosity of a fluid in terms of molecular theory, let us consider a laminar flow in three spatial dimensions in a system of N molecules occupying a volume V with temperature T. If we stipulate that the gradient along the y direction $\gamma = \partial u_x/\partial y$ of the x component of the fluid velocity u_x is fixed, we can write the coefficient of viscosity $\eta(\gamma)$ as

$$\eta(\gamma) = -\frac{\langle P_{xy}\rangle}{\gamma} = \frac{3Nk_{\mathrm{B}}T}{\gamma^2 V}|\lambda_{\max}(\gamma) + \lambda_{\min}(\gamma)| , \tag{SII.5}$$

where P_{xy} is the pressure tensor, and $\lambda_{\max}$ (> 0) and $\lambda_{\min}$ (< 0) are the maximal and minimal values among the Liapunov numbers $\overline{\lambda}_i$ [Evans et al. (1990)]. Here we have assumed $\sum_{i=1}^{6N} \overline{\lambda}_i = 3N(\lambda_{\max} + \lambda_{\min})$. Note that, since we are considering dissipative systems, this sum is negative. Equation SII.5 correctly expresses the molecular viscosity of the system. With respect to nonlinear response, it yields

$$\eta(\gamma) = \eta(0) - a\gamma^{1/2} \quad (a > 0) . \tag{SII.6}$$

The dimension $D(\gamma)$ of the system attractor in phase space decreases as γ increases. Furthermore, writing the difference $\Delta D = D(0) - D(\gamma)$ in terms of the rate of entropy production $\dot{S}$, we have [Hoover and Posch (1994)]

$$\begin{aligned} \Delta D &\equiv D(0) - D(\gamma) \\ &= \dot{S}/k_{\mathrm{B}}\lambda_{\max} = \eta\gamma^2 V/k_{\mathrm{B}}T\lambda_{\max} \\ &= 3N|\lambda_{\max} + \lambda_{\min}|/\lambda_{\max} . \end{aligned} \tag{SII.7}$$

Hence even in a laminar steady state, an orbit in the system's phase space forms a multifractal strange attractor. Therefore, in a nonequilibrium steady state, the phase distribution function is anomalous almost everywhere on the attractor, and this function itself cannot be readily employed. As discussed in Sect. 6.4, it is the distribution of a coarse-grained quantity which becomes useful.

Actually, both the quantity η on the left-hand side and $|\lambda_{\max} + \lambda_{\min}|$ on the right-hand side of (SII.5) are phenomenological quantities that appear through the rate of volume change in phase space, and thus this equation does *not* represent a statistical mechanical relationship that explicitly captures the microscopic mechanism of transport phenomena. When macroscopic layers become turbulent, energy is sent to small vortices due to turbulent viscosity, thereby enhancing the rate of dissipation to molecular thermal motion. Now, turbulence can also be characterized by the Liapunov spectrum, and thus it should be possible to use (SII.5) to describe turbulent viscosity also. However, in terms of statistical mechanics, the molecular viscosity can be obtained from the explicit extraction (by use of the projection operator) of the microscopic thermal motion, and the turbulent viscosity can be obtained from the renormalization of the fragmentation cascade of vortices into progressively smaller vortices. It is difficult to imagine that appropriate statistical mechanical expressions for the molecular and turbulent viscosities can be obtained through dimensional analysis without the use of statistical mechanical methods that employ this kind of projection operator and renormalization.

Summary of Part II

In Part II we have seen how the form and structure of chaos exhibited by nonlinear dynamical systems is determined by the many types of coexisting invariant sets and the unboundedly complex, yet beautifully self-organized invariant manifolds that characterize phase space. Chaotic behavior, whose intricate self-organization exhibits a myriad of different forms, while incorporating the quality of indeterminism, provides in its many varied realizations the principle source of the complexity and variety exhibited by the ever-changing natural world.

In order to obtain a statistical description of chaos and to understand its transport phenomena, we have studied the fluctuations undergone by various physical quantities when the evolution of a system is considered in terms of motion along its chaotic trajectories. We have also seen how the geometric, qualitative dynamical system theory we have presented provides the proper framework for the statistical physics of chaos. For example, the coarse-grained expansion rate $\Lambda_n(X)$ of neighboring, length n orbits expresses the stretching and folding undergone by segments of the unstable manifold W^{u}. In the coarse-grained limit ($n \to \infty$), this quantity reveals the qualitative features of the geometric structure of chaos. In this way, in qualitative terms, the form and structure of chaos are determined by

(1) the presence or absence of tangency structure in the unstable manifold W^{u}, in which chaotic orbits reside (see Sects. 8.3, 8.4 and 10.2);
(2) the manner in which invariant sets contained within the chaotic region change as a result of bifurcation; and
(3) the presence or absence of both normal and accelerator mode tori in the chaotic sea of conservative systems (see Sects. 10.3 and 10.3).

The statistical nature of this form and structure is characterized by such quantities as the slopes s_α , s_β , s_γ , and s_δ of the linear segments of the spectrum $\psi(\Lambda)$, the corresponding slopes of $f(\alpha)$, the dynamical scaling relations (9.4), (9.13) and (10.5), and the values of the various scaling exponents, including ζ , η , β , and δ.

Critical attractors and chaotic attractors introduce a variety of self-similar structures within complexity, including the 'self-similar nested structure' discussed in Chap. 8. These structures exist because the complexity out of which they arise is due to the repeated application of simple maps (laws of motion).

The fundamental process involved in chaos is the stretching and folding of W^{u}, as described by Figs. 6.3 and 6.5. The special feature of $\psi(\Lambda)$ is its ability to accurately capture the action of this mixing process. In the $q < q_\alpha = 2$ hyperbolic phase, the function $f(\alpha)$ can also be expressed in terms of $\psi(\Lambda)$ [see (8.76)].

One may ask how entropy fits into this discussion. To understand this point, let us first consider the uniform partitioning of phase space into numbered boxes of length l. Next, defining the joint probability $p_l(i_0, \ldots, i_{n-1})$ to be that for which a given orbit is in the i_kth box at time $t = k$ for $k = 0, \ldots, n-1$, we can define, as a generalization of the fractal dimension (8.32), the qth order *Renyi entropy*,

$$K_q \equiv \frac{1}{1-q} \lim_{l \to 0} \lim_{n \to \infty} \frac{1}{n} \ln \left[\sum_{i_0} \cdots \sum_{i_{n-1}} \{p_l(i_0, \cdots, i_{n-1})\}^q \right] .$$

Here the sum is over all $(i_0, \ldots, i_{n-1})$ for which $p_l \neq 0$. The total number M of such terms, given by $M = \exp[K_0 n]$, diverges with n. The value of K_q gives a quantitative representation of the complexity of the state of the system in question. Then, with $\gamma(i_0, \ldots, i_{n-1}) \equiv -(1/n) \ln p_l(i_0, \ldots, i_{n-1}) > 0$, if the corresponding probability distribution function $P(\gamma; n, l)$ can be written

$$P(\gamma; n, l) \equiv \langle \delta(\gamma(i_0, \cdots, i_{n-1}) - \gamma) \rangle \propto \exp[-n\{\gamma - h(\gamma)\}]$$

in the $n \to \infty$ limit, $l \to 0$, then $(q-1)K_q = \min_\gamma \{q\gamma - h(\gamma)\}$ results [Sano et al. (1986); see bibliography for Chap. 8]. Here, $h(\gamma) \leq \gamma$, K_1 is the *Kolmogorov–Sinai (KS) entropy,* and K_0 is the *topological entropy.* In the $q < 2$ case of the hyperbolic phase, approximating Λ_n (> 0) by $\Lambda_n(X_0) = \gamma(i_0, \ldots, i_{n-1})$, we obtain $h(\Lambda) = \Lambda - \psi(\Lambda)$ and $K_q = \Phi(q)/(q-1)$. In the $q \to 1$ limit this gives the well-known result $K_1 = \Lambda^\infty$. Thus, for the hyperbolic phase, K_q also can be expressed in terms of $\psi(\Lambda)$.

The thermal motion of molecules is characterized by the Boltzmann entropy. The quantity K_1 is a generalization of this quantity, but it is not possible to characterize chaos on the macroscopic level with just K_1. This is due to the fact that while it is possible at the macroscopic level to ignore fluctuations arising from thermal motion, it is not possible to ignore fluctuations arising from macroscopic chaos. The functional forms of the structure functions $\psi(\Lambda)$, $f(\alpha)$, K_q, etc., associated with these fluctuations characterize the global form of chaos. In Part II of this book, as observables to facilitate our study, we have selected $\psi(\Lambda)$ and $f(\alpha)$, quantities whose physical interpretations are simple and which are able to accurately reveal the geometric structure of chaos. With these quantities we have investigated the statistical physics of chaos.

As physical examples, we have considered only two types of forced oscillators and the Lagrangian turbulence seen in systems of oscillating Bénard convection. However, with these few examples we found that the attractors for many dissipative systems are low-dimensional objects. In addition, we can say that the set of low-dimensional maps we have treated is sufficiently diverse to provide us with a basic understanding of chaos, because the types of bifurcations exhibited by these maps include almost all those which can be considered as *universal* in chaotic systems. Thus, using these maps, we have been able to obtain a picture that describes the statistical nature of the many varieties of chaos. However, there are a number of important topics that we have not treated, including response and control in chaotic systems and the theory of probabilistic processes with regard to fluctuations arising in chaos.

Chaos constitutes a diverse and multifaceted set of phenomena. The aspect of chaos that will be observed in any given instance depends both on the system in question and the observable quantities employed. For this reason, it is not sufficient to undertake research which treats individual systems using only a fixed set of observable quantities. For example, the semi-conductor S-shaped I–V characteristic system displays physical processes that are peculiar to itself, as well as bifurcation phenomena that do not appear in maps of dimension lower than 3. Hence, in regard to systems described by such high-dimensional maps, it is necessary to statistically characterize the geometric nature of the chaos they exhibit and to investigate the nature of the transport phenomena they display. We note here that in a system possessing multiple positive Liapunov numbers ($\overline{\lambda}_1 > \overline{\lambda}_2 \cdots > 0$), we focus on the largest of these, $\Lambda^\infty \equiv \overline{\lambda}_1(X_0)$, and the corresponding fluctuations in the orbital expansion rate $\lambda_1(X_t)$. As the dimension of the systems that we study becomes larger, their geometric descriptions become increasingly difficult to realize, and it thus becomes more and more important to obtain statistical descriptions that reflect the qualitative features of the phase structure.

A. Appendix

A.1 Periodic Points of Conservative Maps and Their Neighborhoods

For a map $F(X)$ $[X = (x, z)]$ under which surface area is conserved $(\partial \dot{x}/\partial x + \partial \dot{z}/\partial z = 0)$, from (6.4) we have $J \equiv \det\{DF(X_t)\} = 1$. Let us consider the time evolution

$$y_{i+Q} = DF^Q(X_i^*) \cdot y_i \tag{A.1}$$

of an orbit $X_i = X_i^* + y_i$ ($|y_i| \ll 1$) passing through the Q neighborhoods of a set of Q-periodic points $X_i^* = F^Q(X_i^*)$ $(i = 1, 2 \ldots Q)$ for one example of such a two-dimensional conservative map, (6.9). Since $\det\{DF^Q(X_i^*)\} = 1$, if we express the trace of $DF(X_t)$ as $s \equiv \mathrm{tr}\{DF^Q(X_i^*)\}$, we can write the eigenvalues of this matrix as

$$\begin{aligned} \nu_\alpha^Q &= (1/2)\{s \pm (s^2 - 4)^{1/2}\} \qquad (\alpha = 1 \text{ for } + \text{ and } 2 \text{ for } -) && \text{(A.2)} \\ &= (1 - 2R) \pm 2\{R(R-1)\}^{1/2}\,, && \text{(A.3)} \end{aligned}$$

where we have introduced the residue $R \equiv (2 - s)/4$ [Greene (1979)]. Here we have $\nu_1^Q \nu_2^Q = 1$, and it is either the case that both eigenvalues are real numbers or each is a complex number on the unit circle.

(1) When $0 < R < 1$, the eigenvalues are complex numbers on the unit circle and we can set $\nu_\alpha^Q = \exp[\pm 2\pi i \sigma]$ $(0 < \sigma < 1/2)$. Here, σ is the rotation number characterizing the neighborhoods of the points X_i^*, and we can write $R = \sin^2(\pi\sigma)$. In the present case, the X_i^* are neutral, and, as shown in Fig. A.1a, myriad *elliptic* (circular in the figure) invariant curves characterized by a constant rotation number σ encircle these points. These curves define a Q-fold torus. The X_i^* here are referred to as 'elliptic points'.

(ii) When $R < 0$ or > 1 the eigenvalues are real, and $|\nu_1^Q| > 1 > |\nu_2^Q|$. In this case the X_i^* are unstable, and in their vicinities exist *hyperbolic* invariant curves. These saddles are created, together with the period Q elliptic points, as a result of the destruction of rotation number $\rho = P/Q$ invariant curves due to the resonance effect of nonlinear perturbation. Thus, according to the Poincaré–Birkhoff fixed-point theorem, these saddles and the elliptic points exist in equal numbers and are positioned in alternating sequence. (In

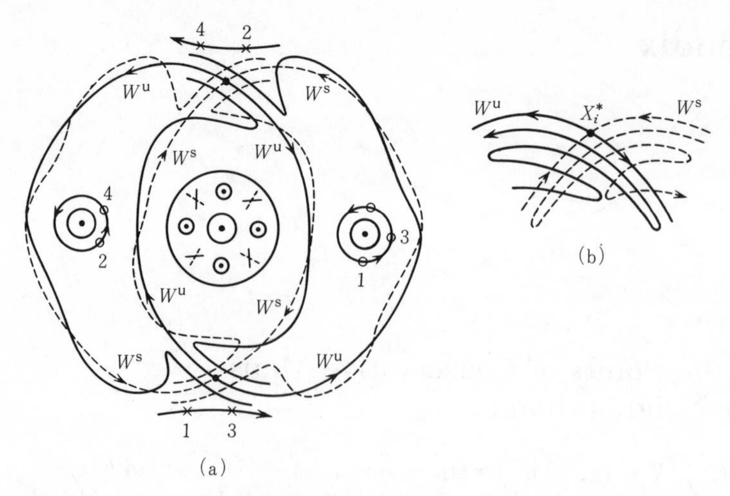

Fig. A.1. **(a)** A schematic of the neighborhoods of the elliptic points together with the saddles and their invariant manifolds W^{u} and W^{s} for the case $Q = 2$. (Note that these are invariant manifolds with respect to F^Q.) There are two sequences of numbers $1, 2, \ldots$ shown here. These represent successive points under F on orbits passing through the neighborhoods of the elliptic points and saddle points. **(b)** Accumulation of primary intersections of W^{u} and W^{s} at X_i^* and the resulting tangle

Fig. A.1, the saddle points are situated near the top and bottom, while the elliptic points are to the right and left sides of the figure.) The invariant manifolds W^{u} and W^{s} of the X_i^* cross to form a set of heteroclinic (or homoclinic) points, thus creating the chaotic sea. Figure A.1 depicts this situation for the case $Q = 2$. As shown in Fig. A.1b, primary intersection points of these various invariant manifolds are created under the action of the maps $F^{\pm}(X)$. These points accumulate at X_i^* along either $W^{\mathrm{s}}(X_i^*)$ or $W^{\mathrm{u}}(X_i^*)$. The area of a given lobe bounded by a given W^{u} and W^{s} is preserved under F, and W^{u} and W^{s} create a limitlessly complex structure referred to as a heteroclinic (or homoclinic) tangle. Chaotic behavior is due to the existence of these complex tangles.

(iii) When $R = 0$ or 1, $\nu_1^Q = \nu_2^Q = 1$ or -1. In this case the motion in the neighborhood of an elliptic point X_i^* cannot be described by a linearization. This situation is termed *parabolic.* These values of R correspond to the disappearance (or, conversely, the appearance in the case of a destabilizing bifurcation) of a certain set of orbits.

As an example of the types of maps to which the present discussion applies, consider the standard map (7.17), whose Jacobian matrix is

$$DF(X) = \begin{pmatrix} 1 - K\cos(2\pi\theta) & 1 \\ -K\cos(2\pi\theta) & 1 \end{pmatrix} . \tag{A.4}$$

For this map we have $s = 2 - K\cos(2\pi\theta)$, and

$$R = (K/4)\cos(2\pi\theta)\ . \tag{A.5}$$

The theory outlined above in application to (A.5) affords an understanding of the fixed points of (7.17).

A.2 Variance and the Time Correlation Function

Let us consider the sum of a quantity u_t (with $\langle u_t \rangle = 0$) along an arbitrary orbit, $\hat{S}_n(X_0) \equiv \sum_{t=0}^{n-1} u_t$. With the time correlation function $C_t \equiv \langle u_t u_0 \rangle$, we can write the variance of this sum as

$$\left\langle \hat{S}_n^2(X) \right\rangle = \sum_{t=0}^{n-1} \sum_{s=0}^{n-1} C_{t-s}\ , \tag{A.6}$$

where $\langle u_t u_s \rangle = C_{t-s}$. This double sum is over each of the lattice points in the square lattice shown in Fig. A.2. If we split this double sum into two sums, one over the terms corresponding to the lower right triangle defined by $t > s$ and one over the upper left triangle defined by $t < s$, we have

$$\left\langle \hat{S}_n^2(X) \right\rangle = nC_0 + \sum_{i=1}^{n-1} \sum_{s=0}^{n-1-i} C_i + \sum_{j=1}^{n-1} \sum_{t=0}^{n-1-j} C_{-j}\ .$$

Finally, because $C_{-j} = C_j$, we can write this as

$$\left\langle \hat{S}_n^2(X) \right\rangle = nC_0 + 2\sum_{t=1}^{n-1} (n-t)C_t\ . \tag{A.7}$$

This gives (6.28), (10.17), and related expressions.

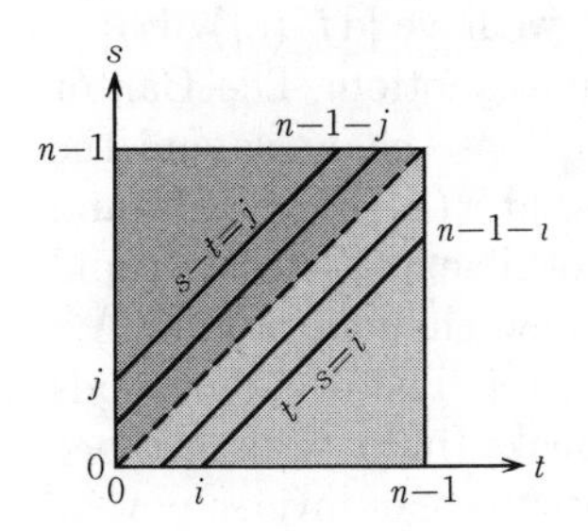

Fig. A.2. The diagonal line $t = s$ and the two triangular lattices $t > s$ and $t < s$

A.3 The Cantor Repellor of Intermittent Chaos

As discussed in Sects. 9.1 and 9.2, Cantor repellors play an important role in a variety of 'crisis' phenomena. We wish here to demonstrate theoretically how they are also important in the context of intermittent chaos and to describe their physical significance in terms of statistical structure functions.

Let us consider the period 3 window corresponding to the quadratic map shown in Fig. 7.10. The Cantor repellor that exists among the three bands which appear for $a > a_{\mathrm{w}}$ also exists in this window, $a_{\mathrm{w}} > a > a_{\mathrm{c}}$, and in the $a < a_{\mathrm{c}}$ region of intermittent chaos. However, within the window this repellor is separated from the attractor, and thus it can only be observed indirectly in transient processes, while in the region of intermittent chaos, its presence is evidenced only by intermittent 'burst' motion.

The time series x_t for a system exhibiting intermittent chaos assumes a form like that shown in Fig. A.3. Here, there appear alternately two types of motion, so-called *laminar motion*, consisting of regular, slow (here period 3) amplitude oscillations, and random, rapid bursts. When $\epsilon \equiv 49(a_{\mathrm{c}} - a) \ll 1$, the average length of an uninterrupted interval of laminar motion is $\tau = 3\pi/\sqrt{\epsilon} \gg 1$, while the average duration of a burst τ_{R} is independent of ϵ. Here, $\tau \gg \tau_{\mathrm{R}}$.

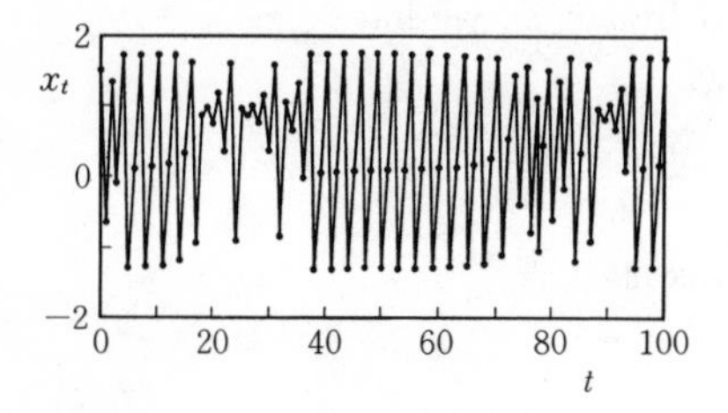

Fig. A.3. The time series x_t (t = 0–100) for intermittent chaos existing just below the period 3 window ($a = 1.749 < a_{\mathrm{c}} = 1.75$) for the quadratic map

We can understand the type of motion exhibited in Fig. A.3 by studying the graph of the map $x_{t+3} = f^3(x_t)$ shown in Fig. A.4. When x_t is inside one of the narrow channels located at $x = c_1, c_2$, and c_3, we have $|\mathrm{d}f^3(x_t)/\mathrm{d}x_t| \sim 1$ and $\Lambda_3(x_t) \sim 0$, and the system exhibits laminar motion. The Cantor repellor corresponds to the steep intervals between these channels and the hopping motion exhibited there. In these intervals, $|\mathrm{d}f^3(x_t)/\mathrm{d}x_t| \gg 1$ and $\Lambda_3(x_t) > 0$, and when x_t is in such a region, the orbit experiences a rapid burst (of average duration τ_{R}) carrying it into a new channel. Figure A.5 exhibits the power spectrum $S(\omega)$ of such a time series. The neighborhoods of $\omega = 0$ and $2\pi/3$ possess series of equally spaced peaks (inter-peak distance $\Delta\omega = 2\pi/\tau$). The envelope of each of these series assumes an inverse power-law form.

As a model of such intermittent chaos, let us consider the SOM map,

$$f(x) = \begin{cases} ax + 0.2 & (0 \le x \le c) \\ a^{-1}(x - 0.8) + 1 & (c \le x \le 0.8) \\ b^{-1}(1 - x) & (0.8 \le x \le 1)\,, \end{cases} \tag{A.8}$$

with $1 > a > 0$, $b = 0.2$ and $c = 0.8/(1 + a)$. The graph of this map appears in Fig. A.6a. For $a > 0.6$ it displays chaotic motion in the interval $I = [0, 1]$.

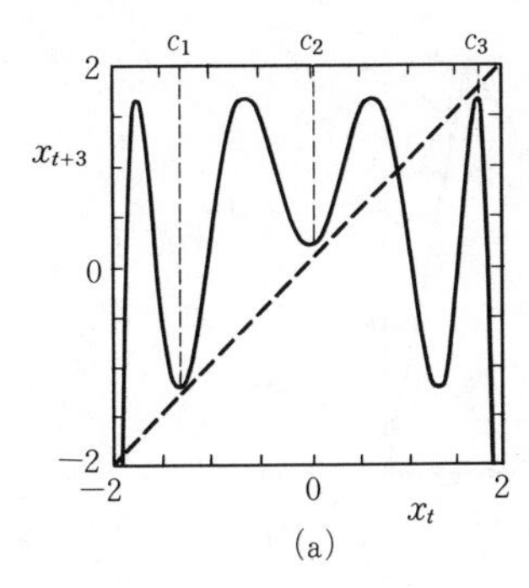

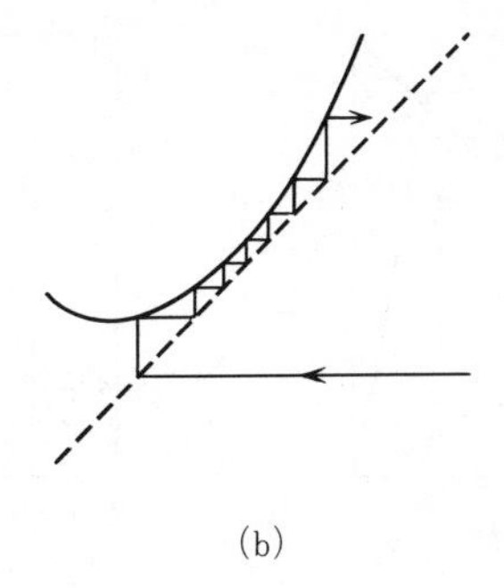

Fig. A.4. (**a**) The graph of $x_{t+3} = f^3(x_t)$ for the quadratic map just prior to the tangent bifurcation ($a = 1.72$). (**b**) An orbit of this map passing through a narrow channel [close-up of the neighborhood of one of the c_i in (a)]

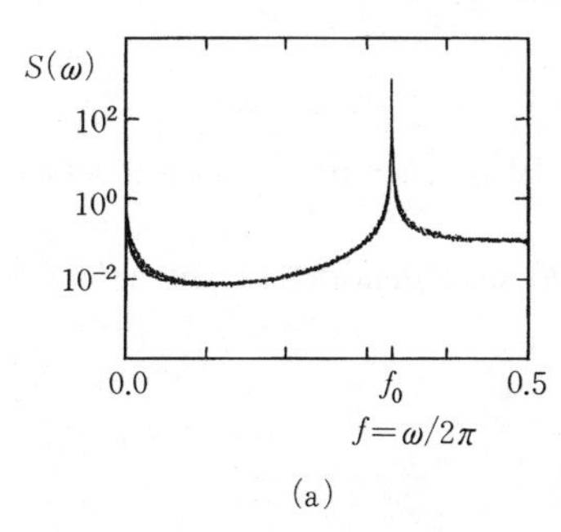

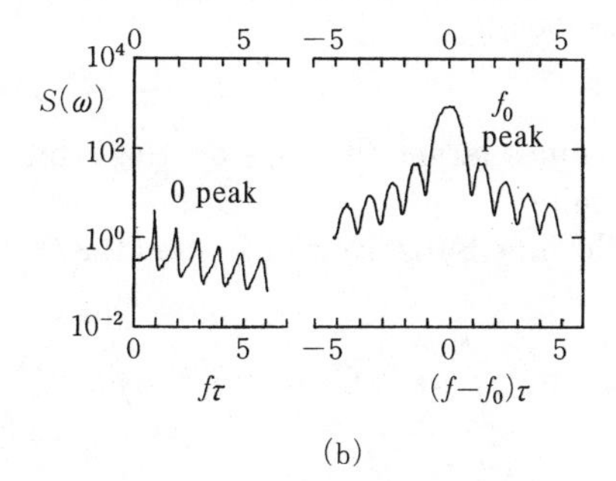

Fig. A.5. (**a**) The power spectrum $S(\omega)$ of intermittent chaos exhibited by the quadratic map with $\epsilon = 4.9 \times 10^{-5}$ ($f_0 = 1/3; T = 2^{15}, N = 400$). (**b**) A close-up of the peaks at $\omega = 0$ and $2\pi f_0$ [Mori et al. (1988)]. The peaks are seen to each be composed of equally spaced series of smaller peaks. The envelope of each series assumes the inverse power-law form $1/|\omega - m\omega_0|^{\zeta_m}$ ($\zeta_0 \approx 1.35, \zeta_1 \approx 1.85$), where $m = 0$ for the 0 curve and $m = 1$ for the f_0 curve, and $\omega_0 = 2\pi f_0$. Under certain conditions, the values $\zeta_0 \approx 1$ and $\zeta_1 \approx 2$ have been obtained analytically

Passing through the $a = 0.6$ tangent bifurcation, the system enters the non-chaotic window lying just below this value, with the point $x^* = 0.2/(1-a)$ becoming an absorbing fixed point. Defining $\epsilon \equiv (a - 0.6)/0.6$, let us consider the chaotic attractor existing for $0 < \epsilon \ll 1$. This attractor consists of the narrow channel situated at $x = c$ together with a Cantor repellor containing an infinite number of saddles, such as the cycle (broken-line loop) shown in Fig. A.6a. An orbit x_t with initial point $x_0 < c$ will display regular, laminar motion continuing for some time [of average length $\tau = (2/\ln 0.6) \ln \epsilon$] as it passes through the channel, and after reaching the value 0.8 it will move in the vicinity of the Cantor repellor, exhibiting a burst of random motion (of average time τ_{R}) until x_t exceeds the value $1 - bc$. This Cantor repellor is an invariant set that continues to exist when the value of a drops below 0.6.

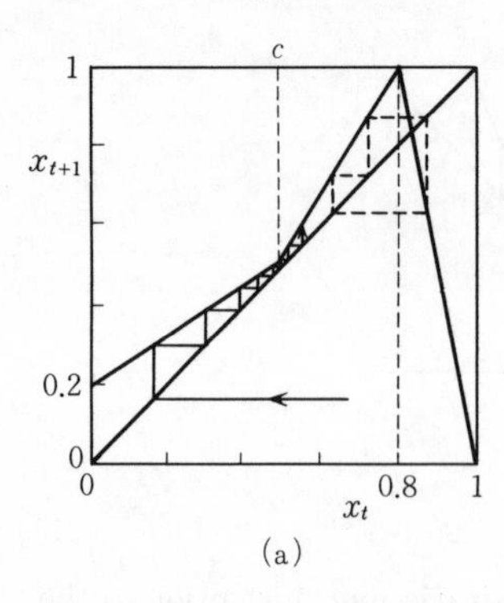

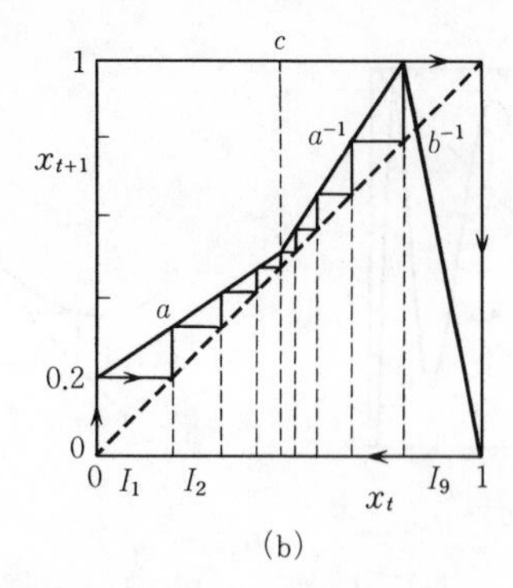

Fig. A.6. (**a**) A graph for the case $0 < \epsilon \equiv (a - 0.6)/0.6 \ll 1$ together with an orbit representing laminar motion (*solid line*) and a period 3 unstable periodic orbit (*broken line*). (**b**) The Markov decomposition for $a = a_4 \approx 0.665 \quad (2m + 1 = 9)$ [Shobu et al. (1984)]

When x_t becomes larger than $1 - bc$, the orbit will be cast into a new $x_t < c$ laminar domain.

Now, following Fujisaka and Inoue (1987), let us consider the linear operator

$$H_q G(x) \equiv \int \mathrm{d}z \, \delta(f(z) - x)|f'(z)|^{1-q} G(z) = \sum_j \frac{G(z_j)}{|f'(z_j)|^q} , \tag{A.9}$$

where the sum $\sum_j$ is over all points z_j in the attractor for which $f(z_i) = x$ is satisfied. Inserting $1 = \int \mathrm{d}x' \delta[f(x) - x']$ into the partition function (8.17) and employing (A.9) n times with the identity $\mathrm{d}f^n(x_0)/\mathrm{d}x_0 = f'(x_{n-1}) \dots f'(x_0)$, we can write $Z_n(q) = \int \mathrm{d}x H_q^n p(x)$. Here, $Z_n(1) = 1$ and $H_1 p(x) = p(x)$. Defining ν_q as the eigenvalue of the linear operator H_q with the largest absolute value, expanding $p(x)$ in eigenfunctions of this operator we obtain, for $n \to \infty$, $Z_n(q) \sim |\nu_q|^n$. Thus the q-potential (8.2) becomes

$$\Phi_\infty(q) = -\ln |\nu_q| \tag{A.10}$$

in the same limit. Note that $|\nu_1| = 1$. Next, let us determine this largest eigenvalue ν_q.

Following Mori et al. (1989), let us consider the largest positive root a_m of the $(m + 1)$-degree polynomial equation

$$a^{m+1} + a^m - 5a + 3 = 0 \quad (m = 3, 4, \cdots) . \tag{A.11}$$

Note that $a_i > a_{i+1}$, $a_\infty = 0.6 + 0$ and $a_3 \approx 0.743$. When $a = a_m$ for some m, the map (A.8) satisfies

$$f^m(0) = c, \quad f^{2m}(0) = 0.8 , \tag{A.12}$$

and, as in Fig. A.6b, we can divide the interval $I = [0, 1]$ into small subintervals $I_i \equiv [f^{i-1}(0), f^i(0)]$ $(i = 1, 2, \ldots, 2m+1)$ within each of which $f(x)$ is monotonic. Note that $f(I_i) = I_{i+1}$ $(i \leq 2m)$ and $f(I_{2m+1}) = I$. In other words, at $a = a_m$ (for $m = 3, 4 \ldots$), (A.8) is Markov. Furthermore, within each small interval, I_i, $f(x)$ is linear, and the eigenvalues of H_q become piecewise constant functions in the Markov partition, while H_q can be expressed as a $(2m + 1) \times (2m + 1)$ matrix:

$$(H_q)_{ij} = \begin{cases} |f_j'|^{-q} & [f(I_j) \supseteq I_i] \\ 0 & \text{(otherwise)} . \end{cases} \tag{A.13}$$

Here f_j' is the slope of $f(x)$ in the sub-interval I_j. The corresponding eigenvalue equation becomes

$$\nu^{2m+1} - b^q \left[\sum_{i=m}^{2m} a^{q(2m-i)} \nu^i + \sum_{i=0}^{m-1} a^{qi} \nu^i \right] = 0 . \tag{A.14}$$

In the $m \to \infty$ limit (i.e., at $a = 0.6 + 0$), the largest eigenvalue ν_q is given by the largest real root of this equation. Hence,

$$\nu_q = \begin{cases} 1 & (q \geq q_\delta \equiv 0.72716 \cdots) \\ a^q + b^q & (q \leq q_\delta) , \end{cases} \tag{A.15}$$

where $q = q_\delta$ is the solution of $a^q + b^q = 1$. Thus, from (A.10), (8.3), and (8.8), at $a = 0.6 + 0$ we have

$$\Phi_\infty(q) = \begin{cases} 0 & (q \geq q_\delta) \\ -\ln(a^q + b^q) & (q \leq q_\delta) \end{cases} \tag{A.16}$$

$$\Lambda_\infty(q) = \begin{cases} 0 & (q > q_\delta) \\ (a^q \ln a^{-1} + b^q \ln b^{-1})/(a^q + b^q) & (q < q_\delta) \end{cases} \tag{A.17}$$

$$\psi(\Lambda) = \begin{cases} (1 - q_\delta)\Lambda & (0 \leq \Lambda \leq \Delta) \\ \text{[right-hand side of (8.11)]} & (\Delta \leq \Lambda \leq \Lambda_{\max}) \\ +\infty & (\Lambda < 0, \Lambda > \Lambda_{\max}) \end{cases} \tag{A.18}$$

Here $\Lambda^\infty = \Lambda_{\min} = 0$, $\Lambda_{\max} = \ln b^{-1} \approx 1.6094$ and $\Delta \equiv \Lambda_\infty(q_\delta - 0) \approx 0.8517$. These functions are shown in Fig. A.7 (*solid curves*). In this system the Renyi entropy can be written $K_q = \Phi_\infty(q)/(q - 1)$, and at $a = 0.6 + 0$ we have

$$K_q = \begin{cases} 0 & (q \geq q_\delta) \\ \{1/(1 - q)\} \ln(a^q + b^q) & (q \leq q_\delta) . \end{cases} \tag{A.19}$$

Note that $K_0 = \ln 2$ and $K_1 = \Lambda^\infty = 0$.

Adding the broken line extensions shown in Fig. A.7 to the functions $\Phi_\infty(q)$, $\Lambda_\infty(q)$ and K_q in the $q < q_\delta$ region and $\psi(\Lambda)$ in the $\Lambda > \Delta$ region, we obtain functions identical to (8.13) for the Baker transformation. These structure functions characterize Cantor repellors, and thus the function $\psi_{\mathrm{R}}(\Lambda)$ corresponds to the Cantor repellor spectrum $\psi_{\mathrm{R}}(\Lambda)$ of Fig. 9.3. Now, let us focus on the shaded square region in Fig. A.8, corresponding to $a = 0.6$. Almost all orbits contained within this region escape under the

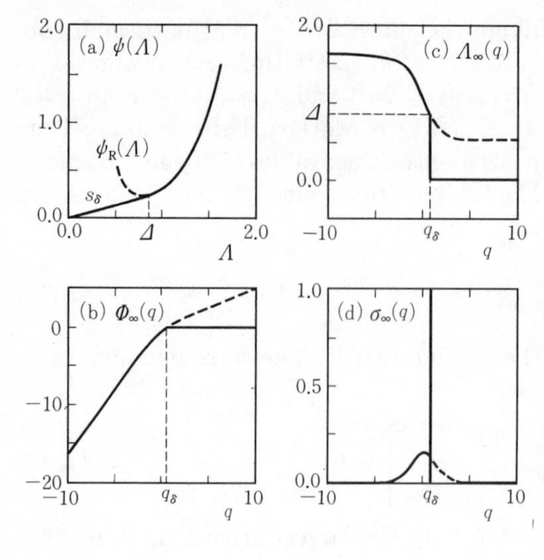

Fig. A.7. Structure functions for intermittent chaos at $a = 0.6 + 0$. Here $s_\delta = 1 - q_\delta$, $q_\delta = 0.72716\ldots$, and $\Delta = \Lambda_\infty(q_\delta - 0) = 0.8517\ldots$. The *broken lines* represent extensions corresponding to the repellor [Mori et al. (1989)]

action of the tent map with slopes a^{-1} and b^{-1} existing here and are eventually absorbed into the fixed point $x^* = 0.5$. However, within this square there remains an invariant Cantor set (a repellor) [see Devaney (1989)]. The chaotic orbits on this repellor determine the form of the structure functions discussed above [Yoshida and So (1988)]. This Cantor repellor is the invariant set that corresponds to the burst phenomenon.

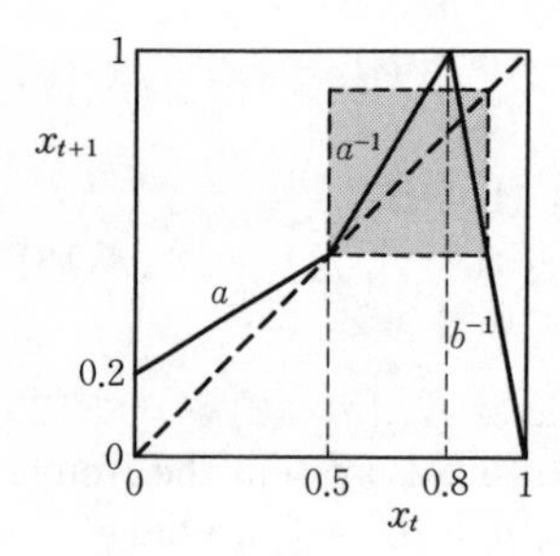

Fig. A.8. A graph of the SOM map at the bifurcation point $a = 0.6$. The Cantor repellor continues to exist at this value, contained within the shaded square shown here. In the corresponding interval of x_t, the graph assumes the form of a tent map whose peak protrudes through the top of this square

The functions $\Lambda_\infty(q)$ for $q > q_\delta$ and $\psi(\Lambda)$ at $\Lambda = 0$ correspond to the $2m+1$-periodic orbits described by Fig. A.6b. The corresponding Liapunov number is $\Lambda_{2m+1}(x_0) = \ln b^{-1}/(2m+1)$. Note that $\lim_{m\to\infty} \Lambda_{2m+1}(x_0) = 0$. These periodic orbits are invariant sets that correspond to laminar motion. The slope $s_\delta = 1 - q_\delta$ linear segment of $\psi(\Lambda)$ connects the point at $\Lambda = 0$ corresponding to laminar motion to the portion of the spectrum, $\psi_R(\Lambda)$, corresponding to burst motion. Fluctuations that occur in intermittent chaos can be understood in terms of the q-phase transition that occurs between the laminar ($q > q_\delta$) and burst ($q < q_\delta$) regimes.

Figure A.9 describes the bending of the slope s_δ linear piece of $\psi(\Lambda)$ due to modulation of ϵ. The four curves appearing here correspond to (from the right) $\tau = 70, 90, 110$ and 130. With the scaling function $B(y)$ ($y < y_{\mathrm{R}} \equiv \Lambda_{\mathrm{R}} - \tilde{\Lambda}$), these curves become almost identically coincident, confirming the similarity relation (9.4). In this figure, $\eta \approx 1.00$, $\Lambda_{\mathrm{R}} \equiv \Delta$, and $y_{\mathrm{R}} \approx 0.27$.

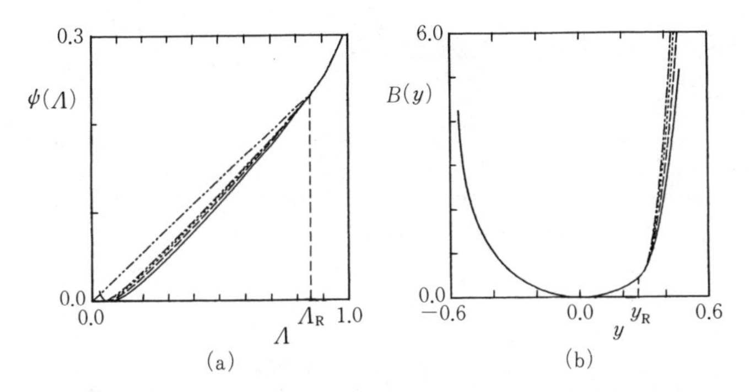

Fig. A.9. (**a**) Pieces of the spectra $\psi(\Lambda)$ corresponding to the values $\tau = (2/\ln 0.6)\ln\epsilon = 70, 90, 110$, and 130. (**b**) The scaling function $B(y)$ ($y = \Lambda - \tilde{\Lambda}$) in (9.4) [Kobayashi et al. (1989)]

Structure functions similar to those discussed here are also seen in the case of the quadratic map just below $a = a_{\mathrm{c}}$. The numerically generated forms of these functions are given in Fig. A.10. These figures correspond to those in Fig. A.7. The functions $\psi(\Lambda)$ and $B(y)$ in Fig. A.11 correspond to those in Fig. A.9.

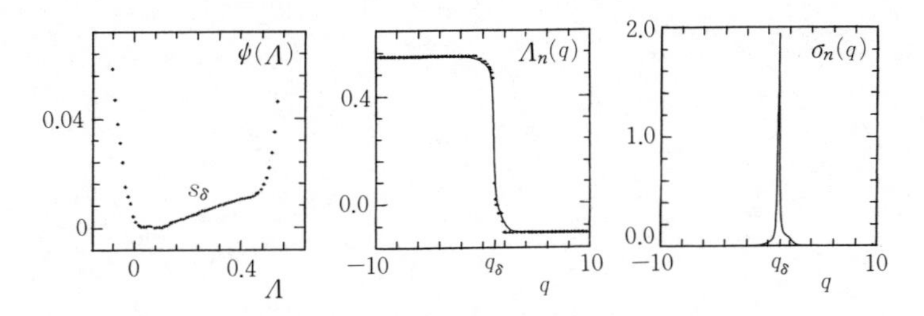

Fig. A.10. The structure functions for intermittent chaos of the quadratic map with $\epsilon = 4.9 \times 10^{-3}$ ($n = 300$, $N = 0.9 \times 10^6$). Here, $q_\delta \approx 0.925$ and $s_\delta = 1 - q_\delta \approx 0.075$. Although it cannot be discerned from this figure, the spectrum $\psi(\Lambda)$ possesses a linear segment of slope $s_\alpha = -1$ to the left of its minimum

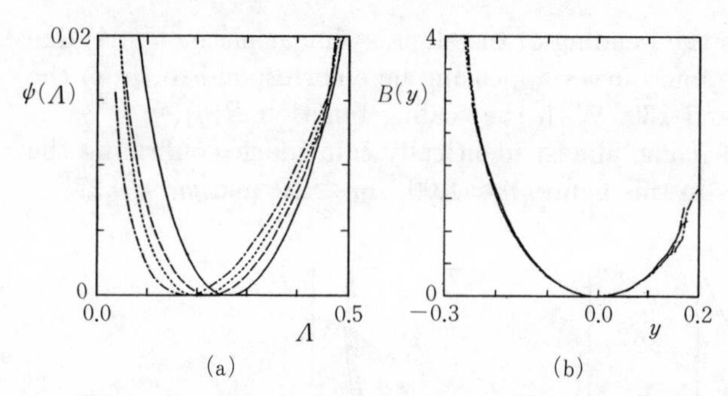

Fig. A.11. (**a**) Four curves depicting the bending of the slope s_δ linear piece of $\psi(\Lambda)$ for the quadratic map. These curves correspond to (from the right) $\tau = 70, 90, 110$, and 130. (**b**) The scaling function $B(y)$ $(y = \Lambda - \tilde{\Lambda})$. Here $\eta \approx 1.06$

Bibliography

Part I – General Reading

The works listed below are of general interest in regard to dissipative structures:

Cross, M.C., Hohengerg, P.C.,(1993) Pattern formation outside of equilibrium. Rev. Mod. Phys. **65** 851–1112,

Glansdorff, P., Prigogine, I. (1971) Thermodynamics of Structure, Stability, and Fluctuations. Wiley, New York

Haken, H. (1983) Synergetics, 3rd edn. Springer, Berlin, Heidelberg

Kuramoto, Y. (1984) Chemical Oscillations, Waves, and Turbulence. Springer, Berlin, Heidelberg

Manneville, P. (1990) Dissipative Structures and Weak Turbulence. Academic Press, London

Nicolis, G., Prigogine, I. (1977) Self-organization in Nonequilibrium Systems. Wiley, New York

Nicolis, G., Prigogine, I. (1989) Exploring Complexity, an Introduction. Freeman, New York

The monograph by Glansdorff and Prigogine (1971) contributed greatly to the early development in the study of dissipative structures during the first half of the 1970s. The point of view presented in that work – that dissipative structures can be understood in terms of an extension of irreversible thermodynamics – was very natural at the time. As a result, it had a strong impact on many physicists whose interest turned to this field. Nicolis and Prigogine (1977), also a product of the 'Brussels school', is something of a continuation of Glansdorff and Prigogine (1971). This work is devoted largely to the analysis of bifurcation phenomena and concentration fluctuations based on nonequilibrium chemical reaction models. This book contains no treatment of fluid phenomena. The book by Nicolis and Prigogine (1989) contains a beginning-level treatment, covering a broad range of topics.

The point of view taken by H. Haken is often compared to that of Prigogine. The former is perhaps more pragmatic, in that it does not adhere to the nonequilibrium thermodynamics that forms the basis of the latter. In Haken's approach, the *slaving principle* is used both as a guiding principle and as a technical 'weapon' in viewing a variety of phenomena at a single level.

An understanding the dynamics of oscillating fields forms the central theme of Kuramoto (1984). This book advocates an approach to the description of nonequilibrium patterns using amplitude equations and phase equations. This was established as a standard approach to such problems during the 1980s. Part I of the present book is similar in its approach, but here the description of oscillating fields is given relatively less weight.

Chapter 1

Chandrasekhar, S. (1961) Hydrodynamic and Hydromagnetic Stability. Clarendon, Oxford

Field, R.J., Berger, M. (1985) Oscillations and Travelling Waves in Chemical Systems. Wiley, New York

Newell, A.C., Whitehead, J.A. (1969) Finite bandwidth, finite amplitude convection. J. Fluid. Mech. **38**, 279–303

Normand, C., Pomeau, Y. (1977) Convective instability: a physicist's approach. Rev. Mod. Phys. **49**, 581–623

Tyson, J.J. (1976) The Belousov–Zhabotinskii Reaction. (Lecture Notes Biomath., vol. 10) Springer, Berlin, Heidelberg

Bénard convection experiments and the linear stability theory of Rayleigh are described in detail in Chandrasekhar (1961). Also, there is an interesting review given in a clear physical context on the emergence of Bénard convection in Normand and Pomeau (1977).

The Newell–Whitehead (NW) equation is derived in Newell and Whitehead (1969). This is a groundbreaking paper in the development of research on dissipative structures. However, the theoretical foundation of the reduction method that these authors present is not particularly clear. We believe that Chap. 5 of the present book makes this theoretical foundation understandable in physical terms.

The reaction mechanism responsible for the Belousov–Zhabotiskii (BZ) equation and the construction and mathematical analysis of the Oregonator are described in detail in Tyson (1976) and in Field and Berger (1985).

Chapter 2

Bodenschats, E., Pesch, W., Kramer, L. (1988) Structure and dynamics of dislocation in anisotropic pattern-forming systems. Physica D **32**, 135–145

Busse, F.H., Clever, R.M. (1979) Instabilities of convection rolls in a fluid of moderate Prandtl number. J. Fluid Mech. **91**, 319–336

Kai, S. (1991) Liquid crystal patterns. Chap. 4 in Pattern Formation. Asakura Shoten, Tokyo (Japanese)

Lega, J., Janiaud, B., Jucquois, S., Croquette, V. (1992) Localized phase jumps in wave trains. Phys. Rev. A **45**, 5596–5604

Manneville, P. (1990) Dissipative Structures and Weak Turbulence. Academic Press, London

Nozaki, K., Bekki, N. (1984) Exact solutions of the generalized Ginzburg–Landau equation. J. Phys. Soc. Japan **53**, 1581–1582

Nozaki, K., Bekki, N. (1985) Formations of spatial patterns and holes in the generalized Ginzburg–Landau equation. Phys. Lett. A **110**, 133-135

Siggia, E., Zipperius, A. (1981a) Dynamics of defects in Rayleigh–Bénard convection. Phys. Rev. A **24**, 1036–1049

Siggia, E.D., Zipperius, A. (1981b) Pattern selection in Rayleigh–Bénard convection near threshold. Phys. Rev. Lett. **47**, 835–858

Manneville (1990) gives a detailed description of the qualitative nature of solutions to the NW equation and similar equations. Focusing on the Boussinesq equation, an analysis of the linear stability of stationary rolls is given in Busse and Clever (1979). This analysis reveals the presence of a number of instabilities that cannot be described with the NW equation and yields a detailed stability diagram. Using an equation resulting from the incorporation of the effect of long-scale horizontally oriented planar flow into the NW equation, the paper by Siggia and Zipperius (1981b) contains a more realistic stability diagram, including features such as the oscillatory instability of stationary rolls and the skewed varicose instability.

Some very nice experimental studies have been made of liquid crystal convection systems by Japanese researchers. Kai (1991) contains a good (Japanese) explanation of this research. The description of the structure and motion of topological defects in the Ginzburg–Pitaevskii equation has its origin in Bodenschats et al. (1988). Theory regarding defect solutions of the NW equation is discussed in Siggia and Zipperius (1981a). The discovery of hole solutions (Nozaki–Bekki solutions) of the complex Ginzburg–Landau equation was reported in Nozaki and Bekki (1984). The experimental observation of such phenomena was first described in Lega et al. (1992).

Chapter 3

Kobayashi, R. (1993) Modeling and numerical simulations of dendritic crystal growth. Physica D **63**, 410–423

Meron, E. (1992) Pattern formation in excitable media. Phys. Rep. **218**, 1–66

Ouyang, Q., Swinney, H.L. (1991) Transition from uniform state to hexagonal and striped Turing patterns. Nature **352**, 610–612

Pelcé, P., Sun, J. (1991) Wave front interaction in steadily rotating spirals. Physica D **48**, 353–366

Rinzel, J., Keller, J.B. (1973) Travelling wave solutions of a nerve conduction equation. Biophys. J. **13**, 1313–1337

Skinner, G.S., Swinney, H. (1991) Periodic to quasiperiodic transition of chemical spiral rotation. Physica D **48**, 1–16

Winfree, A.T. (1974) Sci. Am. **230** No. 6, 82

Winfree, A.T. (1980) The Geometry of Biological Time. Springer, Berlin, Heidelberg

A comprehensive report regarding excitable reaction–diffusion media is presented in Meron (1992). This work also contains an almost exhaustive bibliography that includes essentially all important related works up to 1992. An analytic periodic pulse solution to the McKean model was first found by Rinzel and Keller, as reported in Rinzel and Keller (1973). Numerical results concerning steady rotating solutions of the two-dimensional interface equation arising in this model and the explicit form of spiral waves are given in Pelcé and Sun (1991). Spiral waves can be found not only in BZ reaction systems but also in connection with varieties of living systems. A description of a wide variety of such systems is presented in Winfree (1980).

In recent years, employing the diffusive infusion of a reagent into a gel, a new experimental approach allowing for the study of reaction–diffusion patterns under steady nonequilibrium conditions has been developed. Using this method Skinner and Swinney (1991) have investigated the compound rotational motion of spiral waves. A similar experimental method applied to the study of nonequilibrium chemical reaction systems has allowed, for the first time, the experimental observation of Turing patterns. In this regard, we cite Ouyang and Swinney (1991).

For discussion and analysis of a simple model (the phase-field model) describing the shape formation of interfaces, see Kobayashi (1993).

Chapter 4

Coulet, P., Ioos, G. (1990) Instabilities of one-dimensional cellular patterns. Phys. Rev. Lett. **64**, 866–869

Cross, M. (1983) Phase dynamics of convection rolls. Phys. Rev. A **27**, 490–498

Fauve, S., Bolton, E.W., Brachet, M.E. (1987) Nonlinear oscillatory convection: A quantitative phase dynamics approach. Physica D **29**, 202–214

Kuramoto, Y. (1984) Phase dynamics of weakly unstable periodic structures. Prog. Theor. Phys. **71**, 1182–1196

Manneville, P. (1988) The Kuramoto–Sivashinsky equation: A progress report. in: Systems Far from Equilibrium, ed. by Wesfreid, J.E. et al. Springer, Berlin, Heidelberg, 265–280

Sakaguchi, H. (1992) Localized oscillation in a cellular pattern. Prog. Theor. Phys. **87**, 1049–1053

Sano, M., Kokubo, H., Janiaud, B., Kato, K. (1993) Phase wave in cellular structure. Prog. Theor. Phys. **90**, 1–34

There is a relatively detailed description concerning phase dynamics in Kuramoto (1984). A phenomenological derivation of the type of phase equation discussed in Part I of the present book is given there. An outline of our understanding of the Kuramoto–Sivashinsky equation, as well as a bibliography of related works, is given in Manneville (1988).

By extending phase dynamics models by incorporating the effect of long length scale horizontally oriented planar flow, it is possible to investigate phenomena such as the oscillatory instability of stationary rolls and the skewed-varicose instability. With regard to the former, see Fauve et al. (1987). The latter is discussed in Cross (1983).

In recent years interest has grown in the area of multiple-field dynamics in systems in which amplitude equations and phase equations are coupled. There are many possible types of models existing in such situations, depending on the type of periodic pattern forming the background and the nature of the instability arising therein. In the one-dimensional case, the forms of several multiple-field equations are derived phenomenologically in Coulet and Ioos (1990).

Theoretical considerations regarding coupled-phase instabilities and localized oscillatory solutions in multiple-field systems are given in Sakaguchi (1992). A comprehensive summary of experimental results obtained in investigations of the oscillating grid phase in liquid crystal convection appears in Sano et al. (1993).

Chapter 5

Chen, L.Y., Goldenfeld, N., Oono, Y.,(1996) Renormalization group and singular perturbations – Multiple scales, boundary layers, and reductive perturbation theory. Phys. Rev. E **54** 376–394 (This work approaches problems of the type considered in Part I from a renormalization group point of view.)

Guckenheimer, J., Holmes, P. (1986) Nonlinear Oscillations, Dynamical Systems, and Bifurcations and Vector Fields, 2nd edn. Springer, Berlin, Heidelberg

The point of view regarding reduction presented in this chapter has not previously been properly developed. The formalism given here is in fact close to center manifold theory as applied to bifurcation phenomena. However, the treatment presented in this chapter, consisting of a perturbative determination of the invariant manifold and the form of the evolution law on this manifold, yields, as a perturbative effect, a theory of much broader application than traditional reduction theory. The establishment of the mathematical foundation of this treatment has not yet been realized. Considering the advantage of this more general form of reduction theory, as an example, the reductive structure corresponding to the reduction of one system of PDEs into another can be very naturally interpreted. For discussions of invariant manifold theory in the context of bifurcation phenomena, see Guckenheimer and Holmes (1986).

Supplement I

Kuramoto, Y. (1984) Chemical Oscillations, Waves and Turbulence. Springer, Berlin, Heidelberg
Winfree, A.T. (1980) The Geometry of Biological Time. Springer, Berlin, Heidelberg

A more detailed explanation of the content of this chapter appears in Kuramoto (1984). We cite Winfree (1980) in regard to the far-reaching relationship between living systems and oscillation phenomena.

Part II – General Reading

There are a large number of explanations and references regarding chaos. However, while there are many works concerning the emergence of chaos and related critical phenomena, there are few that describe its structure and statistical nature. The works listed below are of general interest in the study of chaotic behavior. For the reader who may wish to study more advanced problems related to the topics considered in Part II of this book, we provide references containing material that is supplemental to the discussion given in each chapter.

Baker, G.L., Gollub, J.P. (1990) Chaotic Dynamics, An Introduction. Cambridge University Press, New York
Bergé, P., Pomeau, Y., Vidal, C. (1984) Order Within Chaos. Hermann, Paris
Cvitanović, P. (1989) Universality in Chaos, 2nd edn. Hilger, Bristol
Devaney, R.L. (1989) An Introduction to Chaotic Dynamical Systems. Benjamin/Cummings, Redwood City, CA
Guckenheimer, J., Holmes, P. (1986) Nonlinear Oscillations, Dynamical Systems, and Bifurcations and Vector Fields, 2nd edn. Springer, Berlin, Heidelberg
MacKay, R.S., Meiss, J.D. (1987) Hamiltonian Dynamical Systems. Hilger, Bristol
Rasband, S.N. (1990) Chaotic Dynamics of Nonlinear Systems. Wiley, New York

For an introduction to chaotic dynamics, we refer readers to Baker and Gollub (1990) and Bergé et al. (1984). The former exclusively treats examples of oscillators, and describes the manner in which chaos appears and why chaos constitutes a fundamental form of physical phenomena. The latter treats a number of dynamical systems, and presents a discussion regarding the variety of chaotic phenomena and the wide range of physical settings in which it can be observed. Baker and Gollub includes a printout of computer programs, allowing the reader to observe numerically generated chaotic behavior first hand. Rasband (1990) contains an elementary introduction to classical topics in chaos, as understood in 1985. For a beginning-level study of the geometrical concepts and methods involved in chaos employing low-dimensional maps, see Devaney (1989). However, the structural stability theory discussed here must be extended in order to be applicable to physical systems. A well-known dynamical systems text written at the graduate level is Guckenheimer and Holmes (1986). Cvitanović (1989) and MacKay and Meiss (1987) are useful as presentations of results and selected papers up to 1985. However, the statistical physics approach to chaos outlined in Part II of this book has been developed since 1985, and, at present, there is no other book in which such theory is described.

Chapter 6

Baker, G.L., Gollub, J.P. (1990) Chaotic Dynamics, An Introduction. Cambridge University Press, New York

Bergé, P., Pomeau, Y., Vidal, C. (1984) Order Within Chaos. Hermann, Paris

Eckmann, J.-P., Ruelle, D. (1985) Ergodic theory of chaos and strange attractors. Rev. Mod. Phys. **57**, 617–656

Fujisaka, H. (1983) Statistical dynamics generated by fluctuations of local Lyapunov exponents. Prog. Theor. Phys. **70**, 1264–1275

Guckenheimer, J., Holmes, P. (1986) Nonlinear Oscillations, Dynamical Systems, and Bifurcations and Vector Fields, 2nd edn. Springer, Berlin, Heidelberg

Morita, T., Hata, H., Mori, H., Horita, T., Tomita, K. (1988) Spatial and temporal scaling properties of strange attractors and their representations by unstable periodic orbits. Prog. Theor. Phys. **79**, 296–312

Oseledec, V.I. (1968) A multiplicative ergodic theorem Liapunov characteristic numbers for dynamical systems. Trans. Moscow Math. Soc. **19**, 197–231

Ottino, J.M. (1989) The kinematics of mixing: stretching, chaos and transport. Cambridge University Press, New York

Rasband, S.N. (1990) Chaotic Dynamics of Nonlinear Systems. Wiley, New York

Sinai, Y.G. (1981) Randomness in non-random systems. Nature No. 3, 72–80

Takahashi, Y.,(1980) Chaos, periodic points and entropy. Butsuri **35**, 149–161 (Japanese)

Takahashi, Y., Oono, Y. (1984) Towards the statistical mechanics of chaos. Prog. Theor. Phys. **71**, 851–854

Wiggins, S. (1992) Chaotic transport in dynamical systems. Springer, Berlin, Heidelberg

Elementary descriptions of the chaos of dissipative dynamical systems discussed in Sect. 6.1 are given in Baker and Gollub (1990), Bergé et al. (1984), and Rasband (1990). A mathematical treatment is given in Guckenheimer and Holmes (1986). The monograph by Ottino (1989) presents an explanation of experiments involving mixing of fluids in the Lagrangian turbulence of incompressible fluids from a modern point of view (that described in Sect. 6.2). Wiggins (1992) represents an attempt to understand the mixing and diffusion of chaotic orbits in the chaotic sea in terms of mappings of lobes. There presently exist few treatments of the statistical description of chaos discussed in Sects. 6.3 and 6.4. Takahashi (1980), Sinai (1981), and Eckmann and Ruelle (1985) represent pioneering works in connection with the attainment of such a description. The coarse-grained expansion rate of neighboring orbits, $\Lambda_n(X)$, obtained as a generalization of the concept of the eigenvalue of a periodic point, is the fundamental physical quantity used in the description of the stretching and folding undergone by unstable manifolds W^{u} along which chaotic orbits traverse. The first discussions of the fluctuations experienced by $\Lambda_n(X)$ appear in Fujisaka (1983) and Takahashi and Oono (1984). In Part II of this book, this theory is extended to two-dimensional nonhyperbolic systems.

Chapter 7

Cvitanović, P. (1989) Universality in Chaos, 2nd edn. Hilger, Bristol

Devaney, R.L. (1989) An Introduction to Chaotic Dynamical Systems. Benjamin/Cummings, Redwood City, CA

Feigenbaum, M.J. (1979) The universal metric properties of nonlinear transformations. J. Stat. Phys. **21**, 669–706

Grassberger, P., Badii, R., Politi, A. (1988) Scaling laws for invariant measures on hyperbolic and non-hyperbolic attractors. J. Stat. Phys. **51**, 135–178

Grebogi, C., Ott, E., Yorke, J.A. (1983) Crises, sudden changes in chaotic attractors, and transient chaos. Physica D **7**, 181–200

Greene, J.M. (1979) A method for determining a stochastic transition. J. Math. Phys. **20**, 1183–1201

Hata, H., Horita, T., Mori, H., Morita, T., Tomita, K. (1988) Characterization of local structures of chaotic attractors in terms of coarse-grained local expansion rate. Prog. Theor. Phys. **80**, 809–826

Horita, T., Hata, H., Mori H., Morita, T., Tomita, K., Kuroki, S., Okamoto, H. (1988) Local structures of chaotic attractors and q-phase transitions at attractor-merging crises in the sine-circle maps. Prog. Theor. Phys. **80**, 793–808

Hénon, M. (1976) A two-dimensional mapping with a strange attractor. Commun. Math. Phys. **50**, 69–77

Horita, T., Hata, H., Mori, H. (1990) Cascade of attractor-merging crises to the critical golden torus and universal expansion-rate spectra. Prog. Theor. Phys. **84**, 558–562

Jensen, M.H., Kadanoff, L.P., Libchaber, A., Procaccia, I., Stavans, J. (1985) Global universality at the onset of chaos: results of a forced Rayleigh–Bénard experiment. Phys. Rev. Lett. **55**, 2798–2801

MacKay, R.S., Tresser, C. (1986) Transition to topological chaos for circle maps. Physica D **19**, 206–237

Šarkovskii, A.N. (1964) Coexistence of cycles of a continuous map of a line into itself. Ukr. Math. Z. **16**, 61–71

Tominaga, H., Mori, H. (1991) Crossover to the $f(\alpha)$ spectra between critical and chaotic regime, and universal critical scaling laws for two-dimensional fractality. Prog. Theor. Phys. **86**, 355–369

The Hénon map and the standard map discussed in Sects. 7.1 and 7.2 are representative two-dimensional maps corresponding to dissipative and conserved systems, respectively. The study of the phase structure of such maps was begun in Hénon (1976), while that of their geometrical natures was begun in Greene (1979). Research regarding the geometrical structure of bifurcations in chaos that results from collision with saddles can be said to have originated with Grebogi et al. (1983). In the paper by Feigenbaum (1979), the universality present in the period-doubling 2^n-bifurcations of one-dimensional quadratic maps, discussed in Sect. 7.3, is obtained explicitly by use of the renormalization transformation $\mathcal{T}$. Subsequent papers providing rigorous demonstrations of the results of this paper along with a number of generalizations are included in Cvitanović (1989). Šarkovskii's theorem and the mathematics of other one-dimensional chaotic systems are described in Devaney (1989). The important features of the circle map discussed in Sect. 7.4 are the appearance of a critical golden torus and the phase-unlocked chaos resulting from the presence of the quasi-periodic route to chaos. The paper by Horita et al. (1990) contains a treatment following the point of view whereby this route is understood in terms of the cascade of a certain phase-unlocked chaotic sequence leading from the chaotic region. In this paper, an overall picture of the quasi-periodic route is developed.

Chapter 8

Eckmann, J.-P., Procaccia, I. (1986) Fluctuations of dynamical scaling indices in nonlinear systems. Phys. Rev. A **34**, 659–661

Ellis, R.S. (1985) Entropy, Large Deviations, and Statistical Mechanics. Springer, Berlin, Heidelberg

Fujisaka, H., Inoue, M. (1987) Statistical-thermodynamics formalism of self-similarity. Prog. Theor. Phys. **77**, 1334–1343

Grassberger, P., Badii, R., Politi, A. (1988) Scaling laws for invariant measures on hyperbolic and non-hyperbolic attractors. J. Stat. Phys. **51**, 135–178

Halsey, T.C., Jensen, M.H., Kadanoff, L.P., Procaccia, I., Shraiman, B.I. (1986) Fractal measures and their singularities: the characterization of strange sets. Phys. Rev. A **33**, 1141–1151

Hata, H., Horita, T., Mori, H., Morita, T., Tomita, K. (1988) Characterization of local structures of chaotic attractors in terms of coarse-grained local expansion rate. Prog. Theor. Phys. **80**, 809–826

Hata, H., Horita, T., Mori, H., Morita, T., Tomita, K. (1989) Singular local structures of chaotic attractors and q-phase transitions of spatial scaling structures. Prog. Theor. Phys. **81**, 11–16

Horita, T., Hata, H., Mori H., Morita, T., Tomita, K., Kuroki, S., Okamoto, H. (1988a) Local structures of chaotic attractors and q-phase transitions at attractor-merging crises in the sine-circle maps. Prog. Theor. Phys. **80**, 793–808

Horita, T., Hata, H., Mori, H., Morita, T., Tomita, K., (1988b) Singular local structures of chaotic attractors due to collisions with unstable periodic orbits in two-dimensional maps. Prog. Theor. Phys. **80**, 923–928

Jensen, M.H., Kadanoff, L.P., Libchaber, A., Procaccia, I., Stavans, J. (1985) Global universality at the onset of chaos: results of a forced Rayleigh–Bénard experiment. Phys. Rev. Lett. **55**, 2798–2801

Mori, H., Hata, H., Horita, T., Kobayashi, T. (1989a) Statistical mechanics of dynamical sytsems. Prog. Theor. Phys. Suppl. **99**, 1–63

Mori, N., Kobayashi, T., Hata, H. Morita, T., Horita T., Mori, H. (1989b) Scaling structures and statistical mechanics of type-I intermittent chaos. Prog. Theor. Phys. **81**, 60–77

Morita, T., Hata, H., Mori, H., Horita, T., Tomita, K. (1987) On partial dimensions and spectra of singularities of strange attractors. Prog. Theor. Phys. **78**, 511–515

Morita, T., Hata, H., Mori, H., Horita, T., Tomita, K. (1988) Spatial and temporal scaling properties of strange attractors and their representations by unstable periodic orbits. Prog. Theor. Phys. **79**, 296–312

Ott, E., Grebogi, C., Yorke, J.A. (1989) Theory of first order phase transitions for chaotic attractors of nonlinear dynamical systems. Phys. Lett. A **135**, 343–348

Ruelle, D. (1978) Thermodynamic Formalism. Addison-Wesely, Reading, MA

Sano, M., Sato S., Sawada, Y. (1986) Global Spectral Characterisitics of Chaotic Dynamics. Prog. Theor. Phys. **76**, 945–948

Shenker, S.J. (1982) Scaling behavior in a map of a circle onto itself; empirical results. Physica D **5**, 405–411

Takahashi, Y. (1980) Chaos, periodic points and entropy. Butsuri **35**, 149–161 (Japanese)

Takahashi, Y., Oono, Y. (1984) Towards the statistical mechanics of chaos. Prog. Theor. Phys. **71**, 851–854

Tomita, K., Hata, H., Mori, H., Morita, T. (1988) Scaling structures of chaotic attractors and q-phase transitions at crises in the Hénon and the annulus maps. Prog. Theor. Phys. **80**, 953–972

The statistical thermodynamics formalism of dynamical systems discussed in this chapter was first introduced from a mathematical point of view. To understand the motivation behind the development of this formalism, see, for example, Takahashi (1980) and Ellis (1985). The concrete treatment given in this chapter of the multifractal dimension $D(q)$, the two spectra $f(\alpha)$ and $\psi(\Lambda)$, and related quantities for physical dynamical systems follows the theoretical work contained in Halsey et al. (1986), Morita et al. (1988), and Grassberger et al. (1988), and the line of reasoning presented in Mori et al. (1989a), in which a number of subjects are treated in a unified manner. The attempt to unify the geometrical and statistical descriptions of chaos presented in this chapter has its origin in the papers by Horita et al. (1988a), Hata et al. (1988), Tomita et al. (1988), and Mori et al. (1989b), which seek a statistical characterization of the bifurcations arising from fluctuations in the coarse-grained expansion rate of orbits, and the paper Horita et al. (1988b), which deals with the theory regarding the slope s_β resulting from collision with a saddle.

The first experimental results regarding the spectrum $f(\alpha)$ of the critical golden torus for the circle map, discussed in Sect. 8.2, appear in Jensen et al. (1985). This is the experimental study of the critical golden torus exhibited in fluid systems undergoing Bénard convection represented by Fig. 7.5. As indicated in Hata et al. (1989), the spectrum $f(\alpha)$ corresponding to a chaotic attractor in a two-dimensional map possesses a linear segment corresponding to that in $\psi(\Lambda)$. However, there is yet no experimental verification of this finding.

Chapter 9

Grebogi, C., Ott, E., Romeiras, F., Yorke, J.A. (1987) Critical exponents for crisis-induced intermittency. Phys. Rev. A **36**, 5365–5380

Guckenheimer, J., Holmes, P. (1986) Nonlinear oscillations, dynamical systems, and bifurcations and vector fields, 2nd edn. Springer, Berlin, Heidelberg

Hata, H., Horita, T., Mori, H. (1989) Dynamic description of the critical 2^∞ attractor and 2^m-band chaos. Prog. Theor. Phys. **82**, 897–910

Horita, T., Hata, H., Mori, H., Tomita, K. (1989) Dynamics on critical tori at the onset of chaos and critical KAM tori. Prog. Theor. Phys. **81**, 1073–1078

Kobayashi, T., Mori, N., Hata, H., Horita, T., Yoshida, T., Mori, H. (1989) Critical scaling laws of dynamic structure functions for type I intermittent chaos. Prog. Theor. Phys. **82**, 1–6

Mori, N., Kuroki, S., Mori, H. (1988) Power spectra of intermittent chaos due to the collapse of period-3 windows. Prog. Theor. Phys. **79**, 1260–1264

Mori, H., Hata, H., Horita, T., Kobayashi, T. (1989a) Statistical mechanics of dynamical sytsems. Prog. Theor. Phys. Suppl. **99**, 1–63

Mori, N., Kobayashi, T., Hata, H. Morita, T., Horita T., Mori, H. (1989b) Scaling structures and statistical mechanics of type-I intermittent chaos. Prog. Theor. Phys. **81**, 60–77

Mori, H., Okamoto, H., Tominaga, H. (1991) Energy dissipation and its fluctuations in chaotic dynamical systems. Prog. Theor. Phys. **85**, 1143–1148

Murayama, T., Tominaga, H., Mori, H., Hata, H., Horita, T. (1990) q-phase transitions and dynamic scaling laws at attractor-merging crises in the driven damped pendulum. Prog. Theor. Phys. **83**, 649–654

Tominaga, H., Mori, H. (1991) Crossover to the $f(\alpha)$ spectra between critical and chaotic regime, and universal critical scaling laws for two-dimensional fractality. Prog. Theor. Phys. **86**, 355–369

Tomita, K., Hata, H., Mori, H., Morita, T. (1988) Scaling structures of chaotic attractors and q-phase transitions at crises in the Hénon and the annulus maps. Prog. Theor. Phys. **80**, 953–972

Tomita, K., Hata, H., Horita, T., Mori, H. Morita, T., Okamoto, H., Tominaga, H. (1989) q-phase transitions in chaotic attractors of differential equations at bifurcation points. Prog. Theor. Phys. **81**, 1124–1134

Yoshida, T., So, B.C. (1988) Spectra of singularities and entropies for local expansion rates. Prog. Theor. Phys. **79**, 1–6

The treatment in Sects. 9.1 and 9.2 of two types of forced oscillators involving band merging and attractor merging as representative bifurcations follows Tomita et al. (1989) and Murayama et al. (1990). The generalization of the energy dissipation rate follows Mori et al. (1991). In this paper, a quantity representing the spread and randomness displayed by the motion of orbital points on the attractor is defined, and it is found that the variance of the spread corresponding to a single time step changes dramatically as a result of crisis. The demonstration that the self-organized criticality can be straightforwardly understood in terms of the time sequence of the coarse-grained orbital expansion rate, presented in Sect. 9.3, is based on Horita et al. (1989) and Hata et al. (1989). The fact that, in the neighborhood of the emergence point of chaos for a two-dimensional map, $\psi(\Lambda)$ possesses a simple dynamical similarity is shown in Tominaga and Mori (1991). With this similarity, it is shown that the two-dimensional fractality of $f(\alpha)$ disappears as the emergence point of chaos is approached.

The discussion in Sect. A.3 of the Appendix concerning the time series of type I intermittent chaos and the fact that it is clearly characterized by a regular laminar state and a random burst state, as reflected by the characteristic forms of $\psi(\Lambda)$, $f(\alpha)$ and the power spectrum, is based on Mori et al. (1989b), Hata et al. (1989), Mori et al. (1988), and Kobayashi et al. (1989).

Chapter 10

Aref, H. (1984) Stirring by chaotic advection. J. Fluid Mech. **143**, 1–21

Horita, T., Hata, H., Mori, H., Tomita, K. (1989) Dynamics on critical tori at the onset of chaos and critical KAM tori. Prog. Theor. Phys. **81**, 1073–1078

Horita, T., Mori, H. (1992) Long-time correlations and anomalous scaling laws of widespread chaos in the standard map, in: From Phase Transition to Chaos, ed. by Gyorgi et al. World Scientific, Singapore, 290–307

Ichikawa, Y.H., Kamimura, T., Hatori, T. (1987) Stochastic diffusion in the standard map. Physica D **29**, 247–255

Ishizaki, R., Horita, T., Kobayashi, T., Mori, H. (1991) Anomalous diffusion due to accelerator modes in the standard map. Prog. Theor. Phys. **85**, 1013–1022

Ishizaki, R., Horita, T., Mori, H. (1993) Anomalous diffusion and mixing of chaotic orbits in Hamiltonian dynamical systems. Prog. Theor. Phys. **89**, 947–963

Lichtenberg, A.J., Lieberman, M.A. (1982) Regular and Stochastic Motion. Springer, Berlin, Heidelberg, 50–51

MacKay, R.S., Meiss, J.D., Percival, I.C. (1982) Transport in Hamiltonian systems. Physica D **13**, 55–81

Meiss, J.D. (1992) Symplectic maps, variational principles, and transport. Rev. Mod. Phys. **64**, 795–848

Mori, H., Hata, H., Horita, T., Kobayashi, T. (1989) Statistical mechanics of dynamical sytsems. Prog. Theor. Phys. Suppl. **99**, 1–63

Ouchi, K., Mori, H. (1992) Anomalous diffusion and mixing in an oscillating Rayleigh–Bénard flow. Prog. Theor. Phys. **88**, 467–484

Solomon, T.H., Gollub, J.P. (1988) Chaotic particle transport in time-dependent Rayleigh–Bénard convection. Phys. Rev. A **38**, 6280–6286

Wiggins, S. (1992) Chaotic Transport in Dynamical Systems. Springer, Berlin, Heidelberg

In addition to Wiggins (1992), we cite the review by Meiss (1992) as a recent work that treats conserved dynamical systems, the subject of this chapter. Although the long-time correlation of chaotic orbits due to the self-similar nested structure of islands of tori has been investigated numerically by many people, no corresponding theory exists as yet. In Sect. 10.2, we explain the long-time correlation of the local expansion rate of neighboring orbits and the anomalous statistical structure of the coarse-grained expansion rate (in the case in which the islands are not 'too large') in the widespread chaotic sea of the standard map. This discussion is based on Horita and Mori (1992). The special feature of the sea of widespread chaos is the existence of islands of accelerator-mode tori and the resulting anomalous enhancement of diffusion of chaotic orbits. This feature is studied in detail numerically in Ichikawa et al. (1987). Following Ishizaki et al. (1991) and Ishizaki et al. (1993), we explain here the appearance of the long-time velocity correlation and anomalous statistics of the coarse-grained velocity arising as a result of the existence of these islands. The mixing and diffusion undergone in fluid systems found experimentally in the context of Bénard convection by Solomon and Gollub (1988) has an anomalous nature that is similar to that existing in the widespread chaos of the standard map. Furthermore, following Ouchi and Mori (1992), it is shown in this chapter that the two cases can be treated using similar methods.

A very wide variety of chaotic systems exist in Nature. Unfortunately, because of the limited length of this book, we were able to treat only four types of physical systems, two types of forced oscillators and two types of Bénard convection systems. However, despite being small in number, this set of low-dimensional maps is sufficiently diverse, as it collectively exhibits essentially all of the universal bifurcations found in chaotic systems. While we have concentrated on low-dimensional systems in this book, chaotic behavior is also exhibited by more complicated systems, including Eulerian-turbulent fluids, nonlinear nonequilibrium materials systems, systems of an infinite number of degrees of freedom characterized by spatio-temporal chaos existing within regions exhibiting nonuniform spatial patterns, and quantum-mechanical systems. It is important to extend the concepts and methods developed to treat low-dimensional systems in order to make the study of such more complicated systems possible. However, at this point, the fundamental point of view necessary for this generalization is still unclear, and at present no works exist to which we can refer to begin such an endeavor.

Supplement II

Evans, D.J., Cohen, E.G.D., Morriss, G.P. (1990) Viscosity of a simple fluid from its maximal Lyapunov exponents. Phys. Rev. A **42**, 5990–5997

Fujisaka, H., Yamada, T. (1985) A new intermittency in coupled dynamical systems. Prog. Theor. Phys. **74**, 918–921

Fujisaka, H., Fukushima, K., Inoue, M., Yamada, T. (1996) Transformations in chaos and on–off intermittency. Butsuri **51**, 813–820 (Japanese)

Hata, H. (1997) private communication

Hata, H., Miyazaki, S. (1997) Exactly solvable maps of on–off intermittency. Phys. Rev. E **55**, 5311–5314

Hoover, W.G., Posch, H.A. (1994) Second-law irreversibility and phase-space dimensionality loss from time-reversible nonequilibrium steady-state Lyapunov spectra. Phys. Rev. E **49**, 1913–1920

Ishizaki, R., Mori, H. (1997) Anomalous diffusion induced by external force in the standard map. Prog. Theor. Phys. **97**, 201–211

Ott, E., Sommerer, J.C. (1994) Blowout bifurcations; the occurrence of riddled basins and on–off intermittency. Phys. Lett. A **188**, 39–47

Platt, N., Spiegel, E.A., Tresser, C. (1993) On–off intermittency; a mechanism for bursting. Phys. Rev. Lett. A **70**, 279–282

Sarman, S., Evans, D.J., Morriss, G.P. (1992) Conjugate-pairing rule and thermal-transport coefficients. Phys. Rev. A **45**, 2233–2242

Wang, X.-J. (1992) Dynamical sporadicity and anomalous diffusion in the Lévy motion. Phys. Rev. A **45**, 8407–8417

The works cited in the opening paragraph of Sect. SI.1 are Fujisaka and Yamada (1985), Platt et al. (1993), and Ott and Sommerer (1994). Fujisaka et al. (1994) also gives an explanation of on–off intermittency. Ott and Sommerer (1994) and Hata and Miyazaki (1997) present investigations of the phase-space structure of these systems. However, our understanding of such structure is not yet complete.

Wang (1992) is a recent mathematical work concerning the the Lévy process discussed in Sect. SII.2. The analysis presented in this section of anomalous diffusion induced by an external force follows Ishizaki and Mori (1997). This type of anomalous diffusion should be widely observable as a universal feature of conservative dynamical systems containing tori. However, this diffusion is singular in the external forcing term Γ, and thus no corresponding linear response exists.

The phenomenological relationship between the transport coefficients and Liapunov numbers, discussed in Sect. SII.3, is treated in Evans et al. (1990) and Sarman et al. (1992). Considerations concerning the dimension of the attractor ΔD given here are based on Hoover and Posch (1994). The relations among these quantities are of great interest, but in order to obtain a statistical-mechanics formalism of molecular and turbulent transport phenomena, it will be necessary to consider their underlying physical mechanisms.

Appendix

Devaney, R.L. (1989) An Introduction to Chaotic Dynamical Systems. Benjamin/Cummings, Redwood City, CA

Fujisaka, H., Inoue, M., Prog. Theor. Phys. **78**(1987), 268–275

Greene, J.M. (1979) A method for determining a stochastic transition. J. Math. Phys. **20**, 1183–1201

Kobayashi, T., Mori, N., Hata, H., Horita, T., Yoshida, T., Mori, H. (1989) Critical scaling laws of dynamic structure functions for type I intermittent chaos. Prog. Theor. Phys. **82**, 1–6

Mori, N., Kuroki, S., Mori, H. (1988) Power spectra of intermittent chaos due to the collapse of period-3 windows. Prog. Theor. Phys. **79**, 1260–1264

Mori, N., Kobayashi, T., Hata, H. Morita, T., Horita T., Mori, H. (1989) Scaling structures and statistical mechanics of type-I intermittent chaos. Prog. Theor. Phys. **81**, 60–77

Shobu, K., Ose, T., Mori, H. (1984) Prog. Theor. Phys. **71**, 458–473

Index

Accelerator mode periodic orbits 250, 254
Accelerator mode periodic points 256
Accelerator mode tori 250, 255, 256, 265
Activator 61, 67
Activator–inhibitor model 67
Active functional elements 20
Algebraic structure functions 233, 242
Amplitude equation 11, 34, 35, 107
Analysis 192
Anisotropic fluids 26
Anisotropy 27, 63
Annulus map 157
Anomalous diffusion 250, 258
– resulting from external force 262
Anomalous scaling laws 247
Anomalous scaling relation 253
Aperiodic orbits
– neutral 148, 163
Apex orbit *see* Vertex orbit
Archimedes spirals 39, 54
Arnold tongues 175, 227
Aspect ratio 7
Asynchronization 121, 261
Attractor 131, 132, 151
– chaotic 133
– formation 224
– strange 131, 195
Attractor destruction 178, 183
– in the Hénon map 155, 192
– in the quadratic map 164, 187
Attractor merging 177, 183
– in the annulus map 225
– in the circle map 189
– in the forced pendulum 224, 227

Back 46, 56
Baker transformation 185
– statistical structure functions of 186
Balanced equations 106
Band chaos 151, 163, 174
– phase-locked 176
Band merging 152, 183, 218
– in the forced pendulum 218
– in the Hénon map 152, 192
– in the quadratic map 164, 165
Basin 133
– boundary of 155, 224
Belousov–Zhabotinskii reaction 14, 52
Bénard convection 5, 22, 134, 202, 254, 263
Bending 142
Benjamin–Feir instability 38, 41
BF instability *see* Benjamin–Feir instability
Bifurcating solution 12
Bifurcation 12, 151, 217
– in oscillating laminar flow 254
– in the annulus map 158, 225
– in the circle map 174
– in the forced pendulum 218, 224, 227
– in the Hénon map 153, 192
– in the quadratic map 161
– in the standard map 159, 243
– universality of 215
Bifurcation parameter 11, 151, 217
Bifurcation point 13, 89, 98, 153, 154
Bifurcations
– universality of 217
Bistable 44
Boussinesq equation 8
Breaking of time-reversal symmetry 245
Burgers equation 77
Burst 220, 244, 272
Burst state 261

Cantor repellor 218, 226, 271
– slope resulting from 218

Cascade 151, 217
– of 2^m-band bifurcations 235
– of 2^n bifurcations 165
– of F_m attractor merging 237
Cell 131
Chaos 129
– and critical phenomena 230
– bifurcations of 151, 217
– statistical stability of 145
– statistical structure of 148
– universality of 237
Chaotic orbit 129, 133, 136
Chaotic sea 136, 241, 243, 262, 265
Circle map 161, 174, 189
Circular wave 52
Climb 29
Coarse-grained mixing rate 148
Coarse-grained orbital expansion rate 144, 183, 265
Coarse-grained velocity 251, 252
– statistical structure 253
Coefficient of viscosity 263
Common tangents 219, 226
Complex Ginzburg–Landau equation 34, 35, 38, 39, 41, 108, 109
Complex GL equation *see* Complex Ginzburg–Landau equation
Compound rotation 58
Conservative dynamical system
– phase space structure of 269
Conservative map 159, 269
Conservative system 265
Continuous symmetry 86
Core 54
Coupled chaotic oscillators 261
Coupled excitable oscillators 119
Coupled phase equations 87
Coupled phase instability 88
Crisis
– attractor-destruction 155, 187
– attractor-merging 177, 225
– band-merging 165, 218
Crisis curve 175, 180
Critical attractor 183, 199, 217, 230, 233, 265
Critical coupling strength 123
Critical golden torus 158, 178, 202, 232
Critical mode 10, 35, 88, 107
Critical torus 135
Critical 2^∞ attractor 163, 200, 230
Crystal growth 62, 67

Defect solution 31, 38
Diffusion 160, 250, 254
Director 26
Dissipative dynamical system 131, 263
– phase space structure of 129
Dissipative structure 6, 59, 107
Dynamic similarity relations 221
Dynamical self-similarity 241
Dynamical similarity relations 182, 236, 277

Eckhaus instability 23, 72
Electro-hydrodynamic instability 26
Elliptic points 270
Energy dissipation rate 132, 156, 222
Entrainment 79
Entropy 266, 275
– production of 264
– topological 218
Excitability 19
Excitable media 48
Excitable system 19, 47
Excitation wave 48
External forcing term 262

Feigenbaum universal constants 168
Feller's theorem 247, 263
First visiting time 255
FitzHugh–Nagumo equation 19, 48
FKN mechanism 15
F_m attractor merging 217
– similarity relations of 182, 237
F_m time series 232, 242
Folding 131, 138, 145, 174, 225, 266
– caused by type I tangency structure 245
– tangency structure of 203, 212
Fractal 131, 238
Fractal dimension 198, 200
Front 46, 56
Fully extended chaos *see* Widespread chaos

Ginzburg–Pitaevskii equation 28
Glide 29
GP equation *see* Ginzburg–Pitaevskii equation
Group entrainment transition 122

Hausdorff dimension 198
Hénon map 151, 155, 157
– attractor destruction in 155, 192
– band merging in 153, 192
– bifurcation in 154, 192

Heteroclinic points 155, 208, 270
Hole solutions 39
Homoclinic points 153, 209, 218, 228, 270
Hopf bifurcation 34, 88
Hyperbolic dynamical system 142, 187
Hyperbolic phase 211
Hyperbolic structure 142

Indeterminism 144
Inertial terms 8
Information dimension 198
Inhibitor 61, 67
Instability of chaotic orbits *see* Orbital instability
Interface 45
Interface dynamics 47, 115
Intermittency
– on–off 261
Intermittent chaos 272
– structure functions of 275
Intermittent hopping *see* Intermittent transferring
Intermittent transferring 220, 227
Invariant manifold 94, 265
Invariant sets 127, 133, 144, 265
– similarity of 172
Inverted bifurcation 12, 36
Involute 53
Irreproducibility 144
Islands around islands 138, 241, 245, 256

Josephson map 262

KAM torus 159, 241
Keener–Tyson model 18
Kink 45
Kolmogorov–Sinai entropy 266
KS equation *see* Kuramoto–Sivashinsky equation
Kuramoto–Sivashinsky equation 78, 85

Lagrangian turbulence 254
Laminar motion 272
Laminar state 261
Last KAM torus 160, 241
Liapunov number 139, 140
Limit cycle 19
Linear segments of structure functions 189, 212, 214, 245
Linear stability 8, 23, 37, 98
Liquid crystal convection system 27, 91
Living systems 119
Lobes 136, 255
Local dimension 195, 197
Local pattern 88
Local structure 40, 86
Logistic map 161
Long-time correlation 148, 244, 252, 258
Long-time velocity correlation 251
Loss of memory 144
Low-dimensional maps 155

McKean model 48
Mean-field coupled model 122
Meandering 58
Memory loss
– within a band 152
Mixing 131, 138, 144, 160
– anomalous 257
– in widespread chaos 243
– within a band 152
Mixing time 255
Molecular viscosity 264
Mullins–Sekerka instability 66
Multifractal dimension 198
Multiplex string-like structure 131, 178, 225

Natural probability measure 195, 207
Navier–Stokes equation 8, 25, 82
Neural network 119
Neurons 119
Neutral 140
Neutral orbits
– aperiodic 148, 163
Neutral points 139
– periodic 135, 269
Neutral solution 10, 35
Newell–Whitehead equation 13, 21
Nonequilibrium open systems 6, 62
Nonequilibrium steady state 263
Nontrivial solution 12
Normal bifurcation 12, 36
Normal rolls 27
Nozaki–Bekki solutions 39
Nullcline 18
NW equation *see* Newell–Whitehead equation

Oblique rolls 27
Observables 146, 266
On–off intermittency 261

One-dimensional circle map 174
One-dimensional quadratic map 161
Orbital expansion rate 143, 147, 183
Orbital instability 139, 144, 146, 246, 247
Oregonator 16
Oscillating fields 34, 112
Oscillator network dynamics 121
Oscillatory instability 24, 60, 135

Pacemaker 81, 91
Partial dimension 199
Partition function 183, 196
– generalized 199
Period doubling 132, 161
Periodic points
– neutral 135, 269
Perturbation theory of weakly nonlinear oscillators 95
Phase dynamics 69, 84, 112
– of an oscillator field 119
– of interfaces 84
Phase equation 72, 78, 84, 112
Phase field model 62, 67
Phase instability 23, 38, 88
Phase locking 174, 177
Phase model 121
Phase singular point 38
Phase structure 148
Phase turbulence 41, 82
Phase waves 40, 78
Phase–amplitude equation 89
Piecewise-linear model 44
Poincaré map 130, 135, 156
Poincaré section 135
Prandtl number 6
Probability 204, 206
Probability measure 195
– conservation of 210
Projection operator 264
Propagating plane waves 37
Pulse 48, 84

q-average 184, 222
q-band chaos 176
q-phase transition 189, 194, 213
q-point cycles 175
q-point saddles 175
q-potential 184, 222
q-variance 184, 222
Quadratic map 161
Qualitative analysis 192
Quasi-periodicity route 177, 180

Randomness 131
Rayleigh number 6
Reaction–diffusion equation 20, 107, 108
Reaction–diffusion system 19, 43, 47
Renormalization 264
Renormalization transformation 167, 181
Renyi entropy 266, 275
Reproducibility 146
Residue 269
Riddled structure 261
Rotating spiral waves 38
– compound 58
– in excitable media 52
Rotation number 157, 174, 224, 227

Saddle 133, 140, 151
– collision with 151, 194, 205
Saddle-node bifurcation 104, 161
Šarkovskii order 172
Scaling 72
Scaling fields 168
Self-similar structure 265
– hierarchical 165
– nested 131, 196, 231, 242
Self-similarity
– of hierarchical structure 243
Shock wave solution 79
Similarity
– in 2^n-band bifurcations 236
– among bands 173
– in 2^n bifurcations 165
– in 2^n-band bifurcations 171
– in F_m-attractor merging 182
– of invariant sets 172
– within a band 173, 236
Similarity law 127
Similarity relation *see* Similarity law
Similarity relations
– for F_m attractor merging 237
Simple bifurcation 99
Singular spectrum 195
Skewed varicose instability 24
Slaving principle 13
Slope
– of linear segments 203, 205, 219, 265
Solvability condition 97
Spatial inversion symmetry 70, 84, 87
Spatial modulation 12, 70, 109
Spatial translational invariance 67, 70
Spatio-temporal intermittency 41
Stability 140

– statistical 133, 146
Stability of periodic solutions 21
Stable manifold 137, 141, 151, 262
Standard map 159, 241
Statistical structure 148, 252
Statistical structure functions 183
Steadily propagating solution 45
Steady-state traveling wave solution 63
Strange attractor 131, 195
Stretching 131, 138, 142, 144, 145, 266
Subcritical bifurcation 12
Supercritical bifurcation 12
Synchronization 37, 39, 79, 119, 121, 261

Tangency mapping 142
Tangency orbit 155
Tangency structure 142, 245, 265
– folding of 203
Tangent bifurcation 161
Tangent map 140
Tangle 138, 270
Target pattern 79, 92
Thermal conductivity 263
Time coarse-graining 146
– limit of 184
– statistical law of 146
Time correlation 252, 254
Time correlation function 147, 271
Time-reversal symmetry
– breaking of 245
Topological defect 29
Topologically conjugate 187
Tori 265
– islands of 159, 243
Torus
– Q-fold 269
– wrinkled 178
Translational symmetry 22
Trigger wave 48
Trivial solution 8, 12, 98
Turbulent viscosity 264
Turing instability 59, 109
Turing pattern 59
Turnstiles 138, 251, 254
Two-dimensional fractality 161, 173, 238
Two-phase coexistence 190, 194
2^m time series 230
2^n bifurcations 161
– similarity in 165
2^n-band bifurcations 161, 176
– similarity of 171
2^n-band splitting 165, 217

U region 177, 228
Unstable manifold 137, 141, 151, 259, 262, 265
Unstable orbits 144, 269
– aperiodic 133
– periodic 129, 133

van der Pol equation 97
Variance 136, 147, 235, 271
Variational principle 184, 197, 222
Vertex orbit 164, 171, 200
Viscosity 264

Weakly coupled oscillators 119
Widespread chaos 177, 178, 224, 243, 254
Williams domain 27
Wrinkled torus 225

Zig-zag instability 24, 72, 75

Druck: Mercedesdruck, Berlin
Verarbeitung: Buchbinderei Lüderitz & Bauer, Berlin